D1238536

Transactions of the American
Philosophical Society
Volume 86, Part 2

TRANSACTIONS

of the

American Philosophical Society

Held at Philadelphia for Promoting Useful Knowledge

VOLUME 86, Part 2

Ptolemy's Theory of Visual Perception:

An English Translation of the *Optics* With Introduction and Commentary

A. Mark Smith

THE AMERICAN PHILOSOPHICAL SOCIETY

Independence Square, Philadelphia

1996

Library of Congress Cataloging in Publication Data

Smith, A. Mark
  Ptolemy's Theory of Visual Perception

Bibliography, index

  1. Ptolemy, Claudius   2. Optics, visual perception   3. Classical
  literature   4. Translation, Latin   5. History of Science

ISBN: 0-87169-862-5   94-78521
US ISSN 0065-9746

CONTENTS

PREFACE

The English translation that follows is based upon Albert Lejeune's critical Latin text of 1956, which has recently been reprinted along with a French translation and supplementary annotations.[1] With very few exceptions, in fact, I have kept it parallel to that text in terms of both structure and interpretive intent. My translation thus reflects not only various editorial emendations suggested by Lejeune but also various critical slants brought to bear by him on difficult passages. Indeed, without Lejeune's keen insight to guide me, I would have been hard put to unravel certain linguistic knots in the Latin text. I have also taken advantage of his excellent French translation, drawing upon it freely as both a source of inspiration for, and a check against, my own translation.

At this juncture one might naturally ask why, given the ready availability of Lejeune's French version, an English translation of the *Optics* is warranted at all. At least three reasons spring to mind. The first and most obvious is that many potential readers, even within the academic community, are effectively restricted to English as far as linguistic competence is concerned. Second, because Lejeune's translation—along with its scholarly additions—is simply appended to rather than integrated with the original edition of 1956, it is extremely awkward to follow in conjunction with its critical appurtenances.[2] Third, despite his best efforts to address recent scholarship in the new edition, Lejeune's interpretive focus remains fundamentally unchanged in its relative narrowness. My translation and its accompanying annotation, on the other hand, reflect a somewhat broader analytic approach that is more sensitive than

---

[1]Originally published under the title *L'Optique de Claude Ptolémée dans la version latine d'après l'arabe de l'émir Eugène de Sicile* (Louvain: Bibliothèque de l'Université, 1956), Lejeune's edition was republished under the title *L'Optique . . . Sicile: Édition critique et exégétique augmentée d'une traduction française et de compléments* (Leiden, New York, Copenhagen, Cologne: Brill, 1989)—all references to the *Optics* will be from this later edition.

[2]The new edition is structured as follows: 1) original introduction, 2) new supplementary introduction, 3) reprinted Latin text with original annotations faced by French translation with occasional new annotations, 4) various indexes reprinted from the original, 5) original bibliography, 6) a "postface," and 7) a new updated critical bibliography. Since the new edition was published posthumously, this somewhat chaotic structure is surely due to Lejeune's having been prevented by death from properly integrating his new material into the previous edition.

Lejeune's to concerns expressed by recent scholars. Consequently, it is more up-to-date in its critical perspective.

As a translator, I take seriously the notion that any translation, no matter how literal or liberal, constitutes an interpretation—as in fact the Latin term *interpretatio* (= "translation") implies. What is more, I subscribe to the old-fashioned idea that, although a text may not speak for itself, it does speak for its author. I am, in short, a firm believer in authorial intent. I say this mindful of the pitfalls associated with such a hermeneutic position, particularly if it is taken in an uncompromising way. Thus, on the one hand, I reject the denotative extreme that would reduce any text to a bare exposition of fact or observation. Authorial intent is never that simple, as witness poetry, most of whose "meaning" lies in evocation rather than denotation. On the other hand, I reject the connotative extreme that would strip every text of meaning beyond what the reader chooses to import into it. Still, there is no denying that lectorial intent plays a part in any textual interpretation. How, after all, can we help but impose our own conceptual categories and predilections upon what we read? The best we can hope for, then, is not to be too obtrusive in such interpretive intrusions.

These reservations acknowledged, I nonetheless feel secure in assuming that Ptolemy's primary—although not necessarily sole—intent in the *Optics* was didactic and, therefore, that in his exposition he was appealing to common experience and logic as understood within the context of his time. My goal as translator is therefore to convey that intent as clearly as possible. This of course is a goal far easier to set than to meet, particularly in the case of the *Optics*, where the original text lies at so many linguistic removes from the surviving Latin version.[3] Anyone attempting to translate (or interpret) this latest version of the *Optics* is thus faced with a task not dissimilar to that of a paleographer attempting to decipher the bottom layer of a double palimpsest. To make matters worse, the text that is available today was created by a Byzantine Greek whose Latin style Lejeune has characterized with atypical Gallic understatement as "assez barbare."[4] Small wonder, then, that this text is at best difficult to penetrate much of the time and at worst so tortured and confused as to be virtually incomprehensible some of the time.

For this reason I have taken a rather liberal approach to my translation, somewhat more liberal than Lejeune's and probably too liberal by some lights. At times, of course, the style of the Latin

---

[3]See below, p. 5.
[4]*L'Optique*, p. 7*.

exemplar is so convoluted that no other tack seems feasible. There are a few loci, in fact, where my translation is perforce so liberal as to amount to little more than paraphrase. Yet even at places where the text seems clear enough at a superficial level, I have felt compelled by context to take liberties with the Latin. Ptolemy's discussion of the objective grounds of visibility in book 2 serves as an illustration.[5] This discussion revolves about the so-called *res vidende*, a phrase that literally translates to "things to be seen" or, perhaps a little more loosely, "visible objects." As often as not, in fact, that is precisely how it is intended in the *Optics*. Within the context under consideration, however, the reference is not to visible objects themselves but, rather, to the various characteristics that make them visible. It is this point that I have tried to convey by rendering *res vidende* as "visible properties" (Lejeune = "les visibles").[6] These properties, the text continues, are differentiated according to three categories: *que vere videntur* (literally, "those that are truly seen"), *que primo videntur* (literally, "those that are seen first"), and *que sequenter videntur* (literally, "those that are seen afterward"). Here, again, context indicates what is actually intended: the visible properties are being subdivided according to levels of contingency or immediacy. What is "truly seen," on the one hand, is perfectly self-sufficient and immediate: i.e., "luminous compactness" (*lucida spissa* literally, "luminous compact things"), which depends upon nothing else for its visibility. Hence, I have chosen to dub the properties *que vere videntur* as "intrinsically visible" (Lejeune = "vus au sens vrai du mot"). What is "seen first," on the other hand, is color, which is contingent on light for its visibility but which is nonetheless fundamentally visible in itself. It is to convey this sense of fundamental visibility that I have translated *primo videntur* as "primarily visible" (Lejeune = "ce qui est vu [im]médiatement"). Being wholly dependent upon color (and thus luminous compactness) for their visibility, finally, the remaining visible properties— e.g., size, shape, place, and distance—are "seen afterward" (i.e., *sequenter videntur*) insofar as they are not visible *per se* but are only perceived inferentially. That is why I designate these properties as "secondarily visible" (Lejeune = "vu médiatement"). In all these cases it is clear that I (as well as, to some extent, Lejeune) have done violence to literal meaning in order to convey the ulterior sense of the terms at issue.

---

[5] See *L'Optique*, pp. 12–14 for the Latin text, and pp. 71–73 below for the English translation of this account.

[6] I avoided the stock philosophical term *visibilia* in order to avoid confusion and complication.

Like Lejeune in his critical text, I have structured my English translation according to two key subdivisions: books and paragraphs. The constituent paragraphs of each book are numbered in consecutive order precisely as they are in Lejeune's text so that the reader who wishes to compare the English version to the Latin original can do so with ease. Accordingly, all references that I make to the text will be by book and paragraph in the form II, 109 (book 2, paragraph 109). As a further aid to the reader who wishes a handy reference, I have prefaced each book with a detailed topical résumé based upon the excellent general summary provided by Lejeune on pp. 123*–131* of his edition.

Although I have, like Lejeune, also subdivided the text according to propositions, unlike him, I have set these propositional elements clearly apart from the main text. Unlike him, as well, I have made distinctions among these propositional elements in order to render them more accessible to the interested reader. For example, some of the propositions constitute actual mathematical proofs. These I have labeled "Theorems." Others are simply intended to illustrate a given point geometrically. These I have labeled "Examples." Still others describe empirical procedures, in which case I have labeled them "Experiments." In certain instances, when the same general point applies under variable circumstances, I have further subdivided propositional elements into "Cases." All of these propositional elements are numerically designated according to book and consecutive order. Thus, for instance, *EXAMPLE II.6* (the sixth example in book 2) might be followed by *THEOREM II.3*, and so on.

As for diagrams, those that actually accompany the text have been adapted with very few changes from the figures supplied by Lejeune, these in turn being adapted from Gilberto Govi's earlier edition of 1885.[7] Like the paragraphs, they are designated according to book and consecutive order. Thus, for example, the fourteenth diagram of the fourth book is referred to as figure IV.14. Diagrams that accompany the introduction and commentary, on the other hand, are numbered according to consecutive order alone.

A few words, finally, about indexes. I have included two of them in this book. The first is general, covering names as well as certain key words and concepts. The second is intended to serve as a glossary of sorts. It therefore consists of specific technical terms from the Latin text keyed to their various equivalents in the English version

---

[7]*L'Ottica di Claudio Tolomeo, da Eugenio, ammiraglio di Sicilia, scrittore del secolo XII, ridotta in latino sovra la traduzione arabe di un testo greco imperfetto, ora per la prima volta, conforme a un codice della Biblioteca Ambrosiana, per deliberazione della R. Accademia delle Scienze di Torino* (Torino, 1885).

(here, again, I must sing Lejeune's praises for easing my way by having provided an exemplary Latin index for his edition). For instance, in various forms by case and number, the term *actus* appears eleven times in the Latin edition and has been rendered six different ways in my translation: "action" (3 occurrences); "actual/actually" (3 occurrences); "effect" (2 occurrences); "event" (1 occurrence); "inclination" (1 occurrence); and "operation" (1 occurrence). As with *actus*, so with every other Latin term listed, there will be, insofar as feasible, a full listing of English equivalents with their locations in the English text according to page and line number (e.g., "12.3" indicates "the third line on page 12"). Latin, of course, lends itself poorly to word-for-word English translation, so many of the listed "equivalents" will appear improper or bizarre unless they are understood within their appropriate textual environment.

I would like to take this opportunity to acknowledge the generous support of the National Endowment for the Humanities, without which this translation-project would have taken far longer than it did. So, too, I must acknowledge the support of the University of Missouri, Columbia, which has been remarkably forthcoming with supplementary funding. I would also like to acknowledge the courtesy extended to me by the main libraries at the University of California, Berkeley, and California State University, Hayward, as well as by the Bancroft Library. All three collections were invaluable to me in my research. Credit is also due to Jennifer Erickson for her invaluable editorial assistance. Finally, for their help, encouragement, and inspiration, I wish to thank specifically: Lois L. Huneycutt, David C. Lindberg, Marshall Clagett, A. I. Sabra, and Gérard Simon.

INTRODUCTION

1: Ptolemy: A Biographical Sketch

As scant as is the evidence from which to chart the course of Claudius Ptolemy's life, we are at least fortunate enough to have two definite milestones along the way: among the personal observations given by Ptolemy in the *Almagest*, the earliest dates from the eleventh year of Hadrian's reign (127 A.D.), the latest from the fourth year of Antoninus Pius' reign (141 A.D.).[1] Taking the midpoint of this interval (134 A.D.) as an approximate *floruit*, then, we can reasonably assume that Ptolemy was born around 100 A.D., an estimate that squares fairly well with the birthdate derived from late Antique and early medieval sources. As to the place of his birth, we can only conjecture. That it was not Pelusium, as once supposed, is now clear; perhaps it was Ptolemaïs Hermiou in Upper Egypt. Whatever the case, there is no reason to doubt that he was Greco-Egyptian by birth, perhaps also a Roman citizen, as indicated by his otherwise incongruous forename "Claudius." As far as we can tell, his entire working life was spent in Alexandria or its environs, certainly no farther away than nearby Canobus. It is said that he lived to be seventy-eight and survived into the reign of Antoninus Pius' successor, Marcus Aurelius (161–180). These two claims, if true, would lead us to place Ptolemy's death not only somewhere within that span, but probably toward the end.[2]

Suffice it to say, lack of biographical detail makes it all but impossible to date Ptolemy's canon with any precision. We can, however, place certain key works within a rough chronological order on the basis of internal evidence. In all probability the earliest of his

---

[1] See *Almagest* XI, 5 and IX, 7 in G. J. Toomer, trans., *Ptolemy's Almagest* (New York, Berlin, Heidelberg, Tokyo: Springer, 1984), pp. 525 and 450 respectively. Ptolemy also cites a partial lunar eclipse observed in the ninth year of Hadrian's reign (125 A.D.), but there is nothing in the context to indicate that this was a personal observation; see ibid., IV, 9, p. 206.

[2] For a fairly recent account of Ptolemy's life and works, see G. J. Toomer, "Ptolemy," *Dictionary of Scientific Biography*, vol. 11 (New York: Scribner's, 1976), pp. 186–206. See also Franz Boll, "Studien über Claudius Ptolemäus," *Jahrbücher für classische Philologie*, supplementband 21 (Leipzig, 1894), pp. 53–66, and B. L. van der Waerden, "Klaudios Ptolemaios," *Paulys Realencyclopädie der classischen Altertumswissenschaft*, vol. 23.2 (Munich, 1959), cols. 1788–1859, esp. 1788–1791.

extant treatises, the *Mathematike Syntaxis* (or *Almagest* as it came to be known from Arabic sources) had to have been finished after 141, the latest observation recorded in it. If, moreover, the Canobic Inscription imputed to Ptolemy is genuine, then its date (147/148) could be—and, indeed, generally has been—taken as a *terminus ad quem* for the *Almagest*. Recent evidence, in fact, suggests that the *Almagest* was not actually brought to completion until somewhat later.[3]

Whatever its date of composition, though, the *Almagest* serves as a chronological benchmark for the rest of Ptolemy's writings. For instance, citations of the *Almagest* in the *Handy Tables*, the *Planetary Hypotheses*, the *Tetrabiblos*, and the *Geography* indicate that they were all written after it. How long after is far from certain, but there are solid grounds for supposing that the *Planetary Hypotheses* postdates the *Handy Tables* and that both antedate the *Geography*.[4] That the *Optics*, in its turn, postdates both the *Almagest* and the *Planetary Hypotheses* is borne out by two cardinal pieces of evidence. First, unlike the *Almagest*, where the phenomenon of atmospheric refraction is virtually ignored, the *Optics* treats it in a fairly extensive and sophisticated way.[5] In short, there seems to be a conceptual evolution between the *Almagest* and the *Optics* that indicates the former's chronological priority. The second piece of evidence lies in the differing explanations of the so-called Moon Illusion that Ptolemy offers in the *Almagest*, the *Planetary Hypotheses*, and the *Optics*. A fairly clear developmental line in his understanding can be traced through the three: in the *Almagest* he misconstrues the phenomenon entirely, in the *Planetary Hypotheses* he seems to be somewhat confused but on the right track, and in the *Optics* he reduces the phenomenon quite correctly to a mere psychological effect.[6]

---

[3]The text of the so-called Canobic Inscription is appended to some of the medieval manuscripts of the *Almagest* and purports to be a copy of a stele erected in the tenth year of Antoninus' reign by Ptolemy to celebrate his achievements in astronomy. Consisting primarily of astronomical parameters from the *Almagest*, it also contains some discordant figures, one set of which in particular has prompted the recent claim that the Canobic Inscription actually predated the completion of the *Almagest*; for details see N. T. Hamilton, N. M. Swerdlow, G. J. Toomer, "The Canobic Inscription: Ptolemy's Earliest Work," in J. L. Berggren and B. R. Goldstein, eds., *From Ancient Omens to Statistical Mechanics: Essays on the Exact Sciences Presented to Asger Aaboe* (Copenhagen: University Library, 1987), pp. 55–73.

[4]This chronological sequence is deduced from comparison of parameters used in each work as well as from other incidental features, such as the fact that, unlike the other works, the *Geography* is not addressed to a certain Syrus.

[5]It is possibly to atmospheric refraction that Ptolemy refers in *Almagest* IX, 2, p. 421, but the passage is far from clear. Ptolemy's treatment in the *Optics* is to be found in V, 23–30.

[6]The Moon Illusion concerns the apparent enlargement of moon and sun when they are viewed near the horizon. In *Almagest* I, 3, p. 39, Ptolemy explains this apparent enlargement in terms of refraction caused by vapors in the air through which the celestial bodies are viewed. Thus, he concludes, the enlargement is due to the bodies' being seen through a denser medium, just as the images of objects lying in water are enlarged. Two things are wrong with this explanation. First, it assumes incorrectly that the images of the bodies at the horizon are actually enlarged. Second, the explanation on the basis of refraction is backward:

Accordingly, we are fairly safe in concluding that the *Optics* post-dates not only the *Almagest*, but also the *Handy Tables* and the *Planetary Hypotheses*. By how much is open to debate. On the one hand, there is no ostensible warrant for the claim that Ptolemy wrote the *Optics* toward the very end of his life, perhaps as late as 175. On the other hand, Lejeune is probably too conservative in situating it *grosso modo* within the third quarter of the second century.[7] Surely it is reasonable to locate the treatise somewhere within the decade between 160 and 170. But no matter what date is chosen, the fact remains that, far from being a *Jugendwerk*, the *Optics* represents an advanced stage in Ptolemy's intellectual development.

It would be difficult to imagine an environment more conducive to such development than that of Alexandria, even in its relatively lackluster second-century incarnation. Not only was the vaunted Museum still active, but, even more important, the Library remained unsurpassed within the Mediterranean ambit and perhaps the entire world.[8] Moreover, Alexandria had not lost its cachet as a philosophical gathering-point, where Platonists, Aristotelians, Stoics, Epicureans, Skeptics and even Gnostics mingled and argued. The result was a ferment of ideas about, among other things, the nature of the physical world, the nature of sense-perception as an intellectual conduit to that world, and the nature of knowledge as ultimately grounded in sense-perception. It is hardly surprising that, faced with such a welter of conflicting ideas, many thinkers of Ptolemy's day, Ptolemy included, tended toward eclecticism in their effort to assimilate those ideas systematically.[9]

---

the images actually ought to be *diminished* insofar as the bodies lie in a rarer medium (ether) rather than in a denser one (the vapor-filled atmosphere). For Ptolemy's account of the same phenomenon in the *Planetary Hypotheses*, see B. R. Goldstein, "The Arabic Version of Ptolemy's *Planetary Hypotheses*," *Transactions of the American Philosophical Society*, vol. 57.4 (Philadelphia, 1967), p. 9. And for his final explanation, see *Optics*, III, 60.

[7]*L'Optique*, p. 26*.

[8]For an excellent overview of Alexandria as an intellectual and cultural center, particularly in the pre-Roman period, see Peter M. Fraser, *Ptolemaic Alexandria*, 3 vols. (Oxford: Clarendon Press, 1972). According to Fraser's account, Alexandria's intellectual heyday was the third century B.C., after which a slow decline set in until the mid-second century B.C., when Ptolemy Euergetes II undertook a systematic purge of Alexandria's intellectual community, which had a profoundly negative impact on both the Museum and the Library. A significant recovery seems to have occurred in the middle decades of the first century B.C. under Ptolemy Auletes and Cleopatra VII, although the Library was supposedly destroyed in an accidental conflagration in 48 B.C., during Caesar's Alexandrine War. There is still a good deal of uncertainty about the Library itself. Was it the "inner" royal library, which may have been attached to the Museum? Was it housed in a separate, dedicated repository, or was it held in various locations? Whatever the case, it is clear that, by imperial Roman times, the library facilities of Alexandria—whether centralized in the Serapeum or scattered in various smaller holdings—were enormous and enormously varied. As Fraser observes, however, Alexandria did experience a "brain-drain" of sorts during the Roman imperial period, when various intellectuals, particularly those trained in rhetoric and philosophy, sought their fortunes in Rome.

[9]See pp. 17–18 below. By Ptolemy's time, these philosophical schools, among which Christianity could well be included, had achieved the status of uncontestable dogma among their

As rich as was the philosophical heritage to which Ptolemy had recourse in Alexandria, no less rich was the scientific heritage available to him there. For instance, in mathematics, he was heir to the tradition established by Euclid (fl. c. 300 B.C.), himself an Alexandrian, whose great codification of plane and solid geometry, the *Elements*, provided a firm developmental basis for the likes of Archimedes (fl. c. 250 B.C.) and Apollonius of Perga (fl. c. 210 B.C.). In optics, too, it was Euclid who blazed the trail, laying the grounds for Ptolemy's later analysis with his pioneering studies of optics and catoptrics. In astronomy, as well, Ptolemy had at hand the works of distinguished Greek predecessors, such as Eratosthenes (fl. c. 235 B.C.) and Hipparchus (fl. c. 150 B.C.), but he also had access to a wealth of ephemerides dating back to the late Babylonian period.[10] More than just observational data, these tabular sources also provided Ptolemy with important computational techniques that bore not just on his astronomical work but also, as we shall later see, on his optical work.[11] Finally, in anatomy and physiology, Ptolemy was heir to the tradition of Herophilus (died c. 260 B.C.) and his somewhat younger confrère, Erasistratus, a tradition exemplified by Ptolemy's cantankerous but brilliant contemporary, Galen. On a less theoretical level, meanwhile, Ptolemy was heir to a tradition of applied science that was represented par excellence by his Alexandrian predecessor, Hero (fl. c. 60 A.D.), who wrote a variety of studies on practical matters, including a short treatise on mirrors. So, too, the well-developed technological resources of Alexandria were critical to Ptolemy's scientific pursuits. Without a cadre of highly skilled craftsmen upon which to draw, he would have found it difficult, if not impossible, to construct the scientific instruments upon which he relied in so much of his work.

That Ptolemy's *oeuvre* reflects the gamut of these intellectual and technological resources is evident from the extraordinary range of topics covered in his extant writings. Astronomy or cosmology is of course amply represented not only by his chef d'oeuvre, the *Almagest*, but also by related works, such as his tabular epitome, the

---

proponents, in great part because their primary focus was not intellectual (ways of thinking) but practical (ways of living). Moreover, as they became more dogmatic, they became more systematic, particularly Platonism, which by this time was in its "Middle" form and soon to achieve its "Neo" form under the third-century scholar Plotinus. Somewhat ironically, however, the process of systematization was matched by a strong eclecticism, so that ideas and analytic approaches from other philosophical schools were freely assimilated and adapted. As a result, philosophical discourse had developed a sort of conceptual and terminological *lingua franca* that transcended the bounds of any of its particular philosophical sources.

[10]The earliest observation cited by Ptolemy dates from 720 B.C.; see *Almagest* IV, 6, p. 191.

[11]See pp. 48–51 below.

Handy Tables, and the *Planetary Hypotheses,* in which he attempts to "physicalize" the mathematical models constructed in the *Almagest.*[12] Within the same general domain of interest we might also include the *Geography* and *Tetrabiblos,* the former dealing with cartography, the latter with astrology, and both focusing on what could be termed practical astronomy. Somewhat farther afield we find the *Harmonics,* Ptolemy's treatise on music, whose concern with sound provides a thematic link with the *Optics,* where the focus is on the sister-sense, sight.[13] Farthest afield of all, finally, we have Ptolemy's brief philosophical essay, *Peri kriteriou kai hegemonikou* (*On the Criterion of Truth and the Governing Faculty*) in which he deals with the problem of knowledge and certitude.[14]

Although some of these writings, foremost among them *On the Criterion* and the *Almagest,* reveal a decidedly speculative or theoretical bent, most of them incline toward application or *praxis* rather than theory. This inclination is clear, for instance, in the way Ptolemy approaches mathematics, not as a subject for analysis in its own right but as an analytic tool. The clearest token of his preference for *praxis* over theory, however, is his reliance on instruments not only for observational precision but also for experimental certification. In this regard, of course, the *Optics* serves as a paradigm, reflecting not only the depth of Ptolemy's methodological commitment but also the scope of his philosophical concerns.

2. The *Optics:* A Biographical Sketch

The Greco-Arabic Phase: The only surviving text of Ptolemy's *Optics* is a badly mangled, twelfth-century Latin version of an Arabic translation presumably drawn from the Greek original. Lacking any trace of that Greek original or its Arabic counterpart, though, we can only suppose a direct connection between the two. Indeed, for all we know, there may be a lost Syriac intermediary. Suffice it to say, then, that we are hard put to reconstruct the *Optics'* early textual history at all, much less with any certainty. Granted, there are

---

[12]In addition, Ptolemy wrote three minor works bearing on astronomy: a brief guide for use of the *Handy Tables,* the *De analemmate,* and the *Planisphaerium.*

[13]In the sixteenth chapter of his *On the Criterion of Truth and the Governing Faculty,* Ptolemy singles out sight and hearing as the senses that contribute most to our intellectual grasp of the physical world and thus to our ultimate goal of living well; see note below for full reference. Cf. Aristotle, *De sensu et sensato* 1.437ª4–17.

[14]Contrary to Boll, "Studien," and Friedrich Lammert, who edited the *On the Criterion* in *Opera omnia,* vol. 3, pt. 2 (Leipzig: Teubner, 1952), Toomer, "Ptolemy," has doubts about the authenticity of this work; cf., however, A.A. Long, "Ptolemy on the Criterion: An Epistemology for the Practising Scientist," in P. Huby, G. Neal, eds., *The Criterion of Truth: Essays Written in Honour of George Kerferd* (Liverpool: Liverpool University Press, 1989), pp. 151–230.

a few scattered clues to help us in the attempt; yet, despite our best efforts, a number of important questions remain unanswered and, in all probability, unanswerable.

At present we know of only three clear attestations to the existence of the Greek version of Ptolemy's *Optics*, the earliest dating from the fourth century, the remaining two from the sixth and eleventh.[15] The most that we can infer from this sparse evidence is that Ptolemy's *Optics* was probably still available in Greek by the Middle Ages. What we cannot determine from it is whether the cited text was in fact the true source of the Arabic version and thus its medieval Latin derivative. Nor for that matter can we even determine whether it was genuinely Ptolemaic. Until the Greek and Arabic texts are unearthed and compared, these issues will never be satisfactorily resolved.

About the Arabic-Latin connection we can be a good deal more certain than we can about the Greek-Arabic one. For instance, there is some evidence, albeit indirect and inconclusive, that an Arabic version of the *Optics* was in circulation by the mid-ninth century at latest.[16] The first direct evidence comes somewhat later, in the form of actual citations in the tenth and eleventh centuries. Among these citations, the most significant are provided by Ibn al-Haytham (d. 1041), author of the monumental *Kitāb al-Manāẓir* ("Book of Optics"). That Ibn al-Haytham was thoroughly familiar with the *Optics* is beyond question; not only did he cite it at numerous reprises, but he devoted several studies to it. Perhaps the most instructive of these is his "Doubts on Ptolemy," a critique of various Ptolemaic tenets, many of them drawn from the *Optics*.[17] This work is particularly revealing because it allows a point-by-point comparison of what Ibn al-Haytham claims on behalf of Ptolemy to what the surviving Latin text asserts. On the basis of clues such as these a fairly clear picture emerges of the Arabic text Ibn al-Haytham had at hand. According to this picture, his and the medieval Latin text are identical in four key respects. First, as far as we can tell from an admittedly limited sample, the content of Ibn al-Haytham's Arabic *Optics* mirrors that of the Latin version. Second, both reflect an *Urtext* consisting of five books. Third, both lack the first book in its entirety. And, finally, in both versions the fifth book is incomplete. If

---

[15]For details on the two earlier citations see pp. 48–49 below. The third citation, from the eleventh century, is to be found in Simeon Seth's *Conspectus rerum naturalium*; as Lejeune observes, however, this latter citation may only be at second hand; see *L'Optique*, p. 27*.

[16]See pp. 54–55 below.

[17]See A. I. Sabra, "Ibn al-Haytham's Criticisms of Ptolemy's *Optics*," *Journal of the History of Philosophy* 4 (1966): 145–149.

not conclusive proof, these similarities at least give us strong reason to believe that Ibn al-Haytham's version of the *Optics* was virtually the same in all important respects as that from which the Latin translation was derived.

The Early Latin Phase: As far as the textual history of the *Optics* is concerned, the transition from Arabic to Latin is like the passage from night into day. For a start, we know by attribution who the Latin translator was: a certain *amiratus* (= "admiral" or "emir") Eugene of Sicily. Better yet, we know a fair amount about him.[18] We know, for instance, that he was a high functionary in Norman-ruled Sicily during the second half of the twelfth century, when that island was nearing its cultural apex as a meeting-ground for Arabs, Byzantine Greeks, and Latins.[19] We also know that he was part of an intellectual élite connected with the court of William I (1154–1166), a group that included the renowned humanist Henry Aristippus, an avid promoter of scientific learning. Grecophone by birth, Eugene was considered by his contemporaries to be adept in Arabic and at least competent in Latin. He was thus remarkably well-placed in terms of both linguistic capacity and cultural environment for the task of rendering the *Optics* from Arabic into Latin. We know, as well, that this was not his only translation. He was also involved in the Latin translation of Ptolemy's *Almagest* from a Greek text supplied by Aristippus. This project was completed around 1160, Eugene's participation in it evidently due to his scientific expertise, which had already gained him a reputation. He is also known to have translated at least two other works, one from Greek into Latin and the other from Arabic into Greek. Unfortunately, despite this relative plethora of details about his scholarly activities, we know next to nothing about their chronology. The best we can do is locate them roughly within the second half of the twelfth century, and this applies to his translation of the *Optics* as well.[20]

According to Eugene's own testimony, his translation was based upon two exemplars of an Arabic version "previously translated from the Greek language."[21] Of these two exemplars, Eugene

---

[18]For a somewhat more complete account than the one that follows, see Lejeune, *L'Optique*, pp. 7*–12*. For details, see C. H. Haskins, *Studies in the History of Medieval Science* (Cambridge, MA: Harvard University Press, 1924), pp. 155–193, and Evelyn Jamison, *Admiral Eugenius of Sicily: His Life and Work* (London: Oxford University Press, 1957).

[19]Jamison estimates that Eugene was born around 1130 and died around 1203; see *Admiral Eugenius*, p. 5.

[20]According to Jamison, Eugene probably translated the *Optics* between 1156 and 1160 but may have revised it later in the 1190s; see *Admiral Eugenius*, pp. 5 and 143.

[21]*Optics*, heading of book II, p. 70 below. As Lejeune, *L'Optique*, p. 28*, observes, this claim on Ptolemy's part would seem to exclude the possibility of a Syriac intermediary between Arabic and Greek versions. But since we have no idea upon what grounds Eugene makes this

continues, he drew his translation from the more recent because it was better. How poor a "better" it must have been is evident from the deformed state of its Latin derivative. At the grossest level, for example, the Latin text is marred by the two major lacunae we already discussed: the absence of book 1 and the truncation of book 5.[22] A closer look at the Latin text reveals less obvious internal lacunae as well. But that is not all. In several instances, entire passages have been transposed willy-nilly. Even in their proper context, certain passages are simply incoherent, and, to add to these problems, there are several cases of maladroit interpolation. It is of course possible that Eugene himself imported these flaws into the Latin text, but, faithful translator that he was, it is far likelier that he simply reproduced them from the Arabic exemplar without any attempt at rectification.[23]

The Later Latin Phase: It is more than a little ironic that, almost as soon as it appeared, Eugene's translation of the *Optics* was rendered obsolete by the Latin version of Ibn al-Haytham's *Kitāb al-Manāẓir*, which may have been completed as early as the late eleventh century.[24] Circulating under the Latin title *De aspectibus* and attributed to "Alhazen" (the Latin transliteration of Ibn al-Haytham's first name, al-Ḥasan),[25] this work represented a critical departure from that of Ptolemy. For one thing, Alhazen's and Ptolemy's theories of light and vision were so fundamentally different as to be mutually exclusive. For another, Alhazen's analysis was far more sophisticated, thorough, and logically compelling than Ptolemy's—hence, the *De aspectibus'* rapid dissemination within Scholastic circles of the thirteenth century. Indeed, by the second half of that century, an entire tradition of optical analysis (the so-called Perspectivist tradition) had developed on its basis. But Alhazen's triumph was necessarily at Ptolemy's expense, and, despite evidence of the *Optics'* use during the thirteenth century, not a single manuscript copy survives from that period.[26]

The thirteen manuscripts that do survive date from the early fourteenth to the early seventeenth century.[27] Roughly speaking, they

---

claim (he may simply have assumed a direct filiation between the Greek and Arabic texts), we cannot take it as proof of anything.

[22]See p. 6 above.

[23]See Lejeune, *L'Optique*, pp. 12*–13*.

[24]It is in fact possible, but by no means certain, that the *Kitāb al-Manāẓir* was translated into Latin by Gerard of Cremona, in which case it must have been completed before his death in 1187.

[25]See A. I. Sabra, *The Optics of Ibn Al-Haytham: Books I–III on Direct Vision*, vol. 2 (London: The Warburg Institute, 1989), p. xii.

[26]Roger Bacon (fl. c. 1260) and Witelo (fl. c. 1275), both of them Perspectivists, demonstrate a first-hand knowledge of Ptolemy's *Optics*; for details, see pp. 57–58 below.

[27]For the most current listing, see David C. Lindberg, *A Catalogue of Medieval and Renaissance Optical Manuscripts* (Toronto: Pontifical Institute of Mediaeval Studies Press, 1975),

fall into two chronological groups, the earlier of which consists of
three manuscripts from the first half of the fourteenth century. The
existence of this particular group is somewhat puzzling. Might it indicate an otherwise hidden surge of interest in the *Optics* during the
early 1300s? Perhaps so, but it probably indicates no more than the
play of chance in the process of transmission. Somewhat more
amorphous than the first group, the second consists of one manuscript from the fifteenth century, eight from the sixteenth,[28] and one
from the seventeenth. The existence of this group is readily explained in terms of the almost slavish classicism of Renaissance
scholars inspired by a Golden-Age view of classical Antiquity. As
part of the classical heritage, the *Optics* was bound to arouse the interest of such scholars, particularly because of its attribution to
Ptolemy, author of the vaunted *Almagest*. So lively was this interest,
in fact, that the idea of publishing a printed edition of the treatise
was entertained seriously during the Renaissance.[29] Yet, in a sense,
this renewed focus on the *Optics* was its undoing. The more carefully it was scrutinized, the more obvious its deficiency by comparison with Alhazen's *De aspectibus*. Kepler's revisionary work of 1604
and 1610 simply provided the coup de grâce.[30] As a result, the *Optics* underwent a radical transformation during the early part of the
seventeenth century: once regarded as a promising source of optical lore, it had now become an historical artifact. As such, it slid into
a desuetude so complete that, by the mid-eighteenth century, it was
thought to be lost by no less a scholar than the historian of mathematics, J. F. Montucla.[31]

Like Poe's purloined letter, however, the *Optics* actually lay hidden in plain sight, lurking in manuscript-collections throughout
Europe, many of them carefully catalogued. Nonetheless, it remained unseen at least in part because scholars such as Montucla
were looking in the wrong direction. They expected, quite reasonably, to find it in Greek or perhaps Arabic, but certainly not Latin.

---

p. 74; see also Lejeune, *L'Optique*, pp. 38*–47*. An additional manuscript, belonging to the private collection of Prince Baldassare Boncompagni, survived until the later nineteenth century
but has since been lost.

[28]This total is raised to nine when Boncompagni's now-lost manuscript is added.

[29]Regiomontanus (d. 1476), the renowned student of Ptolemaic astronomy, actually
planned to publish the *Optics*. Nearly a century later, Georg Hartmann, editor of John
Pecham's *Perspectiva communis* (1542), could only lament that, having only a badly damaged
fragment of the *Optics*, he could not publish it. See Govi, *L'Ottica*, pp. iv and xli–xlii.

[30]Indeed, Kepler's two major optical studies, the *Ad Vitellionem Paralipomena* of 1604 and
the *Dioptrice* of 1611, undermined not just Ptolemy's but also Alhazen's account of light and
vision. In short, with Kepler's contributions the *Optics* was rendered doubly obsolete as a theoretical work.

[31]See Montucla, *Histoire des mathématiques*, vol. 1 (Paris, 1758), pp. 312–314. In fact, not only
did Montucla believe the work was lost, but he had no idea that it had ever been rendered
into Latin.

So its recovery at the very end of the eighteenth century was a matter more of serendipity than of scholarly acumen or careful research. The recovery itself was the work of three scholars working independently—G. B. Venturi, Montucla, and J. J. A. Caussin de Perceval—all of whom stumbled upon separate versions.[32] Within little more than a decade of these discoveries, preliminary studies had been presented to learned societies by Venturi, Caussin, and J. B. Delambre, and by 1822 all three studies had been published.[33] Meantime, serious preparations were underway for the text's publication. At least three transcriptions had been completed, and by the early 1820s there could be little doubt that at least one of them would find its way into print. In fact, none of them did, and interest in the *Optics* soon abated.

By the 1870s, however, there was renewed enthusiasm for the publication-project. Doubts about the work's authenticity had finally been put to rest,[34] and nearly all of the manuscripts currently known to exist had been uncovered and examined.[35] Furthermore, at the prompting of Gilberto Govi, who proposed to carry out the task of editing, the Royal Academy of Sciences of Torino had agreed to underwrite publication. A physicist by education, though, Govi was ill-suited for the task he set himself. Not only did he lack training or experience in paleography, but he knew nothing of the process of critical editing. For him it was enough to copy the "most correct" (i.e., the earliest) manuscript available. Published in 1885, the resulting text was far from critical; indeed, aside from modern

---

[32]Venturi discovered two manuscript versions of the *Optics* in the period 1797–1798, the first in Paris and the second in Milan. Montucla's discovery was made by means of a catalogue-listing for the Bodleian Library and was announced in the second edition of his *Histoire des mathématiques*, vol. 1 (Paris, 1799), p. 314. Caussin made his discovery in Paris in 1798. Both Venturi and Caussin made transcriptions with the intent of eventually publishing them. For details on this recovery and its outcome, see Lejeune, *L'Optique*, pp. 33*–35*, and Govi, *L'Ottica*, pp. v–viii.

[33]Given in 1811 to a meeting of the Istituto Nazionale Italiano, Venturi's study, "Considerazioni sopra varie parte dell'ottica presso gli antichi," was published in his *Commentari sopra la storia e le teorie dell'ottica* , vol. 1 (Bologna, 1814), pp. 31–62. Delambre's study, "Sur l'Optique de Ptolémée comparée à celle qui porte le nom d'Euclide et à celles d'Alhazen et de Vitellion," was first presented to the Institut Royal de France in 1811 and published in *Connaissance des temps pour l'an 1816* (Paris, 1813), pp. 239–256; it was subsequently republished in his *Histoire de l'astronomie ancienne* (Paris, 1817), vol. 2, pp. 411–431. Caussin's exceptionally erudite paper, "Mémoire sur L'Optique de Ptolémée," was presented to the Institut Royal de France in 1812 but was not published until 1822 in *Mémoires de L'Institut Royal de France, Académie des Inscriptions et Belles-Lettres* (Paris, 1822), pp. 1–43.

[34]Th. H. Martin, "Ptolémée, auteur de l'Optique . . . est-il le même que Cl. Ptolémée, auteur de l'Almageste?" *Bullettino di bibliografia e di storia delle scienze matematiche e fisiche* 4 (1871): 466–469.

[35]For a listing of the manuscripts known to exist as of 1870, see B. Boncompagni, "Intorno ad una traduzione latina dell'Ottica di Tolomeo," *Bullettino di bibliografia e di storia delle scienze matematiche e fisiche* 4 (1871): 470–492. Among the manuscripts listed is the one belonging to his private collection, which has since been lost.

punctuation and lettering, it represented little or no improvement over its medieval and Renaissance predecessors. Not that Govi's edition was entirely without merit. After all, he did provide some historical and critical analysis in his brief introduction to the text, and he took great pains to reconstruct a number of the accompanying diagrams, which had been distorted out of all recognition in the course of scribal transmission.

The Modern Phase: Whatever its flaws from the standpoint of critical scholarship, Govi's edition represented a significant advance in one respect: it made the *Optics* accessible to a relatively wide audience in an easily readable version. It was therefore on the basis of this edition that the *Optics* entered into the canon of sources for intellectual history in general and history of science in particular. Yet, despite its availability in Govi's edition, the *Optics* had little or no real impact among specialists, such as Franz Boll, who ignored it entirely in his pathbreaking "Studies" of Ptolemy's *oeuvre* and its philosophical underpinnings. Understandable as a purist reaction to the uncritical nature of Govi's editing, this omission is still somewhat surprising in view of the obvious philosophical implications of Ptolemy's account of visual perception and its physical and psychological grounds.

While Govi's edition had no discernible impact among specialists, it found a ready welcome among generalists, such as George Sarton, who saw in Ptolemy's optical work a prime example of Hellenistic applied mathematics. Unfortunately, the resulting popularized account of the *Optics* was badly skewed toward the more "scientific" aspects of Ptolemy's theory, particularly his experimental approach to reflection in book 3 and refraction in book 5.[36] Because of this highly selective approach, the *Optics* was construed too narrowly as a study of pure physical optics rather than as what it really is: a broad-ranging analysis of vision. This skew is clearly reflected in the selections from the *Optics* included by M. R. Cohen and I. E. Drabkin in their influential *A Source Book in Greek Science*.[37] Predictably enough, Cohen and Drabkin chose precisely those portions of books 3 and 5 that seem most congenial from a modern scientific perspective. Nevertheless, such popularizations were useful in publicizing the *Optics* as an important and, to some extent, exemplary study in ancient mathematical science.

[36] See, for example, Sarton, *Introduction to the History of Science*, vol. 1 (Baltimore: Williams & Wilkins, 1927), pp. 274 and 276–277; Sarton lauds Ptolemy's study of refraction as "the most remarkable experimental research of antiquity" (p. 274).
[37] Cambridge, MA: Harvard University Press, 1948, pp. 268–283.

By the mid-1940s, on the heels of World War II, the time was ripe for a full-scale critical analysis of the *Optics* based on proper philological and historical grounds. Undertaken by the Belgian scholar Albert Lejeune, this analysis unfolded over an eleven-year period in a series of articles and monographs, among which the two most significant were his *Euclide et Ptolémée, deux stades de l'optique géométrique grecque* of 1948[38] and *Recherches sur la catoptrique grecque* of 1957.[39] But the centerpiece of Lejeune's project was the superb critical edition of the *Optics* that he published in 1956. Not only is this edition superior to Govi's, but it is almost incomparably so. Unlike Govi, for example, Lejeune based his edition upon a close comparative reading of all known manuscripts, so that the resulting critical text is a model of its kind.[40] Unlike Govi, as well, Lejeune annotated the Latin text extensively, offering the sort of elucidation that Govi failed almost entirely to provide. And, finally, unlike Govi, Lejeune made a serious effort to put the *Optics* into historical context by examining its various sources as well as its influence. Of particular note in this regard is his close scrutiny of the question of the *Optics'* authenticity on the basis of not only doxographic but also philological evidence.[41] In one key respect, however, Lejeune's edition was deficient: it lacked a modern-language version of the Latin text. Only recently has this gap been filled by the appearance of Lejeune's long-awaited French translation, published by Brill in 1989 as a sort of addendum to the original Latin edition. This translation includes supplementary material intended to update the introduction as well as the annotations and bibliography provided in the original Latin edition of 1956.

Although Lejeune's edition and the interpretive framework he built around it have held up remarkably well so far, they have not escaped criticism entirely. Over the past decade his analysis has been attacked on at least two fronts by Gérard Simon and Wilbur Knorr. In his provocative study, *Le regard, l'être et l'apparence dans l'Optique de l'antiquité*,[42] Simon accuses Lejeune of misunderstanding the true theoretical focus of the *Optics*. Ptolemy was not, as Lejeune would have it, attempting to explain the physics of radiation;

---

[38]Louvain: Bibliothèque de l'Université.

[39]*Mémoires de l'Académie Royale de Belgique: Classe de Lettres et des Sciences Morales et Politiques* 52.3.

[40]Altogether, Lejeune consulted twelve manuscripts for his edition; ironically, the Milan manuscript (Ambrosiana T.100.sup) upon which Govi based his transcription turns out by Lejeune's reckoning to be the closest of all to the original. For details on Lejeune's editing procedures, see *L'Optique*, pp. 38*–122*.

[41]*L'Optique*, pp. 13*–26*; this account is repeated verbatim in *Recherches*, pp. 13–26.

[42]Paris: Seuil, 1988.

he was instead attempting to account for the process of vision. In short, his preoccupation was with sight, not light.[43] Simon goes on to show in convincing fashion how this shift in analytic perspective offers significant new insights into Ptolemy 's approach to both theoretical and practical issues in the *Optics*.

If Simon's line of attack on Lejeune is somewhat oblique, involving relatively subtle issues of interpretation, that of Wilbur Knorr is anything but. In an intriguing and wide-ranging essay entitled "Archimedes and the Pseudo-Euclidean *Catoptrics*: Early Stages in the Ancient Geometric Theory of Mirrors," he confronts Lejeune squarely on the issue of authenticity.[44] For a start, Knorr argues, the evidence in favor is so slight and inconclusive as to render the case for Ptolemy's authorship not just questionable but doubtful. If not Ptolemy, however, then who might the author of the *Optics* actually be? A likely candidate, Knorr suggests, is the second-century philosopher Sosigenes, who is presumed to have written an eight-book treatise on vision. As it now stands, then, the *Optics* must be a remnant, at third linguistic hand, of that eight-book original. Since Lejeune has already responded in detail to this claim, there is no need to deal with it at length here.[45] Suffice it to say that, although it cannot be dismissed out of hand, Knorr's case is not compelling enough to warrant abandoning the assumption of Ptolemy's authorship. Still, at the very least, Knorr has done yeoman service by making crystal-clear just how tenuous the evidential support for that assumption really is.

It should be clear by now that, despite significant advances in scholarship over the past few decades, much remains to be done before we can feel secure in our understanding of the *Optics* within its proper historical context. At a textual level, on the one hand, recovery of the Greek version would go a long way toward resolving issues of authorship and textual intent, particularly if the recovered version were to include book 1 and the concluding portion of book 5.[46] A far more likely path of recovery, however, would be via the Arabic, since so many of the manuscript resources in that language have yet to be consulted and published. At an interpretive level, on the other hand, the time is ripe, or at least ripening, for an intensive analysis of the *Optics* within the context of the

---

[43]For a somewhat more detailed synopsis of Simon's thesis, see my review in *Isis* 83 (1992): 118–119.

[44]*Archives internationales d'histoire des sciences* 35 (1985): 27–105.

[45]See *L'Optique 2*, pp. 133*–138* for Lejeune's rebuttal.

[46]The same could be said, of course, if what were recovered, added to the extant portion of the treatise, were to comprise eight books, as Knorr speculates in behalf of Sosigenes.

entire Ptolemaic corpus. Properly done, the result would be a sweeping comparative study not unlike Boll's but more comprehensive in its interpretive scope.

3: An Overview of the *Optics*

A Survey of Ptolemy's Sources

At first glance, Ptolemy's *Optics* would seem to be little more than a study in what we would today call physical optics. This impression is reinforced by the propositional nature of so much of the treatise, particularly of the final three books, in which reflection and refraction are analyzed both experimentally and geometrically. But such an impression is misleading. The primary intent of Ptolemy's *Optics* is not to give an objective account of light and its physical action at all. It is, instead, to explain visual perception in the most general sense, albeit in terms of a physical efflux from the eyes that is generically related to light. As a result, Ptolemy's *Optics* deals with a much broader set of issues than those covered by physical optics alone. This breadth is reflected in the range of sources upon which Ptolemy drew in constructing his account. Since he provides no specific citations, though, it is impossible to determine whether the sources we identify were direct or indirect, that is, whether Ptolemy's access to them was mediated or not. Indeed, some of his potential sources are so indefinite as to indicate nothing more than probable influences.

The sources themselves can be divided into two main categories: technical, or "scientific," and general, or "philosophical." Among the technical sources, the most easily identifiable is Euclid's *Elements* (c. 300 B.C.), which is the primary basis for Ptolemy's geometrical reasoning in the *Optics*.[47] Not surprisingly, books 1 and 3 of the *Elements* figure most prominently, the latter being particularly noticeable in Ptolemy's treatment of spherical convex and concave mirrors. Book 5, on basic proportionality, and book 6, on the proportionality of plane figures, are also used by Ptolemy, but to a much lesser extent than books 1 and 3. Possible additional sources in mathematics, but in a far more tangential way, are Archimedes' *On the Sphere and Cylinder* (c. 250 B.C.) and Hero of Alexandria's lost commentary on the *Elements* (1st century A.D.?).[48] Ptolemy may

---

[47]Ed. J. L. Heiberg, *Euclidis opera omnia*, vols. 1–5 (Leipzig: Teubner, 1883–1888), and trans. T. L. Heath, *The Thirteen Books of Euclid's* Elements, 3 vols. (second edition 1925; rpt. New York: Dover, 1956).

[48]Archimedes, *On the Sphere and Cylinder*, ed. J. L. Heiberg, *Archimedis opera omnia*, vol. 1 (Leipzig: Teubner, 1910), pp. 1–229; trans. T. L. Heath, *The Works of Archimedes* (Cambridge:

also have owed a conceptual debt to Geminus' missing treatise on mathematical classification (1st century A.D.).[49]

Another group of technical sources, still mathematical in orientation, focuses specifically on optics. The earliest of these is Euclid's *Optics* (c. 300 B.C.), which provides the general framework for mathematical optics on the basis of the visual ray.[50] Whatever use Ptolemy made of this work is so fundamental, however, that there is no way of pinpointing specific propositional borrowings. Such is not the case with the *Catoptrics* commonly attributed to Euclid, but thought to have been compiled much later.[51] Whether genuinely Euclidean or not, this work certainly represents an early stage of evolution in optical analysis, one that predates Ptolemy by a considerable amount.[52] Thus, the clear similarities between propositional elements in the *Catoptrics* attributed to Euclid and in Ptolemy's *Optics* may be taken to indicate either direct or relatively direct borrowing—assuming the *Catoptrics* is genuinely Euclidean—or a shared progenitor—assuming it is not. In addition, Ptolemy may have been influenced by Hero of Alexandria's *Catoptrics* (1st century A.D.?), but two points militate against this conclusion. First, whatever Ptolemy might have derived from this treatise he could as easily have derived from the *Catoptrics* attributed to Euclid. Second, if Ptolemy actually did use Hero's *Catoptrics* as a source, his failure to mention the least-lines demonstration of

Cambridge University Press, 1897), pp. 1–90. On Hero's lost commentary on the *Elements*, see Heath, *Elements*, vol. 1, introduction, pp. 20–24.

[49]On Geminus, see Proclus' commentary on the *Elements*, trans. Glenn R. Morrow, *Proclus: A Commentary on the First Book of Euclid's* Elements (Princeton: Princeton University Press, 1970), pp. 31–35 and 90–92. Although Morrow dates Geminus to the first century B.C., more recent work by Otto Neugebauer has established to the satisfaction of most scholars that his floruit belongs in the first century A.D.; see Neugebauer, *A History of Ancient Mathematical Astronomy* (Berlin, Heidelberg, New York: Springer, 1975), pp. 579–581.

[50]Ed. J. L. Heiberg, *Euclidis opera omnia*, vol. 7 (Leipzig: Teubner, 1895), pp. 1–121; for a French translation of the entire treatise, see Paul Ver Eecke, *Euclide: L'Optique et la Catoptrique* (Paris: Blanchard, 1959), pp. 1–51; parts of the *Optics* have also been translated into English by Cohen and Drabkin, *Source Book*, pp. 257–261.

[51]Ed. J. L. Heiberg, *Euclidis opera omnia*, vol. 7, pp. 283–343; for the French translation, see Ver Eecke, *Euclide: L'Optique et la Catoptrique*, pp. 99–123. Heiberg concluded not only that the *Catoptrics* was not genuinely Euclidean but that it was probably compiled by Theon of Alexandria, who also produced a recension of the *Optics*. Most scholars have followed Heiberg in this assumption, but in his 1985 essay, "Archimedes and the Pseudo-Euclidean *Catoptrics*," as well as in his more recent "Pseudo-Euclidean Reflections in Ancient Optics", *Physis* 31 (1994): 1–45, Wilbur Knorr has argued forcefully (and I think convincingly) against Heiberg and in favor of Euclidean authorship.

[52]In accepting the claim of Theonine origin, Lejeune thought that he could discern three chronological strata in the *Catoptrics*: the first predating Archimedes and probably representing a true Euclidean foundation; the second predating Ptolemy and perhaps harking back to the time of Hero of Alexandria; and the most recent dating to Theon himself. Of the thirty propositions comprising the treatise, Lejeune assigned only five to this third stratum, so even by his reckoning the lion's share of the *Catoptrics* predates Ptolemy. For details, see Lejeune, *Recherches*, pp. 112–151.

the equal-angles law of reflection is puzzling.[53] Archimedes, finally, rounds out this list of possible optical sources. He is presumed to have written a *Catoptrics* that went beyond the study of mirrors to include a rudimentary analysis of refraction.[54]

A third group of technical sources for the *Optics* is so indefinite as to constitute mere influences. Precisely where Ptolemy learned how to use the diopter for observational purposes we cannot tell. Archimedes is of course a possibility, if a distant one.[55] Likewise, we have no idea where Ptolemy gained his knowledge of scenographic effects, but that he understood the principles of illusionism in painting is quite clear from allusions he makes to them in the *Optics*.[56] Finally, it is impossible to determine with any precision where Ptolemy learned the technique of constant second differences that he applied in his analysis of refraction. That such a technique was routinely used in late "Babylonian" astronomy is common knowledge, of course, as is the fact that Ptolemy was conversant with such astronomy.[57] Still, there is no way of pinpointing the text or texts upon which his expertise was based.

The second main category of sources, the so-called general or "philosophical" ones, can be identified from various allusions in the *Optics*. For instance, there is no doubt from his account of color-perception in book 2 that Ptolemy was familiar with (and rejected) Plato's theory of color as outlined in the *Timaeus*.[58] What we cannot determine, of course, is whether that knowledge was mediated. There are fairly clear indications, as well, that Ptolemy knew the theory of *emphasis* ascribed to Democritus, but we cannot tell whether it was Democritus or some secondary authority who was the actual source of that knowledge.[59] Likewise, there are clear Peri-

---

[53]For the text of Hero's *Catoptrics*, see W. Schmidt, ed. and trans. in *Heronis Alexandrini opera quae supersunt omnia*, 2.1 (Leipzig: Teubner, 1900). For an English translation of the first six chapters, see Cohen and Drabkin, *Source Book*, pp. 261–268. The Least-Lines demonstration of the equal-angles law of reflection is found in chapter 4, where Hero proves for any plane or convex mirror that, of every possible incidence-and-reflection ray-couple, the one that subtends equal angles with respect to the surface of reflection on each side of the point of reflection represents the shortest path.

[54]For details, see Lejeune, *Recherches*, pp. 50–53, 142–145, and 176–179. Knorr, "Archimedes and the Pseudo-Euclidean *Catoptrics*," suggests that the *Catoptrics* attributed to Euclid may, for a time (from c. 150 A.D. to c. 400 A.D.), have been misattributed to Archimedes and then re-edited in Euclid's name. For Lejeune's response to Knorr's argument, see *L'Optique*, pp. 357–362. For a recent reiteration of, and elaboration upon, Knorr's claim for the authenticity of the *Catoptrics*, see Ken'ichi Takahashi, *The Medieval Latin Traditions of Euclid's* Catoptrica (Fukuoka: Kyushu University Press, 1992), pp. 13–49.

[55]See Albert Lejeune, "Le dioptre d'Archimède," *Annales de la Société Scientifique de Bruxelles* 61 (1947): 27–47; esp. p. 45.

[56]See, e.g., II, 126–128.

[57]See below, pp. 41–44.

[58]See below, p. 27.

[59]See note 92 below.

patetic elements in the *Optics*, particularly in book 2, but as far as specific Aristotelian treatises are concerned, we can only speculate.[60] The *De anima* and the Pseudo-Aristotelian *Problemata* seem to be the likeliest of such sources, but one cannot rule out the *Parva naturalia* or *Physics*. Last, least definite, and yet perhaps most easily recognized of Ptolemy's philosophical sources are the Stoic ones that underlie his theory of visual perception. Throughout the *Optics* hints abound that Ptolemy conceived of visual perception in terms of a pneumatic system that is subject to a controlling faculty in the brain.[61]

As we conclude this survey of sources and influences, we should bear in mind that it is just that—a survey, and a tentative one at that. On the whole, in fact, the more specific the identification the more tentative it is. We should also bear in mind that some of the sources mentioned (e.g., Plato's *Timaeus*) exerted a negative rather than a positive influence upon Ptolemy, inasmuch as they were reacted against rather than followed by him. Such negative sources are of course no less important than positive ones in any thinker's intellectual formation.

The Methodological Foundations of the Optics

A fair amount has been written about Ptolemy's philosophical leanings, the vast majority of it based upon works other than the *Optics*. Depending upon the treatise, in fact, Ptolemy has been variously portrayed as a Platonist-Pythagorean, an Aristotelian, a Stoic, an Empiricist, and even a Positivist.[62] The resulting impression is of an intellectual schizophrenic whose philosophical allegiances are sworn and resworn whenever the analytic circumstances demand.

---

[60]See below, pp. 18–19.

[61]See below, pp. 28–29.

[62]The following is a mere sampling: (1) for Ptolemy as a Platonist-Pythagorean, see Izydora Dambska, "La théorie de la science dans les oeuvres de Claude Ptolémée," *Organon* 8 (1971): 109–122; (2) for Ptolemy as an Aristotelian, see Silvia Fazzo, "Alessandro d'Afrodisia e Tolomeo: Aristotelismo e astrologia fra il II e il III secolo D.C.," *Rivista di Storia della Filosofia* 43 (1988): 627–649; (3) for Ptolemy as a non-Aristotelian with strong Platonist ethical leanings, see Liba Taub, *Ptolemy's Universe: The Natural Philosophical and Ethical Foundations of Ptolemy's Astronomy* (Chicago: Open Court, 1993); (4) for Ptolemy as a Stoic, see F. Lammert, "Eine neue Quelle für die Philosophie der mittleren Stoa I," *Wiener Studien* 41 (1920): 113–121, and "Eine neue Quelle . . . II," *Wiener Studien* 42 (1920/21): 36–46; (5) for a comparative analysis of Sextus Empiricus and Ptolemy on epistemological grounds, see Gualberto Lucci, "Criterio e metodologia in Sesto Empirico e Tolomeo," *Annali dell'Istituto di Filosofia di Firenze* 2 (1980), 23–52; (6) for Ptolemy as a "naive" empiricist, see Louis O. Katsoff, "Ptolemy's Scientific Method," *Isis* 38 (1947): 18–22; and (7), for Ptolemy as a Positivist, see Pierre Duhem, *To Save the Phenomena: An Essay on the Idea of Physical Theory from Plato to Galileo*, trans. Edmund Doland and Chaninah Maschler (Chicago: University of Chicago Press, 1969), pp. 14–18.

Such an impression is of course inevitable if we suppose that the philosophical schools of his day were so rigidly defined and exclusivist as to preclude any crossover. But, as one scholar has recently observed, crossover was far more common among these schools than exclusivity:

Within the Hellenistic period itself a common philosophical jargon had developed, and terms emanating from one school were frequently employed by another. What came to be shared, moreover, was not just words, or concepts, but something we might call professionalism or expertise. Ptolemy and his contemporaries were writing for audiences who had been similarly educated to themselves and whom they could expect to be familiar with an intellectual tradition characterised by a community of concepts, standard questions and answers, common argumentative methods and objections.[63]

In short, "like most of his contemporaries, Ptolemy [worked] from within a tradition that [was] irreducibly composite . . . but broadly unified in the perspective of its own members."[64] Ptolemy's susceptibility to various philosophical influences was thus a matter not of shifting allegiances but of a consciously eclectic outlook. As a result, his philosophical approach may have been multifaceted, but in a holistic sense. To characterize his thought in terms of particular philosophical strands is therefore to risk compartmentalizing it out of all recognition. This point is worth remembering as we deal specifically with the *Optics* and its methodological foundations.

In 1981, Anna de Pace published a two-part study entitled "Elementi Aristotelici nell'*Ottica* di Claudio Tolomeo."[65] The gist of her argument is that Ptolemy was "Aristotelian" (and therefore not "Platonist") in his approach insofar as he accorded ontological priority to the sensible reality under analysis rather than to the mathematical constructs according to which it was analyzed. Pace's point is well taken. For one thing, as described by her, Ptolemy's scientific methodology is perfectly consonant with the sense-based theory of knowledge outlined in *On the Criterion*.[66] For another, Ptolemy does tailor his geometrical analysis to the sensible phenomena so that his visual-ray theory, unlike Euclid's, takes into full consideration such non-mathematical aspects of sight as variations in visual acuity, color-perception, or even misperceptions due to

---

[63]A. A. Long, "Ptolemy on the Criterion," p. 153.
[64]Ibid., p. 171.
[65]*Rivista critica di storia della filosofia* 36 (1981): 123–138, and *Rivista critica . . .* 37 (1982): 243–276.
[66]See below, p. 28.

false subjective judgments. The result is a fundamentally empirical approach—exemplified in Ptolemy's experiments on binocular vision, reflection, and refraction—that is, if not actually Aristotelian in its foundations, at least compatible with Peripatetic norms.[67]

Yet, although the epistemological foundations of Ptolemy's analysis may be legitimately characterized as "Aristotelian," the structure of that analysis may no less legitimately be characterized as "Platonic." Having already discussed this point in detail elsewhere, I will offer only the briefest summary here.[68] My basic contention is that, like Greek mathematical astronomy, Greek geometrical optics was subject to a particular methodological paradigm that falls under the rubric of "saving the appearances." According to this paradigm, the sensible world consists of appearances or illusions, primary among which are the appearances of irregularity or disorder. Called anomalies, such irregularities are irrational and therefore unreal in Platonic terms, and if they are to be rationalized, or "saved," they must be reduced to perfect regularity so that their underlying reality can be discovered. This end is achieved by applying a fundamental and perfectly simple principle of order. In Greek optics that principle is the visual ray, whose saving grace is absolute rectilinearity.[69]

The theory of vision based on the visual ray principle is as straightforward as it is simple. The eye is reduced to a center-point of sight directly linked by visual rays to point-objects. As long as this visual link remains unbroken and uniform, the object will be seen as it actually exists in physical space. This is the normative case for vision, so any departure from it will constitute an anomaly. The two major optical anomalies are reflection and refraction, both of them arising when the visual ray strikes an optical

---

[67]Even the application of mathematics to physics has its justification in Aristotle; see *Physics* 2.2.194ᵃ6–10, where Aristotle speaks of "the more natural of the branches of mathematics, such as optics, harmonics, and astronomy." "While geometry," he continues, "investigates natural lines but not *qua* natural, optics investigates mathematical lines, but *qua* natural, not *qua* mathematical," trans. R. P. Hardie and R. K. Gaye, in Jonathan Barnes, ed., *The Complete Works of Aristotle*, vol. 1 (Princeton: Princeton University Press, 1984), p. 331.

[68]For the full account, see "Saving the Appearances of the Appearances: The Foundations of Classical Geometrical Optics," *Archive for History of Exact Sciences* 24 (1981): 73–99, and "Ptolemy's Search for a Law of Refraction: A Case-Study in the Classical Methodology of 'Saving the Appearances' and its Limitations," *Archive for History of Exact Sciences* 26 (1982): 221–240.

[69]There is an even more fundamental governing principle, which I call the Principle of Natural Economy. According to the dictates of Natural Economy, the rectilinearity of visual radiation is a result of the need for it to be as efficient as possible: i.e., along the shortest spatial path (note how Hero's Least-Lines proof of the equal-angles law manifests this economy principle). The same principle dictates that celestial motion be circular, since the circle encompasses the greatest possible area for the circumference. For a more detailed account, see my "Extremal Principles in Ancient and Medieval Optics," *Physis* 31 (1994): 113–140.

interface and is broken. If the interface is impermeable, the imping-ing ray will break completely and rebound to cause reflection. If the interface is permeable, the impinging ray will be partially bro-ken and thereby diverted in its passage. In that case, refraction will result. In both cases, however, the appearance to be saved is es-sentially the same: not the object as it actually exists in physical space but a displaced image (and, therefore, an illusion) of it will be seen. To save this appearance is therefore to relate the image to its true, objective counterpart. In other words, the anomaly is resolved when image-location is perfectly determined with respect to both eye and object.

That reflection and refraction share much in common is clear from this account. But they are more than just similar; they are in fact so closely and systematically linked as to constitute special cases of one another in terms not only of cause and effect, but also of resolution.[70] This systematic relationship dictates the fundamen-tal three-tiered analytic structure of mature classical optics. At the lowest level comes *optics* proper, which deals with all aspects of unimpeded visual radiation, so it is within this tier that the very ba-sis for subsequent analysis is established. At the next level comes *catoptrics*, which analyzes reflection of visual rays from mirrors of various shapes. The progression to this level constitutes a full step upward in analytic complexity. At the highest level of complexity, finally, comes *dioptrics*, which treats the phenomena associated with refraction.

This analytic structure is clearly reflected in the organization of Ptolemy's *Optics*. Accordingly, the first two of the five books com-prising the treatise are devoted to the study of *optics* proper. To the extent that we can reconstruct it, book 1 would have dealt with the basic geometry of visual radiation, so its theorematic content would have been commensurately simple. The general analysis of visual radiation continues in book 2, where Ptolemy deals with issues of visual perception. In the course of this examination, he begins a pi-oneering study of binocular vision that extends into the third book. The actual focus of this study is on diplopia, the apparent doubling of the image when the visual axes are displaced from their custom-

---

[70]In both cases, of course, the cause is impedance and breaking/bending by a resisting sur-face, and the effect is image-displacement. Theoretically, at least, the resolution of the anom-aly is basically the same as well: i.e., a simple proportionality between angle of incidence and angle of reflection/refraction. In reflection, that proportionality is one of equality; in refrac-tion it is somewhat more complex; see below p. 44. Bear in mind, incidentally, that by ground-ing the equal-angles law of reflection in the principle of least distance, Hero reduces reflection to a mere special case of unimpeded visual radiation, which also hews to the principle of least distance.

ary convergence. Diplopia therefore represents a true anomaly insofar as it entails image-displacement with respect to one or the other eye. In the third and fourth books, Ptolemy undertakes a detailed analysis of reflection, starting with the basic principles by which image-location is determined. After a relatively brief examination of the simplest case, reflection from plane mirrors, Ptolemy progresses to the more complex case of reflection from spherical convex mirrors. He then turns in book 4 to the even more complex case of reflection from spherical concave mirrors, after which he offers a brief discussion of reflection from composite mirrors. In book 5, finally, he caps his overall study with an examination of refraction, the centerpiece of which is his ingenious but failed attempt to determine the law of refraction.

In the analysis that follows, we will use this topical organization as our basic guide. We will start by looking at Ptolemy's fundamental theory of visual perception, taking into account its physical and mathematical, as well as physiological and psychological, dimensions. We will then turn to a systematic analysis of his account of reflection from plane, spherical convex, and spherical concave mirrors. And we will conclude by examining his account of refraction, our focus in that case being upon his attempt to derive the governing principle of refraction.

Ptolemy's Basic Theory of Vision

Whatever their specific differences, all Greek theories of vision share in common the fundamental premise that without physical contact between eye and visible object vision cannot occur.[71] How this contact is supposed to be established is in fact what differentiates these theories at the most fundamental level. There are two basic alternatives. The first, which is intromissionist, is so designated because it entails the passage of something from the object into the eye. This is the alternative chosen by the atomists, who theorized that atom-thick replicas (*eidola*) are continually sloughed off from the surfaces of objects into the space that surrounds them and thence directly or indirectly into the eye. Serving as visible representations of their generating objects, these replicas, or the physical

---

[71]For an overview of these theories, see David C. Lindberg, *Theories of Vision from Al-Kindi to Kepler* (Chicago: University of Chicago Press, 1976), pp. 1–17; Bernard Saint-Pierre, "La physique de la vision dans l'antiquité: Contribution à l'établissement des sources anciennes de l'optique médiévale" (Ph.D. diss., University of Montréal, 1972); and J. I. Beare, *Greek Theories of Elementary Cognition from Alcmaeon to Aristotle* (Oxford: Clarendon Press, 1906), pp. 1–92.

impressions generated by them, provide the means by which visual perception is effected.[72] Aristotelian visual theory is also intromissionist but without the crass materialism of its atomist counterpart. Instead of a material efflux or mechanical impression, it is a qualitative effect (i.e., color) that reaches the eye. In this case, the requisite physical contact is provided by a continuous transparent medium through which the effect can pass from object to eye.[73]

The second basic alternative, this one extramissionist, is exemplified by the so-called Pythagorean theory, according to which vision is due to a sort of internal fire or flux that issues from the eye to strike external objects. Once touched by the outgoing flux, those objects are visually perceived, so, in a sense, the eye sees the physical world by shedding its own light upon it.[74] A variation of sorts on this extramissionist theme is to be found in the Stoic explanation of sight. In this account, the eye is assumed to exude a specific form of visual pneuma that impregnates the air at the surface of the eye and transforms it into a visual pathway. It is by means of this pathway that the eye picks out external objects visually, much as a blind man "sees" his way with a cane.[75]

There is one final theory that deserves mention here, this one neither wholly intromissionist nor wholly extramissionist. As de-

---

[72]The atomists, who included Leucippus (fl. c. 450 B.C.), Democritus (fl. c. 420 B.C.), and Epicurus (fl. c. 400 B.C.) are all credited with the belief that vision is the result of simulacra entering the eye. According to Theophrastus, Democritus explains the production of such simulacra in terms of physical impressions (*apotyposis* = "molding," or even "stamped impression") created in the air by the atomic replicas sloughed off by visible objects. Upon reaching the eye, these impressions create a representation in the pupil that forms the basis for subsequent visual perception. Democritus' term for this representation is *emphasis*, which connotes reflection or rebound. For a more detailed description, see George M. Stratton, trans., *Theophrastus and the Greek Physiological Psychology before Aristotle* (London: Allen and Unwin, 1917), 49–54; see also Aristotle, *De sensu et sensato* 2.438ª5–13. In *De rerum natura* 2.26–447, Lucretius (fl. c. 100 B.C.) asserts that it is the atom-thick simulacra themselves, not intermediate physical effects, that enter the eye to cause vision.

[73]For Aristotle, transparency is a potential of certain media such as air or water; only when light is present does this potential become actualized. Thus, light serves as a sort of catalyst for vision by preparing the medium for the passage of the qualitative effect to the eye. See *De anima* 2.7 and *De sensu et sensato* 2 and 3. It is worth noting, however, that Aristotle appears to espouse the theory that sight issues from the eye in his account of the rainbow, halos, mock suns, and rods in *Meteorology* 3.2–6.

[74]Alcmaeon (fl. early 5th c. B.C.?) is taken as a representative of the Pythagorean theory. According to Theophrastus, *De sensibus* 26, Alcmaeon asserts that the eye contains a fire and that vision is due to its "gleaming." The exact meaning of this assertion is uncertain, especially since the "gleaming" is equated with transparency, which may indicate that Alcmaeon, like the atomists, imputed sight to a sort of reflection in the pupil. According to Aristotle, *De sensu et sensato* 2.437ᵇ24–438ª3, Empedocles (fl. c. 465 B.C.) also attributed sight to a fire sent out from the eye.

[75]Galen is one of the crucial sources for our understanding of the Stoic theory of vision; the walking-stick analogy, which was later adopted by Descartes, is described and critiqued by Galen in *De placitis Hippocratis et Platonis* 7.7, ed. and trans. Phillip De Lacy, *Corpus medicorum graecorum* V 4, 1, 2 (Berlin: Akademie-Verlag, 1980), pp. 472–475; for Galen's alternative account, which is also based on pneumatic effusion from the eye, see *De placitis* 7.5, pp. 452–455.

scribed by Plato in the *Timaeus*, in fact, this theory appears to be a sort of Pythagorean-Atomist hybrid inasmuch as it posits a double efflux.[76] The first comes from the eye, which emanates a subtle fire that melds with daylight to form a sort of percipient medium extending out to visible objects. The second comes from the objects themselves, which continually shoot off particles of different sizes. On meeting the percipient medium, these particles cause it to dilate or contract, and it is through these dilations and contractions that the perception of color arises.[77]

Evidently Pythagorean in inspiration if not in detail, Ptolemy's account of vision, like Euclid's, is grounded in the visual ray model. According to this model, the eye radiates visual flux in the form of a cone, whose vertex lies within the eye to define the center of sight and whose base defines the visual field. This visual cone can in turn be resolved into a bundle of discrete rays, each one projected outward from the vertex.[78] Ptolemy insists, however, that the physical emanation of visual flux is perfectly continuous in every sense. Unlike Euclid, then, he denies the individual ray's physical reality. For him it is nothing more than a convenient analytic fiction (II, 50).

There are indications in the *Optics* that, as far as the physical cause of visual radiation is concerned, Ptolemy's theory is fundamentally Stoic in its grounding. If so, then the visual flux should be conceived of as a pneumatic emission from the eye and the ray as a line of visual sentiency established in the surrounding air. On the other hand, Ptolemy may have been thinking in more blatantly mechanistic terms of a continuous particulate emanation from the eye.[79] Whatever the physical cause, however, Ptolemy's theory of radiation puts him squarely in the extramissionist camp.

Issuing forth at enormous speed, the visual flux eventually strikes external objects and, in so doing, feels them visually. Thus, while it may not be an actual species of touch, sight is like touch in its basic operation (II, 13 and 67). The visual flux's sensitivity is variable, though. For instance, the farther it extends from its source, the weaker its capacity to sense what it touches, which explains why visible objects are more difficult to see at a distance. Eventually, of

---

[76]See Theophrastus, *De sensibus* 5.

[77]See *Timaeus* 45b–d and 67c–68c.

[78]For Euclid's articulation of this model, see *Optics*, definitions 1 and 2.

[79]For a discussion of possible Stoic elements in Ptolemy's theory, see my "Psychology of Visual Perception in Ptolemy's *Optics*," pp. 195–196. The more blatantly mechanistic side of Ptolemy's account comes out in his explanations of reflection and refraction, both of which depend on a dynamic interaction between the visual flux, treated as particulate, and the reflecting surface, treated as an impervious solid, or the refracting surface, treated as a pervious fluid; see below, pp. 37 and 42–43.

course, those objects become so distant that to reach them the visual flux must extend beyond its capacity to sense at all. When that happens, the objects are no longer seen, not because they fall within the gaps between discrete rays, as Euclid would have it, but because they lie beyond the threshold of minimum visibility for their size (II, 48–51; cf. Euclid, *Optics*, proposition 3). Such diminution in visual sensitivity, and thus in visual acuity, can be understood dynamically in terms of physical projection. As with a projectile, then, so with visual flux, the farther it extends from its source, the more weakly (i.e., slowly) it moves and, consequently, the less forcefully it acts on impact (II, 20).

Distance from the vertex of the visual cone, as a function of ray-length, is not the only factor affecting visual acuity. Another is obliquity with respect to the cone's axis, a factor that Euclid ignores entirely. How obliquity functions in this case is evident from experience: the closer the line of sight gets to the axial limit along the so-called visual axis, the greater the visual acuity along it. It is for this reason that, in order to get a clear visual grasp of any object as a whole, we must scan it with the visual axis. It is also for this reason that peripheral vision is so notoriously indistinct (II, 20).

Not only are ray-length and obliquity with respect to the visual axis the prime determinants of visual acuity; they are also the prime determinants of spatial perception, which is ultimately grounded in the visual apprehension of distance and orientation. Distance, according to Ptolemy, is ascertained through ray-length, each ray being endowed with a sense of its own outward extension from the center of sight (II, 26). Orientation, for its part, is grasped in two ways. On the one hand, the apprehension of left-to-right and top-to-bottom orientation is a function of the directional privilege that each visual ray possesses (II, 26; cf. Euclid, *Optics*, definitions 5 and 6). This directional privilege alerts the center of sight to the relative leftward, rightward, upward, or downward disposition of every ray. On the other hand, the slant of any given object with respect to the center of sight is apprehended through a comparative analysis of the lengths of all the rays striking its surface.[80] In both cases of orientation, though, the primary referent is the visual axis in relation to which left, right, up, down, or slant are ultimately determined.

---

[80]This description of how slant is detected follows from Ptolemy's account of how depth of curvature is discerned and how the perception of simple plane figures, such as circles and squares, is affected by viewing them from an oblique perspective; see II, 63 and 67–73.

The intuitive sense of eye-to-object distance, as well as of direction along the vertical and horizontal, enables the visual flux to define a basic coordinate system within which to map objective reality. The mapping itself is a matter of determining the fundamental spatial characteristics of visible objects on the basis of the surface they present to the visual flux. These characteristics include place, size, shape, and motion. Place, for instance, is determined on the basis of the object's position relative to the visual axis within the plane of the visual cone's base—i.e., along the horizontal and vertical—as well as along the visual axis itself (II, 26–27). Size-determination is somewhat more complex. The most primitive indicator of size is the visual angle, which is formed at the vertex of the visual cone by rays that touch the outer edges of the given object. Alone, however, the visual angle is inadequate to indicate true size. Distance must also be taken into account so that larger objects of whatever shape lying farther from the eye but subtending the same angle as smaller, nearer ones of the same shape will be perceived as larger. But even the addition of distance to visual angle is not enough to ensure true size-perception. Slant must also be factored in. Thus, while two objects of the same shape may lie the same distance from the eye along the visual axis and may subtend the same visual angle, they will not always be perceived to be of the same size. Indeed, if one of them faces the eye directly and the other is slanted with respect to it, the latter will be perceived to be larger, as indeed it is (II, 52–62).

The perception of shape and motion is more straightforward than that of size. Shape-perception, for its part, is doubly determined. Circumferential form is ascertained by means of the rays that strike the outer edges of the visible object's surface, whereas convexity or concavity in the plane of the visual axis is grasped by a sort of feeling that the visual flux has at the common section of the visual cone and the object's surface. This feeling is analogous to that of a hand when it apprehends the convexity of an object, such as a ball, as it grasps it (II, 64–69). Somewhat the same sense of touch underlies the perception of motion in the plane of the visual field, which depends upon the realization of passing aroused in the flux as the object sweeps through it, ray-by-ray, in consecutive order. The directional sense of each ray passed, as well as the sense of speed of passage, is of course what determines the perception of the object's velocity. Meantime, the perception of motion toward or away from the center of sight depends upon the sense of lengthening or shortening that is aroused in the visual rays that intercept the moving object (II, 76–81).

In principle, one eye would suffice for visual perception, but, as Ptolemy remarks, "nature has doubled our eyes so that we may see more clearly and so that our vision may be regular and definite" (II, 28). But if the eyes are doubled, why is it that we generally see one rather than two images? Ptolemy's response is that, under normal circumstances, the two visual axes work in perfect concert to focus on precisely the same spot of whatever object is being viewed. As a consequence, the bases of the two visual cones coincide so that corresponding rays in each cone converge at the appropriate spot on the object's surface (II, 27–28). Diplopia, or double vision, occurs when the visual axes fail to converge upon the object under visual scrutiny (II, 29). To illustrate, Ptolemy devises a simple apparatus consisting of a wooden plank of moderate width and length and a couple of thin pegs, one white and the other black.[81] Ptolemy then describes a series of experiments (II, 30–45) in which the plank is placed horizontally to the eyes so that the midline along its length bisects the line connecting the pupils. The pegs are then placed one after the other along the midline or side by side at varying distances, and the eyes are brought to focus upon different spots. Depending upon the specific conditions, then, one or even both of the pegs will be seen double, sometimes separately and sometimes overlaid. Later on, Ptolemy extends this analysis beyond point-objects (as represented by the pegs) to line-objects and, on that basis, shows how image-displacement, and thus image-location, can be geometrically determined (III, 25–58).

So far we have been looking at Ptolemy's visual theory from a subjective standpoint, examining how the eye acts on visible objects by means of the visual flux. But there are certain objective criteria that must be met if those things are to be effectively visible to the flux.[82] The most fundamental of these criteria is what Ptolemy calls "luminous compactness." What this means is that, by virtue of its physical constitution, the visible object must present a dense front to the impinging rays so as to block or impede them and, in the process, allow them to sense the impingement. That way they are made aware that they are seeing something. Compactness alone is not adequate, though; the object must also be illuminated, either by

---

[81] See II, 30; Ptolemy in fact provides no specific guidelines other than that the ruler should be "short," the pegs "thin," and the distance between the pegs and the ruler's edge "moderate."

[82] That Ptolemy was fully aware of the distinction between subjective and objective factors in vision is evident from his claim in II, 3 that "visible properties exist in two ways, one of which depends upon the disposition of the visible property [itself] and the other upon the action of the visual faculty." Clearly, the first mode of existence is objective, the second subjective.

itself or from elsewhere, if it is to be seen (II, 4). Yet even with these two criteria fulfilled, the object will not be visually perceived, because compactness and luminosity, whether alone or in concert, are necessary but insufficient grounds for true visibility.

It is in fact color, not luminosity, that Ptolemy regards as truly (or, in his terms, "primarily") visible (II, 5). Color, in short, is the proper object of sight (II, 13). As such, it is not a mere perceptual effect arising from particulate interactions but a real and inherent property of physical objects (II, 15–16). In this opinion Ptolemy sides with Aristotle against both Plato and the atomists.[83] Although it is real, however, color cannot subsist by itself; it must be instantiated in physical objects. Hence, it requires compactness not only for its visibility but for its very existence. It also requires illumination, because without light color is only potentially, not actually, visible, even though it may be properly objectified. In fact, for Ptolemy, as for Aristotle, the main function of light is to actuate the visibility of color.[84] As if to emphasize this actuating function, Ptolemy lumps light and visual flux under the same genus (II, 23). Its role as an agent of vision is therefore indirect rather than direct. In other words, light is not itself visible; what we see as light or luminosity is actually brightness, which constitutes a species of whiteness, so it is through the inherent "color" of brightness that light is visually apprehended (II, 69). Ptolemy seems to agree with Aristotle, at least at a general level, that the two primary or polar colors are white and black and that the remaining colors represent various intermediate blends of the two.[85]

As far as visual perception is concerned, the real purpose of color is to define the surfaces of physical objects. Indeed, so interdependent are color and surface that it is quite easy to be misled into equating the two (II, 12). By defining physical surfaces, color fulfills two crucial functions. First, it marks the spatial boundaries of visible objects so that they can be perceived as individuals apart from surrounding objects. As in a painting, therefore, so in physical space, color-separation provides the outlines that set various objects in the visual field off from one another (II, 6–7). Second, by defining the boundaries of physical objects in three dimensions, color

---

[83]For Aristotle's view on color as an inherent quality of objects and as the proper object of sight, see *De anima* 2.6–7. In II, 24 Ptolemy specifically attacks the Platonic theory of dilations and contractions as the cause of color-perception.

[84]In an analogy with distinctly Aristotelian overtones Ptolemy likens light to the "form" of color's "matter"; see II, 16.

[85]II, 24; as examples of intermediate colors, Ptolemy cites red, rose, and blood-red. For Aristotle's account of color as a blending of white and black, see *De sensu et sensato* 3.

permits the spatial characteristics of those objects to be perceived. Every color arouses in the visual flux a distinct passion that constitutes a sort of coloring. This passion ultimately creates a perception of the color that arouses it while, at the same time, giving the flux the requisite clues for determining the shape, size, place, and motion of the objects defined by that color (II, 23). This is why Ptolemy characterizes these spatial properties—which are included in Aristotle's list of common sensibles—as "secondarily" visible, because in fact they are not actually seen at all but are perceptually inferred from color (II, 6).

Evidently, then, Ptolemy conceives of visual perception along roughly Aristotelian lines as a complex process unfolding in three basic phases.[86] The first phase, which is physical in nature, entails the emission of visual flux from the eye and the consequent establishment of physical contact with a given external object. To be visually effective, though, this contact must be made with an object dense enough to impede the visual flux and luminous or illuminated enough to render its inherent color visible. These preconditions met, the second stage, which is essentially sensitive in nature, begins. During this phase, the visual flux suffers the passion of coloring that is aroused as soon as the initial physical contact is established. The apprehension of color that ensues accidentally conveys the object's spatial properties as manifested by its surface. These properties are then visually determined according to certain parameters, such as ray-length, obliquity, and visual angle, that are provided by the flux. From such determinations, finally, perceptual judgments or inferences are drawn during the third and concluding phase of the process. The result is a sort of conceptual conclusion about the object as it actually exists in physical space.[87]

The tripartite scheme just outlined dovetails perfectly with the admittedly sketchy psychological and physiological system that can be pieced together from clues scattered throughout the *Optics*. That system is under the overall control of what Ptolemy refers to as the Governing Faculty (i.e., *virtus regitiva*), which has obvious parallels with the Stoic *hegemonikon* and, like it, seems to be located

---

[86]For an account of the similarities between Ptolemy's and Aristotle's accounts of visual perception, see my "Psychology of Visual Perception," pp. 200–203.

[87]The process just described mirrors in a specific way Ptolemy's general description of sense-perception as a conduit to conceptual understanding in *On the Criterion*, particularly as outlined in chapters 7–12. As Ptolemy himself sums it up in chapter 4: "To the faculty of sense perception belong both sense organs and *phantasia*. The sense organs are the bodily instruments through which contact is made with perceptible things, while *phantasia* is the impression and transmission to the intellect, whose retention and memory of the things transmitted we call conception," trans. A. A. Long, "Ptolemy on the Criterion," p. 187.

in the brain.[88] From there it apparently controls all the senses, including sight, as the center of nervous activity (II, 13). Serving as both initiator and ultimate arbiter of visual perception, the Governing Faculty is the origin of visual flux/pneuma, which it passes on to the center of the eye where the actual source-point (i.e., *principium*) of vision lies.[89] From here the flux is channeled outward through the optical humors and the pupil, which limits the visual field and thereby defines the edges of the visual cone.[90] Since binocular vision assumes two such cones whose axes are both focused on a single point, there must also be a faculty of oversight that keeps the eyes working in concert to maintain this focus. That role is fulfilled by the Apex, which lies beyond the source-points of each eye. Providing the ultimate spatial referent for binocular vision, the Apex lies at the crown of what Ptolemy calls the common axis of sight (III, 35 and 61). It is of course tempting to locate the Apex at the optic chiasma, where the two optic nerves join in front of the brain. As long all three axes—i.e., the two proper axes of the individual cones and the common axis—converge on the same point, binocular vision will yield a single image as it is intended to do.[91]

Once visual contact is established with the external object and the visual flux undergoes the passion of coloring, that passion and all the visual information it conveys are transmitted back to the eye, where they reach the surface of the cornea. Referred to as the viewer (*aspiciens*) by Ptolemy, the cornea "assumes the nature of a convex mirror in terms of its shape and smoothness" (II, 16). Ptolemy's intent in styling the cornea thus is far from certain, but one plausible interpretation is that he regards the cornea as a sort of imaging-device in which an initial visual representation of the object is created.[92] If so, then it is at the cornea that the sensitive

---

[88]Ptolemy adverts to the Governing Faculty at several places in the *Optics*; see, e.g., II, 22–23; II, 76; II, 98; III, 61. For an extended discussion of this faculty and its relationship to the brain and, thus, perception and reasoning, see *On the Criterion*, chapters 13–16, in Long, "Ptolemy on the Criterion," pp. 207–213.

[89]That the optical source-point lies at the center of the ocular globe is never specifically said by Ptolemy, but it is tacitly assumed in II, 20 and IV, 4.

[90]On Ptolemy's estimate for the maximum angle at the vertex of the visual cone (90 degrees) and the consequent estimate of the visual field, see II, 24, n. 42 below.

[91]Note the incoherence of this model insofar as the unitary image in binocular vision seems to depend on the perfect coincidence of the bases of the visual cones, which would entail each eye's having precisely the same field of vision. In other words, binocular vision would do nothing to broaden the field of vision beyond what is comprehended by either eye singly, which is patently false.

[92]Here Ptolemy may be adopting the Democritean notion of *emphasis* in accounting for the initial act of visual imaging in the eye. Note that, in choosing the cornea (which is taken to be equivalent to the pupil), Ptolemy is either ignorant of or in disagreement with Galen's claim that the primary instrument of sight is the crystalline lens, which lies inward from the cornea;

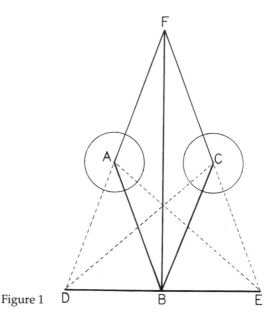

Figure 1

phase of vision begins. Meantime, angular and linear determinations of the spatial characteristics of the object are made at the source-points at the centers of the ocular globes. The resulting visual information is then passed to the Governing Faculty, which exercises its discriminative power (*virtus discernitiva*) in judging this information and, on that basis, making conceptual conclusions about the object under scrutiny.[93]

The visual system just described can be grasped quite easily in diagrammatic form. Figure 1, for instance, illustrates the principles of binocular vision as given by Ptolemy. From the two source-points **A** and **C** in the center of the eyes two visual cones, defined by the broken lines, are projected to the surface of object **DE**. Since the two cones share a common focus and a common base, their respective axes will meet at a single point, **B** in this case, on the object's surface. Likewise, corresponding rays **AE/CE** and **AD/CD** converge respectively on points **E** and **D** on that surface. Beyond points **A** and **C**, meanwhile, lies point **F**, from which common axis **FB** is projected

see *De usu partium* 10.1, trans. Margaret Talmadge May, *Galen on the Usefulness of the Parts of the Body*, vol. 2 (Ithaca, NY: Cornell University Press, 1968), p. 463.

  [93]Beyond the vague reference to "nervous activity," Ptolemy says nothing specific about the optic nerves, much less about their role in providing the pathway for the transmission of visual information to the Governing Faculty in the brain. Still, by his time the anatomy of the visual system was well known, and the nerves had been isolated and traced to their source in the brain; see, e.g., *De usu partium* 10.1, trans. May, p. 463.

to the same focal point **B** as the two common axes. As long as the three axes, **AB**, **CB**, and **FB** converge at the same point on the object, it will appear single. Figure 2 illustrates the perceptual system for a single eye. Acting in its capacity as seat of nervous activity, the Governing Faculty sends visual flux/pneuma to source-point **A** within the eye. From there the flux issues outward through pupil **FG** and cornea **HK** in the form of a cone. Reaching the surface of a properly dense and illuminated object, **B**, the visual cone forms common section **CD** with it, midpoint **E** of that common section being touched by visual axis **AE**. Sensing the color of **B**'s surface, the visual flux within the cone suffers the appropriate passion, which is then transmitted back to the cornea where it is presumably imaged as a point-by-point representation, much as it would be in a mirror. Meantime, on the basis of that passion, the spatial characteristics of the object are gauged in reference to visual axis **AE**. As a result, the object's place (according to the relative rightward and leftward obliquity of **AC** and **AD** with respect to axis **AE**, and according to the distance along **AE**), motion (according to changes in those parameters), shape (according to the relative lengths of **AE** and **AC**, **AE** and **AD**, etc.) and size (according to visual angle **CAD** and

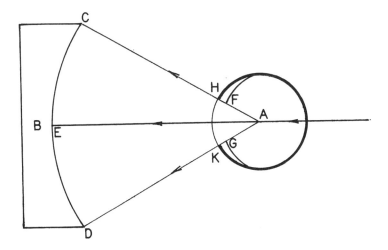

Figure 2

ray-length **AE**) are ascertained. The resulting determinations, in-cluding that of color, are scrutinized and judged by the Governing Faculty, which forms from them a complete perception of the object as it exists in space.

Under proper conditions, the visual process as outlined above will yield a correct perception of physical space and its constituent objects. But conditions are often less than optimal, and, depending on the abnormality, a variety of misperceptions or illusions can arise. These misperceptions can be gathered under two main head-ings: those due to objective factors and those due to subjective causes. On the objective side, if certain threshold conditions are ex-ceeded or not met, the visual faculty will not function as it should. Too much or too little illumination will cause an object not to be seen properly, as will too much or too little distance (II, 18–19). Thus, an overly bright object next to a moderately illuminated one can occlude it, whereas a square object seen at too great a distance can appear rounded (II, 90–91 and 97). Taken to an extreme, of course, a decrease in illumination or an increase in distance can cause the object to disappear entirely from sight. Restrictions in the flow of visual flux also have an adverse effect. Sight with one eye is thus poorer than with two since it lacks the double intensity of vi-sual radiation that comes with normal binocular vision (II, 18). By the same token, sight is impaired in myopia because of a diminished flow of visual flux, whereas in hyperopia that flow is interfered with by the optic humors (II, 85–87). A motion that is too fast, such as that of a rapidly spinning wheel, cannot be properly detected, nor can a motion that is too slow, such as that of the planets (II, 98). Move-ment also affects the perception of color, so that when a particolored wheel is rotated swiftly, the individual colors are no longer dis-cerned. Instead, the wheel takes on an intermediate hue that is blended from all the constituent colors (II, 96). Each of these mis-perceptions is due to a change in the objective conditions of sight rather than in the visual system itself. Even when the misperception stems from a diminution in the flow of visual flux, as in myopia and hyperopia, the cause is external or objective insofar as it is not the flux itself, but its intensity that is changed.

Of the misperceptions or illusions that are subjective in origin, some can be traced to the faculty of sight and some to the faculty of perceptual judgment. There are a few salient examples of the for-mer. For instance, accidental coloring of the visual flux can cause misperception, as happens when things viewed in a bronze mirror take on the cast of the reflecting medium (II, 108–109). Afterimage is another case of accidental coloring in which the initial passion of coloring in the flux outlasts its cause. The continuing passion thus affects the perception of subsequent colors, as happens when we shift our gaze from a brightly colored object to a white field (II, 107). Diplopia, reflection, and refraction are all instances in which an illusion is created on account of some distortion in the visual flux, which is either diverted from its customary convergence or broken. The resulting illusion in all three cases involves a misperception of where the object actually is, since it appears to be where its displaced image lies (II, 114–118). Another example of misperception specific to the visual faculty is the oculogyral illusion in which the viewer spins for awhile, stops, and immediately afterward sees the surrounding environment spin. In this instance, Ptolemy explains, the flux continues to sweep along inertially, even though it has actually stopped. This inertial sweep is then perceptually transferred to the environment, which thus appears to move for awhile (II, 121).

The final group of illusions, those due to false perceptual judgments, includes an interesting assortment of misperceptions, one of which pertains to illusionism in painting. By a judicious use of color-contrasts, or even bare outlining, an artist can fool the eye into perceiving a spatial representation in what is in reality nothing more than a multi-hued plane (II, 10–11, 124, 127–128). Perceptual misjudgment can also arise when we are being conveyed in a vehicle, such as a boat, that is moving so smoothly we do not detect our own motion. In such a case, we often impute the motion perceptually, albeit not intellectually, to the passing scenery (II, 132). Even our recognition of image-reversal in plane mirrors involves a perceptual misjudgment. Thus, when we see ourselves face-to-face in a mirror, we are initially impelled by perceptual misjudgment to assume that our true left is represented at the right side of the image, because when we actually do face someone else, his left side would be seen by right-hand rays (II, 138). The last, and surely the most interesting, of the perceptual misjudgments to be considered here is the so-called moon illusion, in which celestial objects appear larger the nearer they are to the horizon. Although the details of Ptolemy's explanation are somewhat vague, its general grounding in perceptual psychology is clear enough (III, 59).

This brief description of visual illusions brings our examination of Ptolemy's basic account of vision to a close. What is perhaps most striking about that account is how tightly organized it is. Ptolemy progresses from topic to topic in a sequence so carefully and logically ordered that its apparent naturalness masks the exquisite artifice involved. Remarkable, too, is the comprehensiveness of his account. Not only does Ptolemy examine the objective and subjective aspects of sight in great detail, but he extends his analysis to visual illusions, all within the framework of a relatively sophisticated psycho-physiological system. Equally noteworthy is the coherence of his account. This point is best illustrated by the way in which the psycho-physiological framework of analysis is extended to the very typology of visual illusions, some of which are objectively or physically determined, and some of which are subjectively determined, either by false sensation or by false perceptual judgment. Finally, and by no means least important, we cannot help but be struck by how effectively Ptolemy uses common experience to bolster theoretical conclusions. That he is as consummate an observer as he is an organizer is clear at virtually every turning in the *Optics*.

In highlighting the strengths of Ptolemy's account, however, we should not be blind to its shortcomings. For one thing, Ptolemy tends at times to overdetermine the phenomena he explains. A clear instance can be seen in his account of distance-perception. By rooting it in an intuitive grasp of ray-length and thus ignoring the clues provided by surrounding objects, he fails entirely to acknowledge that the perception of distance might be determined relatively rather than absolutely. Similarly, by basing his analysis of shape-perception (i.e., of concavity or convexity) on the absolute measure of ray-length, he ignores the crucial role of color-contrast in the determination. This failure is all the more egregious in view of his later acknowledgement that color-contrasts create the illusion of depth in paintings. So, too, his explanation of size-perception, while not lacking in ingenuity, depends too heavily on absolute determinations, particularly of slant and distance.

Ptolemy's analysis of binocular vision has its share of problems too. To start with, by treating the visual field as a plane rather than a curve, Ptolemy is bound to misconstrue the fusion of optical images.[94] His explanation of axial focus and the resulting convergence of corresponding rays in the two visual cones is thus far more determinate than it should be, requiring the perfect coincidence of both visual bases. As we have seen, such perfect coincidence means that doubling the eyes does nothing to broaden the visual field. Another problem with Ptolemy's analysis of binocular vision has to do with purpose. Ptolemy sees the primary function of binocular vision in terms of visual acuity, so that in bringing two eyes to bear on a given object, the viewer brings twice as much visual flux to bear on it. By focusing on this not implausible explanation, though,

---

[94]This point is noted and discussed in Lejeune, *Euclide et Ptolémée*, pp. 145–147 and 169–72, and Simon, *Le regard*, pp. 144–147.

Ptolemy misses the relatively obvious point that binocular vision is specifically designed to aid in depth-perception.[95] Still, for all its deficiencies, the very attempt on Ptolemy's part to make sense of binocular vision marks a signal advance beyond any of his sources.

Ptolemy's Analysis of Reflection

Ptolemy opens his study of reflection by setting forth three basic principles for determining image-location in mirrors. That these principles are not original to him is clear from their appearance in one form or another in the *Catoptrics* attributed to Euclid.[96] Indeed, as far as basic theoretical elements are concerned, Ptolemy's analysis of reflection offers very little in the way of originality. What sets his analysis definitively apart from that of his predecessors is the observational and organizational skill he brings to bear upon it.

As articulated by Ptolemy, the first principle of image-location "asserts that objects seen in mirrors appear along the extension of the [incident] visual ray that reaches them through reflection" (III, 3). In other words, the image of any visible point appears to lie below the mirror along the imaginary extension of the visual ray passing from the center of sight through the point of reflection. This principle is empirically confirmed by blocking the spot where the image of the visible point is seen in the mirror (i.e., the point of reflection) and thus determining that the image lies in a direct line with the center of sight and that spot (III, 4). The second principle states "that particular spots [on a visible object] seen in mirrors appear on the perpendicular dropped from the visible object to the mirror's surface and passing through it" (III, 3). To verify that the image does indeed lie along this line (the so-called cathetus of reflection), we need only stand a thin, straight rod orthogonal to a plane mirror and see that the rod's image will always lie in a straight line with the reflecting surface (III, 4). It is important to note, however, that the cathetus is normal to the actual reflecting surface only when the mirror is plane; otherwise, it is normal to the surface tangent to the mirror's surface at the point of reflection. In the case of spherical convex or spherical concave mirrors, then, the cathetus passes through the center of curvature (III, 3).

---

[95]One fairly straightforward clue to this function is the fact that, when viewing a rounded object (e.g., a spherical or cylindrical object) with both eyes, we see more of it than we do with one eye. This point, which is not acknowledged by Ptolemy anywhere in the *Optics* in its present form, is specifically addressed by Euclid in propositions 24–26 of the *Optics*.

[96]See *Catoptrics*, definitions 3 and 4, propositions 1 and 16–18.

It follows from these two principles that the image of any visible point lies at the intersection of the extended incident visual ray and the cathetus of reflection (III, 5). But this intersection-point is not yet fully determinate. A third and even more fundamental principle is necessary: namely, "that the disposition of the reflected ray connecting pupil to mirror and mirror to visible object is such that each of the ray's two branches joins at the point of reflection and that both form equal angles with the normal dropped to that point" (III, 3). This principle of equality between the angles of incidence and reflection completes the determination of image-location by specifying precisely where the point of reflection must lie and, therefore, precisely how far below the mirror's surface the image must be situated along the cathetus of reflection. Yet, in order to give that determination as much rigor as possible, Ptolemy makes explicit a crucial underlying assumption that none of his predecessors seems to have recognized. All of the aforementioned ray-segments and normals, he tells us in III, 5, lie in one plane (i.e., the plane of reflection), and this plane is normal either to the plane of the mirror or to the plane tangent to the mirror at the point of reflection.[97]

With the analytic framework thus established, Ptolemy launches into his famous experimental confirmation of the equal-angles law. The procedure is described in III, 8–12. A circular bronze plaque of "moderate" size is marked off into quadrants, each quadrant in turn being divided into 90 degrees. One of the diameters subdividing the quadrants is inscribed into the plaque's face, and to that diameter a convex circular arc and a concave circular arc are inscribed tangent at the centerpoint. The diameter normal to that diameter is also inscribed. Meantime, three strips of clear, polished iron are fashioned in such a way that one of them is straight and the other two are molded precisely to the curvature of the circular arcs inscribed on the face of the plaque. A diopter is attached to the plaque in one of the quadrants so that it can be rotated along the plaque's circumference. The iron mirror-strips are then applied vertically by a pin to the centerpoint of the plaque, the straight one lying along the inscribed diameter and the curved ones lying along their respective arcs. Finally, a colored marker is attached to the plaque in the quadrant next to that of the diopter so that it too can be rotated along the plaque's circumference.

Now that the apparatus is properly set up, the diopter is placed at some determined point along the plaque's circumference within

---

its quadrant, and through it a line-of-sight is established with the central pin-cum-marker holding the mirror. The colored marker in the neighboring quadrant is then rotated along the plaque's circumference until its image coincides with the marker at the center— i.e., until its image lies directly behind the pin along the line-of-sight established through the diopter. When the arcal distance of the diopter with respect to the normal is compared with that of the colored marker, the two will invariably be equal, no matter which kind of mirror (straight, convex circular, or concave circular) is installed. In short, the law of equal angles holds for all three kinds of mirror and thus, presumably, for all others.

So much for the theoretical principles of reflection and their empirical verification. What about the physical explanation? Ptolemy appeals to the same dynamic model of projection that he employs in his account of unimpeded visual radiation. In the case of reflection, however, the ray-projectile is assumed to strike a perfectly resistive surface so that it is forced to rebound. The angle of rebound is, of course, dependent on the angle of incidence, "for projectiles are scarcely obstructed by objects they strike at tangents, whereas they are obstructed to a considerable extent by objects that resist them [directly] along the line of projection" (III, 19). No matter how tangential the incidence, though, the resulting impact does weaken the projection. That, Ptolemy assures us, is why images appear dimmer in reflection than the original objects would at the same distance.[98] In choosing this dynamic model, Ptolemy has a clear precedent, if not an actual source, in Aristotle's *Problemata* and, more to the point, Hero of Alexandria's *Catoptrics*.

If, indeed, reflection is due to physical impact and rebound, then surely the ray ought to feel the impact and thereby "see" the reflecting surface rather than the image in it. It does not, apparently, because, when the visual flux impinges on an opaque body, its "power" enters into that body (II, 4). In order to see it, then, the flux must somehow become fixed onto the visible object it strikes. But, far from becoming fixed onto the reflecting surface, the visual flux rebounds cleanly from it, so it fails to sense not only the surface itself but also its own breaking at that surface (III, 15). Unable to perceive that the flux has broken, the visual faculty is deluded into believing that the object it is actually seeing along the reflected ray

---

[98]III, 22 and 64; Ptolemy and his successors for the next fourteen centuries or so were convinced that reflection naturally weakens the radiated effect because the reflecting capacity of the mirrors they had available was so inferior to that of mirrors developed since the late Renaissance.

is being seen along a perfectly unbroken incident ray (III, 16). The object therefore appears to the visual faculty to lie below the mirror's surface. But this appearance is entirely illusory, because there is in fact nothing "behind" the mirror to see. So the images seen by reflection are mere figments of perceptual misjudgment; like the sound of the proverbial falling tree, they have no independent existence apart from the perceiver.[99]

In the paradigm case of reflection from plane mirrors, image-location is perfectly symmetrical with respect to the object and the eye. That is to say, if image, object, and eye are all taken as points, the distance from eye to object along the reflected ray-couple is precisely the same as the distance from eye to image along the incident ray and its imaginary extension below the mirror.[100] Now, if we assume that objects with spatial extension consist of innumerable visible points (Ptolemy uses lines for analytic convenience), then it follows that each point's image is symmetrically located with respect to both its generating point-object and the eye. In that case, the composite image must also be symmetrically located with respect to both the generating composite object and the eye. It will therefore constitute a perfect likeness of that object in terms of all its spatial characteristics, from apparent distance, through apparent shape and size, to basic orientation. To demonstrate this point, Ptolemy concludes his analysis of plane mirrors with a series of five propositions (III, 68–96) showing that images projected in plane mirrors suffer no distortion whatever in terms of these characteristics.

Having accounted for image-formation in the first and simplest kind of mirror, Ptolemy is prepared to address the two remaining types of basic mirror, spherical convex and spherical concave. Unlike plane mirrors, of course, both of these types distort the images formed in them, yet of the two, the spherical convex is simpler because the distortion it creates is both more regular and less radical. Or, as Ptolemy puts it, "just as plane mirrors have a greater capacity than other kinds for preserving the integrity [of images], so also convex mirrors have a greater capacity than concave mirrors for preserving [such] integrity" (III, 98). Accordingly, Ptolemy undertakes the analysis of spherical convex mirrors first. He begins by

---

[99]Simon makes this same point rather elegantly in *Le regard*, p. 164: "[The image] is invariably distinguished as an *appearance* without reality [so that], far from being conceived as a device apart from the viewer, obeying objective physical laws, the mirror is no more than a snare for sight, a creator of phantasms" (my translation, his italics).

[100]Likewise, by extension, the distance along the cathetus of incidence from the center of sight to the reflecting surface is equal to that along the cathetus of reflection from the object-point to the reflecting surface, as well as from that surface to the image-point.

demonstrating that in such mirrors the image, which is located at the intersection of the incident ray and the cathetus of reflection, always lies behind the reflecting surface (III, 99–103). It is nonetheless possible under certain conditions for the incident ray and the cathetus of reflection to meet on or above that surface, in which case the surface itself will block the intersection-point from view along the incident ray or its extension. Even so, Ptolemy hastens to add, the visual faculty will perceive the image as if it lay behind the surface (III, 107–109). This seemingly absurd claim reveals a crucial assumption on Ptolemy's part: as long as the reflected visual ray reaches the visible object, that object must somehow be perceived, even when the image-location precludes such a possibility.[101] In such cases, the visual faculty adjusts the anomalous perception to its norms—hence the perceptual replacement of the image behind the mirror where it ought to be. (This point will be more clearly illustrated below when we discuss image-formation in concave mirrors.) In the next two theorems (III, 110–120), Ptolemy demonstrates that, although the object-image distance is less than the eye-object distance in spherical convex mirrors, the image appears smaller than the object. He then goes on to show how the curvature of the mirror affects the apparent shape of the image, so that, for instance, a plane object (represented by a straight line), will always appear convex, and so forth (III, 121–126). In the final theorem (III, 127–130), he demonstrates that the right-to-left and top-to-bottom orientation of the image in convex spherical mirrors is the same as it is in plane mirrors.

In progressing from plane and convex spherical to concave spherical mirrors, the main subject of book 4, Ptolemy takes the analysis to a much higher level of complexity. For a start, unlike plane or convex spherical mirrors, concave spherical mirrors can produce multiple images of the same point to a given center of sight, depending on where the object-point and eye are located. Lacking a general method for determining precisely what combination of reflections, if any, will result from what arrangement of point-object and eye, Ptolemy undertakes a case-by-case analysis that occupies 21 theorems. In the first theorem (IV, 3–5), the center of sight and the center of curvature are assumed to coincide, so that the center of sight also constitutes the object-point. Under these conditions, reflection will occur from every point on the reflecting surface. This

---

[101]To quote Simon again: "For Ptolemy *a visual ray that reaches an object after reflection from a mirror always sees something of it*, because it is in the nature of the visual ray to see what it reaches" (*Le regard*, p. 164—my translation, his italics).

theorem is followed by a group of seven (IV, 6–25), in which the center of sight and the object-point are assumed to lie upon the sphere's diameter on either side of the center of curvature. Depending upon whether the eye-to-centerpoint and object-to-centerpoint distances are equal, reflection will occur either from every point on the great circle whose plane is normal to the aforementioned diameter or from every point on a smaller circle parallel to that great circle. A third group of theorems, consisting of thirteen altogether (IV, 22–61 and 84–96), covers all cases in which the center of sight and the object lie upon a chord of the sphere on either side of the radius normal to that chord. In this case, reflection can occur only within the plane containing the chord and radius; and, depending on the relative placement of eye and object, as well as on which of the two arcs cut off by the chord faces the eye, a maximum of three reflections can take place.

Not only do spherical concave mirrors differ from plane and spherical convex mirrors in terms of the number of point-images they can project, but they also differ in terms of where those images can be projected. As Ptolemy lays the problem out in Theorem IV.22 (IV, 63–65), there are four distinct possibilities according to the relationship between the ray of incidence and the cathetus of reflection. On the one hand, the two lines can intersect behind the mirror or between the mirror and the center of sight, in which case the images produced will be perceived clearly.[102] On the other hand, the two lines can intersect at or behind the center of sight, or they can fail to intersect at all. Whichever the case, the image produced will appear at best quite nebulous and at worst so indistinct as to be virtually imperceptible. That any image whatever is perceived in these instances requires some explanation, however, and Ptolemy gives us a by-now familiar one: just as before, when the intersection-point lay above the surface of a convex spherical mirror, so now, the visual faculty is forced to transfer the image to a more suitable location. Accordingly, when there is no intersection-point, the image will seem to coalesce with the mirror's surface and take on its color, whereas, when the intersection-point lies beyond or at the center of sight, the image will be perceived to lie in front of the mirror, between the eye and the reflecting surface.[103]

---

[102]The first kind of image is what we today call "virtual"; the second we call "real." For Ptolemy, of course, this distinction would have no meaning whatever, since no image can be real, and, given his theoretical commitment, it is doubtful that he would have changed his mind even had he seen that real images can be projected onto a screen.

[103]IV, 69–70; clearly, Ptolemy is struggling here to make sense of a visual experience that is inconsistent with its geometrical analysis: we actually do see a sort of nebulous image when

To verify these assertions, Ptolemy proposes two simple experiments, both of them requiring the apparatus used earlier for confirming the equal-angles law (i.e., the circular plaque with pivoting
diopter and colored marker). In the first experiment (IV, 71–73), the
concave mirror-strip is attached to the centerpoint, the incident line-
of-sight is established through the diopter to the centerpoint, and
the colored marker is moved to and fro along the corresponding line
of reflection. The images thus formed will be as described in the preceding theorem according to precisely where that marker is situated
along the line of reflection. The second experiment (IV, 74–80) employs the same apparatus. A thin rod is placed normal to the mirror
and the eyes are situated so that the normal from the midpoint of
the line connecting them coincides with the rod along its length. After describing the appearance of the image under these conditions,
Ptolemy goes on to explain its actual formation. In the course of that
explanation, he is compelled at times to adjust the geometry to fit
the appearance, in particular, the appearance of image-continuity
that is contravened by the pure geometrical explanation.

The various determinants of image-formation having been established, Ptolemy can now bring his study of spherical concave mirrors to a close by analyzing image-distortion according to the two
definite image-locations described earlier: i.e., behind the mirror
and between the mirror and the eye. Of the seven theorems devoted
to this analysis, two deal with the apparent distance of the image either behind or in front of the mirror (IV, 109–119), two with apparent size (IV, 120–129), one with apparent shape (IV, 130–141), and
two with top-to-bottom and right-to-left orientation (IV, 142–151).
Having thus completed his account of spherical concave mirrors,
Ptolemy concludes the fourth book of the *Optics* with a brief and
somewhat *ad hoc* account of composite mirrors. Accordingly, after
explaining image-distortion in convex and concave conical mirrors
(IV, 165–170), he discusses the projection of multiple images from
various juxtapositions of plane mirrors, either ranged about the
viewer (IV, 171–173) or placed aslant to one another in odd or even
combinations (IV, 175–182).

---

the intersection lies at or behind the eye. As for the case when there is no intersection, Ptolemy
assumes that the image coalesces with the mirror's surface and takes on its color in order to
account for the fact that we actually see nothing. Since, however, we must see something-
whenever the reflected ray strikes a visible object, as indeed it does in this case, Ptolemy is
forced to invent a ghost appearance. In such instances, when experience seems to contravene
"objective" analysis, we must explain things "not according to particular aspects, but according to the essential and true nature of the phenomenon" (IV, 109). In short, we must adjust the salvation to fit the actual appearances.

In spite of this decidedly anticlimactic conclusion, Ptolemy's overall study of reflection in books 3 and 4 of the *Optics* represents a significant achievement in terms of both analytic structure and empirical verification. On those grounds, certainly, it constitutes a marked improvement over its sources. True, his account of image-formation in concave mirrors is somewhat off the mark by modern standards, but, as Lejeune quite rightly observes, the deficiency of that account is due to Ptolemy's theoretical commitment to the visual ray model and, therefore, to an analytic structure based first and foremost on the eye as source of radiation.[104] To demand a better account on his part is therefore to demand that he change his theoretical commitment from the visual ray model to the light-ray model of modern physical optics. This point bears on what may seem to be a puzzling omission in his study of mirrors. Nowhere does Ptolemy mention burning mirrors, an omission that may seem all the more puzzling in view of the last proposition in the *Catoptrics* attributed to Euclid, where it is falsely asserted that solar rays can be focused by concave spherical mirrors.[105] But since the problem of such focus involves light rays rather than visual rays, and since image-formation is therefore not at issue, the topic of burning mirrors is wholly irrelevant to Ptolemy's purpose in the *Optics*.

Ptolemy's Analysis of Refraction

As we noted earlier, Ptolemy is well aware of the systematic relationship between reflection and refraction, and he does not hesitate to seize upon the implications of this relationship in framing his account of refraction in book 5 of the *Optics*. As concerns physical explanation, for instance, he draws freely upon reflection for his basic model of refraction, seeing in each a special case of the other (V, 1). Thus, whereas the reflecting interface blocks the ray/projectile's passage entirely, the refracting interface blocks it only partially, thereby slowing it down and diverting it toward the normal (V, 36).[106] In both cases, however, Ptolemy tacitly assumes that the dynamic effect at the interface is instantaneous so that,

---

[104]*Recherches*, pp. 77 and 108.

[105]See *Catoptrics*, prop. 30; this false claim about focusing is made in the second part of the theorem.

[106]Hero of Alexandria uses virtually the same explanatory model as Ptolemy, treating refraction as a case of imperfect rebound from imperfectly solid bodies; see *Catoptrics*, chapter 3. Suffice it to say, however, that this model applies only in the case of passage from an optically rarer into an optically denser medium; otherwise the ray is diverted *away* from the normal (V, 2).

whether the ray is forced to rebound or is merely diverted, it suffers no aftereffect whatever. That is why, once refracted, the ray proceeds in a perfectly straight line through the optically denser medium rather than continually curving toward the normal as would a real physical projectile hurled obliquely into water or some other resisting medium.

Reflection and refraction are systematically linked by more than physical cause. As Ptolemy points out in V, 3, they also share the same basic principles of image-location. In both cases, therefore, the image is located where the continuation of the incident ray intersects the cathetus dropped to the reflecting/refracting surface from the object. In both cases, as well, the ray with all its appurtenant normals lies in a single plane orthogonal to the plane of reflection/refraction. And, most important of all, in both cases "the angles . . . bear a certain consistent quantitative relation to one another with respect to the normals" (V, 2). In reflection, of course, that relationship is one of equality. But what might it be in refraction? To answer this question, Ptolemy proposes a series of three experiments to ascertain the precise amount of bending when the ray passes into various refracting media.

The basic apparatus for all three experiments consists of the circular bronze plaque, divided into quadrants, that was used before in the study of mirrors, and a watertight semicylindrical vessel of the same curvature. The first experiment (V, 8–11) is designed to measure refraction of the visual ray from air to water. The plaque is stood upright in the semicylindrical vessel, and water is poured into the vessel until the waterline coincides with one of the plaque's two incised diameters. The other incised diameter is thus normal to the waterline. A tiny marker is attached to the centerpoint of the plaque, another marker is placed at the normal on the arc above the waterline, and a line of sight is established between the two markers along that normal. Then a third marker is placed upon the arc below the water so that it appears to fall in line with the other two. Under these conditions, that third marker will also lie on the normal, so no refraction will have occurred. Next, the marker above the water is placed at an angle of 10 degrees from the normal. The angle thus marked represents the angle of incidence i. Then, while sighting along the line between this marker and the one at the centerpoint, the observer moves the third marker along the opposite quadrant below the water's surface until it appears to fall in line with the other two. Its angle, which represents the angle of refraction r, is then measured. The same procedure is repeated at ten-degree intervals for i all the way to 80 degrees. The resulting measurements according to Ptolemy are as follows: $i = 10°$, $r = 8°$; $i = 20°$, $r = 15.5°$; $i = 30°$, $r = 22.5°$; $i = 40°$, $r = 29°$; $i = 50°$, $r = 35°$; $i = 60°$, $r = 40.5°$; $i = 70°$, $r = 45.5°$; and $i = 80°$, $r = 50°$.[107]

In the next two experiments, a glass semicylinder whose diameter is somewhat less than that of the plaque is attached to the plaque so that its plane surface coincides with one of the incised diameters. In the first of the two experiments (V, 14–18), the refraction is from air to glass. The marker used to measure i is again placed at 10-degree intervals from the normal, a line of sight is established between it and the marker at the plaque's centerpoint, and the marker in the quadrant below the glass semicylinder is moved until it appears to line up with the other two. Its arc along that quadrant with respect to the normal is then measured.[108] As recorded by Ptolemy, the resulting comparative measurements are as follows: $i = 10°$, $r = 7°$; $i = 20°$, $r = 13.5°$; $i = 30°$, $r = 19.5°$; $i = 40°$, $r = 25°$; $i = 50°$, $r = 30°$; $i = 60°$, $r = 34.5°$; $i = 70°$, $r = 38.5°$; and $i = 80°$, $r = 42°$. In the third and final experiment (V, 20–21), the refraction being measured is from water to glass. The plaque, with its attached glass semicylinder, is placed upright in the water-filled semicylindrical vessel so that the plane surface of the glass semicylinder and the waterline coincide. Then, assuming that the vessel itself is constructed of glass, the observer places the marker on the quadrant below the water at 10 degrees from the normal, establishes a line-of-sight along it to the marker at the plaque's center, and then moves the marker in the opposite quadrant above the glass semicylinder until the three markers appear to line up. The arcal distance from the normal of the marker above the glass semicylinder is then measured. The same procedure is repeated at ten-degree intervals. The results by Ptolemy's tabulation are as follows: $i = 10°$, $r = 9.5°$; $i = 20°$, $r = 18.5°$; $i = 30°$, $r = 27°$; $i = 40°$, $r = 35°$; $i = 50°$, $r = 42.5°$; $i = 60°$, $r = 49.5°$; $i = 70°$, $r = 56°$; and $i = 80°$, $r = 62°$.

Much has been made of these experiments and the relative accuracy or inaccuracy of their results. But to compare these results with those based upon the sine-relationship between i and r in order to determine how close Ptolemy came to the modern law of refraction is to miss the point entirely. In all three experiments Ptolemy was not collecting raw data with a mind uncluttered by presupposi-

---

[107]Although I have used decimal format here for the sake of convenience, the actual tabulations provided by Ptolemy would doubtless have been in sexagesimal format; for details, see V, 11, note 9 below.

[108]As in the previous case, of course, so in this and the subsequent one, when the line of incidence lies along the normal itself, no refraction occurs.

tions. On the contrary, like any practicing scientist worth his salt, he was tabulating his results with a working hypothesis in mind. Having already discussed this point at length in my article "Ptolemy and the Search for a Law of Refraction," I will offer only a synopsis here. First, even the most cursory glance at the results for r listed above will reveal an obvious pattern: they are all rounded off according to increments of half a degree. Second, when we examine the progression of r's in all three cases, we see immediately that those progressions continually regress as i passes from $10°$ to $80°$. Thus, for example, in the air-to-water series, the progression runs 8; $8 + (8 - .5 = 7.5) = 15.5$; $15.5 + (7.5 - .5 = 7) = 22.5$, etc. Underlying this series is an ideal constant progression of $r = 8°$ for every $i = 10°$, but that ideal progression is countered by a constant regression of half a degree for every ten-degree increment of i. Accordingly, while the ideal holds for the first 10 degrees of incidence (i.e., $r = 8°$ when $i = 10°$), the second eight-degree increment of r that would be expected for the next 10 degrees of incidence is reduced by half a degree to $7.5°$, and so on to $4.5°$ when $i = 80°$. As Govi observed over a century ago, this algorithm reduces to a rule of constantly diminishing first differences subject to decrease by constant second differences, a rule that Otto Neugebauer has shown to have been applied to astronomical calculations by the late Babylonians.[109] Thus, the working hypothesis that Ptolemy had in mind was borrowed from astronomy and consisted in the supposition that the apparently irregular relationship between i and r in refraction could ultimately be reduced to regularity through a simple saving principle: constant second differences.

As both Govi and I have shown, a precise law of refraction can be derived quite easily from Ptolemy's obviously doctored results. But two things are worth noting about this law, whether it takes Govi's form ($r = ai - bi^2$) or mine ($r = R - (n^2d_2 - nd_2)/2$). First, it is not the sine-law, nor does it approximate it. Second, even had Ptolemy intuited it in either of the two versions given above, he could not have formulated the law properly because of the limitations of Euclidean proportionality theory, in terms of which he would necessarily have had to express it. In short, without a proper algebraic framework within which to construct it, he could not possibly have arrived at the requisite formulation. *Faute*

---

[109]For Govi's discussion, see *L'Ottica*, pp. xxiv–xxxii; for the application of this type of algorithm to late Babylonian (i.e., Seleucid) astronomy, see Otto Neugebauer's discussion of the linear zigzag function in *The Exact Sciences in Antiquity* (second edition, 1957; rpt. New York: Dover, 1969).

de mieux, then, the law that Ptolemy eventually does provide amounts to the weak generalization that, as *i* increases, so does the difference between *i* and *r* (V, 34).

Even before reaching this inconclusive conclusion, however, Ptolemy departs abruptly from the main line of inquiry to take up the issue of atmospheric refraction and its effect on astronomical observation (V, 23–30). The first point he establishes in this regard is that atmospheric refraction is caused by the bending of the visual ray away from the normal as it passes from the denser terrestrial atmosphere into the rarer ether of celestial space. Therefore, whenever a celestial body is observed, it will invariably appear to lie closer to the viewer's zenith than it actually is, because the incident line-of-sight along which it appears is necessarily higher than the refracted ray along which visual contact is actually made with it. By extension, then, the closer to the zenith the body actually lies, the less the refraction, until it diminishes to nothing along the zenith itself (V, 25–26). Ptolemy's second point follows from the first. Given that the effect of atmospheric refraction is greatest at the horizon and diminishes continually toward the zenith, then the arc that any celestial body makes as it moves from rising to setting during the course of a night will be perturbed accordingly (V, 27–29). Ptolemy is far from blind to the practical implications of this analysis. In order to estimate the effect of refraction for bodies such as the sun and moon whose distances are already well-determined, we need only know how high the interface between atmosphere and ether is (V, 30). Unable to determine this height, however, Ptolemy sadly concludes that the sought-after determination of atmospheric refraction is impossible.[110]

Resuming his normal line of inquiry after this brief excursus into the subject of atmospheric refraction, Ptolemy offers an experimental confirmation of the principle of reciprocity that culminates with the vague "law" of refraction mentioned earlier (V, 32–34).[111] This he follows with an ineffectual stab at demonstrating that the angular relationship between *i* and *r* in refraction cannot be the same as it is in reflection (V, 38–45). Then, hewing faithfully to

---

[110]In "Archimedes and the Pseudo-Euclidean *Catoptrics*," pp. 98–99, Knorr claims that, given certain parameters with which he was well acquainted, Ptolemy could easily have estimated the effect of atmospheric refraction at the horizon to be somewhere between roughly 15' and 30'. Even so, Ptolemy might well have regarded the resulting estimate as too tentative for practical application.

[111]According to the principle of reciprocity, the relationship between *i* and *r* in refraction through a given interface is reversed when the direction of passage is reversed: i.e., the original *i* = the new *r*, and vice-versa.

the pattern of analysis already established for reflection, Ptolemy takes up the problem of image-location for point-objects seen through plane, spherical convex, and spherical concave refracting interfaces (V, 46). He begins with a puzzling series of four propositions designed to ascertain if and when the cathetus of incidence (i.e., the normal dropped from the center of sight) and the refracted ray might intersect below the refracting interface (V, 47–54). As it turns out, the two lines can intersect under two specific conditions: first, when the refracting surface is convex and the ray is refracted toward the normal—i.e., when it passes into a denser medium (V, 49–50); and second, when the refracting surface is concave, the center of curvature lies between that surface and the center of sight, and the ray is refracted away from the normal (V, 53–54). What makes this entire analysis so puzzling is that whether the cathetus of incidence and the refracted ray meet or not is wholly irrelevant to the determination of image-location.

The relevant intersection, of course, is between the cathetus of reflection and the continuation of the incident ray, and it is to the conditions under which this intersection occurs that Ptolemy devotes the next three theorems. In the first case, when the refracting surface is plane, the two lines always meet no matter which way the ray is refracted (V, 57). In the second case, when the refracting surface is convex, it is possible for the two lines not to intersect when the ray is refracted toward the normal (V, 58–59). In the final case, when the refracting surface is concave, two distinct outcomes are possible (V, 60–61). On the one hand, when the center of sight lies between the center of curvature and the refracting surface, the intersection will always occur no matter which way the ray is refracted. On the other hand, when the center of curvature lies between the center of sight and the refracting surface, the two lines will not meet when the ray is refracted away from the normal. When the two lines fail to intersect, as is the case in this last instance, the image "is shifted to the common intersection of the normal and the refracting surface" with which it coalesces (V, 63). In other words, as in reflection, so in refraction, even if no image can actually be seen, one has to be invented as long as the refracted ray makes contact with the visible object.

As in reflection, so also in refraction, images are subject to distortion, depending on the shape of the refracting interface. In refraction, however, the refracting interface's shape is not the only factor involved; even plane surfaces create distortion, and the direction of refraction is a crucial determinant. To make sense of all this empirically, Ptolemy suggests the construction of three hollow glass vessels, one roughly cubical in shape, another cylindrical, and yet another roughly cubical with one of its sides indented so as to form a concave semicylinder (V, 67). We start with the cubical vessel. Filling it with water, we stand a ruler vertically in the water so that it is partly submerged. Then, looking at the ruler straight on, we see that, although the portion of it above water stands directly in line with the submerged portion, this latter portion appears nearer and larger (V, 68).

Explaining this appearance on the basis of point-objects and point-images, Ptolemy demonstrates that, whenever the visual ray passes into a denser medium, the intersection of the extended incident ray and the cathetus of refraction, which defines image-location, lies closer to the refracting surface, and thus to the eye, than the intersection of the refracted ray and the visible point (V, 70–72). Conversely, if the visual ray passes into a rarer medium, the situation is reversed so that the image appears commensurately more distant than its generating object (V, 74). As Ptolemy goes on to show in the next two theorems (V, 75–78), this holds not only for point-objects but also for objects with extension (as represented by lines), which appear to enlarge or shrink according as the apparent distance diminishes or increases. Nonetheless, Ptolemy concludes in V, 79–82, such distortions in apparent distance and size do not extend to shape; flat objects appear flat under the conditions of refraction, as indeed convex and concave objects appear convex and concave under such conditions. Ptolemy is in fact wrong. Shape does get distorted to some extent, even when the refracting interface is plane, but his failure to realize this is understandable in view of his ignorance of the correct law of refraction as well as his dependence upon reflection as the basic model for refraction.[112]

As expected, the analysis now shifts to refraction through convex interfaces, starting with a description of what is seen when the ruler is placed upright and moved to various positions in the water-filled cylindrical container (V, 83). Again, as expected, Ptolemy follows the verbal account with two theorems showing precisely how point-images are formed in refraction through a convex surface for both directions of radial passage (V, 84–87). However, before the second theorem is completed, the treatise comes to an abrupt halt. Even so, given the clear pattern of analysis established to this point, we can easily predict that Ptolemy's next step would be to address the problem of size-distortion in the case of refraction through convex

---

[112]In other words, since reflection from plane mirrors does not distort shape, then, by extension, refraction through plane surfaces should not do so either.

interfaces. This time, perhaps, bearing in mind the effect of reflection from convex mirrors, he would also take into account the slight distortion in shape that accompanies such size-distortion. And, finally, turning to the case of refraction through a concave interface, Ptolemy would follow the same sequence of steps, using the indented vessel to provide the observational basis for a theorematic account of image-location and distortion.

There is of course no way of knowing whether the fifth book or, for that matter, the treatise as a whole would have ended here with the analysis of refraction through concave surfaces. Given his focus on visual appearances, Ptolemy certainly should have, and in fact may have, taken into account such basic "meteorological" phenomena as rainbows and haloes, covered by Aristotle in his *Meteorologica*.[113] Precisely where such an analysis would have fit in the *Optics* is a matter of pure speculation. It might have been included in the now-lost first book as part of Ptolemy's supposed discussion of color, or it might have been relegated to a grab-bag sixth book, where he would have addressed a variety of ancillary topics, including burning mirrors.[114] Whatever the case, the text of the *Optics* that has come down to us, however deficient it may be, reveals a maturity of insight—particularly in the analysis of refraction—that puts it among the most remarkable scientific works of Antiquity.

4: The Historical Influence of the *Optics*

Late Antiquity

We have already noted in our discussion of its textual history that Ptolemy's *Optics* is cited in very few contemporary or near-contemporary sources. In fact, we know of only two references

---

[113]See n. 117 below. In his *Apology*, Ptolemy's contemporary, Apuleius of Madaura, suggests that, aside from accounting for reflection and its various appearances, the optical theorist also ought to account for "why hollow mirrors, if they are held opposite to the sun, ignite the tinder set near; [and] why it happens that arcs are seen in clouds in various ways, (or why) two suns (are seen) with rivalling appearance," trans. Wilbur Knorr, "Archimedes and the Pseudo-Euclidean *Catoptrics*," p. 31. It seems to have been traditional from late Antiquity onward, however, to treat both burning mirrors and meteorological phenomena separately from optics proper.

[114]Actually, if Ptolemy had followed Aristotle in attributing the rainbow to reflection, then he should have analyzed it somewhere in his study of reflection—i.e., in book 3 or 4. Since it is not to be found in either of these, and since both of them appear to be complete, it follows that, if he addressed the problem of the rainbow at all, he must have done so in another context, as suggested. The same holds of course for the analysis of burning mirrors. For details on the analysis of rainbows from Aristotle to late Antiquity, see Carl B. Boyer, *The Rainbow: From Myth to Mathematics* (New York: Yoseloff, 1959), pp. 38–65. For the earliest known treatise on burning mirrors, perhaps roughly contemporary with Archimedes, see G. J. Toomer, *Diocles on Burning Mirrors* (Berlin/Heidelberg/New York: Springer, 1976).

from Antiquity that can be safely classified as definite. The earlier of these—assuming that it actually does hark back to the fourth century—is found in Damianos' *Optical Hypotheses*, where Ptolemy's *Optics* is said to contain an experimental confirmation of the recti-linear emission of visual rays.[115] The second reference occurs in Simplicius' sixth-century commentary on Aristotle's *De caelo*. There he alludes to Ptolemy's discussion of elemental motion in the *Optics* as well as in the now-lost *Elements*.[116] Unfortunately, these citations are as inconclusive as they are sparse. Neither the experiment mentioned by Damianos nor the account of elemental motion cited by Simplicius crops up in the surviving text of the *Optics*. We are thus compelled to assume, somewhat glibly perhaps, that both authors are referring to the missing portion of the treatise, probably to book 1.[117]

Such a marked absence of testimony seems more than a little surprising. Indeed, as Knorr observes, "this is a meager record for a treatise which we should have expected to become the definitive textbook in its field, just as [the] *Syntaxis, Tetrabiblos, Harmonics* and *Geography* did in theirs."[118] Yet, unlike those four compendia, the *Optics* apparently sank into oblivion almost as soon as it was produced. Why? Perhaps the easiest way of accounting for this slide into oblivion is to explain it away, that is, to argue that the failure of the *Optics* to spark contemporary interest is more apparent than real. After all, negative evidence is notoriously misleading, so the absence of manuscripts and contemporary testimony may indicate nothing beyond the failure of such evidence to survive. On the other hand, if we suppose with Knorr that the *Optics* was written not by Ptolemy but by a contemporary of lesser stature (e.g., Sosigenes) and only later misascribed to Ptolemy, then we could hardly expect it to have enjoyed the kind of success that the genuine Ptolemaic compendia, such as the *Almagest*, did. It is even possible that the *Optics* represents a forgery of sorts, written in the late fifth or early sixth century at the earliest and deliberately misattributed to

---

[115]See *Optical Hypotheses*, chap. 3, ed. and trans. Richard Schöne, *Damianos Schrift über Optik, mit Auszügen aus Geminos* (Berlin, 1897), p. 4. Although Damianos has traditionally been dated to the fourth century, Knorr, "Archimedes and the Pseudo-Euclidean *Catoptrics*," pp. 89–96, argues for a later dating, certainly not before the fifth century, and probably not before the sixth century, if even that early.

[116]The relevant text from Simplicius is given in *L'Optique*, p. 271.

[117]In his sixth-century commentary on Aristotle's *Meteorology*, Olympiodorus adverts to Ptolemy's claim that the rainbow comprises seven colors; see Lejeune, *L'Optique*, p. 272, for the passage. Since this claim is nowhere articulated in the *Optics* as now extant, and since Olympiodorus does not cite any specific source for it, we can hardly take Olympiodorus as a definitive witness to the existence of the *Optics*.

[118]"Archimedes and the Pseudo-Euclidean *Catoptrics*," p. 97.

Ptolemy.[119] While not implausible, however, these explanations are far from convincing. Not only do we have no compelling reason at this point to deny the authenticity of the *Optics*; we have no compelling reason to doubt that its career was as blighted as the evidence (or lack of it) indicates.

Another way of accounting for the *Optics'* failure to capture a contemporary audience is by recourse to the character of the treatise itself. To put it simply, the *Optics* may have been regarded in its day as more comprehensive and technically demanding than it was worth. At first glance, this argument might seem specious, given that the *Almagest*, which seems to have enjoyed a much wider dissemination than the *Optics*, is by far the more daunting of the two. But technical difficulty is not the decisive factor in this case; utility is. In other words, what ultimately destined the *Optics* to failure was that whatever practical benefits it offered were considered insufficient to warrant the effort involved in its mastery.

The other four textbooks mentioned offer an illuminating contrast. Despite its extraordinary length and complexity, for instance, the *Almagest* attracted readers because, in an age that was so thoroughly star-struck as late Antiquity, its practical application to astrology was obvious. Mastery of astronomy in exacting detail was indispensable to the proper casting of horoscopes and, consequently, to the proper regulation of life. The *Almagest* was therefore assured a warm reception by astrologers, whose paramount concern was accurate prediction of celestial cycles.[120] That same concern would also have ensured the success of Ptolemy's interpretive guide to astrology, the *Tetrabiblos*, as well as of his *Geography*, for, as Ptolemy asserts in the second book of the *Tetrabiblos*, some knowledge of geography is necessary for "universal [prognostication], which relates to whole races, countries, and cities."[121] And the *Geography*'s appeal would have extended even further to include not only specialists but also armchair dilettantes titillated (as many of us are today) by geographical descriptions.[122] Finally,

---

[119]The dating here is of course contingent on pushing Damianos' floruit forward as suggested by Knorr.

[120]Suffice it to say, astrologers formed by no means the only, nor perhaps even the primary, circle of interest for the *Almagest* during late Antiquity; indeed, the commentaries by Theon of Alexandria and Pappus show little inspiration on the astrological side. That the study of astronomy was considered worthwhile for its own sake is clear from at least the time of Plato, and the simple demands of time-reckoning, etc., made it imperative to know some astronomy. Still, as far as practical application is concerned, astrology provides the most obvious case for the need to master astronomy.

[121]*Tetrabiblos* II, 1, ed. and trans. F. E. Robinson (Cambridge: Harvard University Press, 1940), pp. 116–117.

in regard to the *Harmonics*, we should bear in mind two interrelated points. First, during classical Antiquity music was considered to be a practical discipline insofar as it exemplified the application of arithmetic to physical reality. Second, for that very reason it formed an integral part of the "liberal arts" canon from fairly early on, as indeed did astronomy, which exemplified the application of geometry to physical reality.[123] Thus, by Ptolemy's era, the study of music, like that of astronomy, had the full sanction of an educational regime that was designed to be as general as possible in its scope and appeal.

Unlike astronomy or music, however, optics was never fully integrated into the liberal arts scheme. True, there were those, such as Vitruvius, who regarded the study of optics as an essential part of a well-rounded education, at least as far as architecture is concerned.[124] As a specific discipline, however, optics was considered at best peripheral to the basic educational regime of late Antiquity. Consequently, it lacked the curricular sanction that astronomy and music enjoyed, even though some knowledge of optics seems to have been requisite to the formation of a properly educated gentleman.[125] On the other hand, like astronomy and music, optics was recognized as having practical ramifications during late Antiquity.

---

[122]By Ptolemy's time the study of geography had been long established on the basis of the Greek *periplus*, which was designed primarily for navigation. With Hellenistic and Roman expansion, however, there was an increasing fascination with faraway places and customs, to which Roman geographers in particular responded; see, e.g., George Stahl, *Roman Science* (Madison: University of Wisconsin Press, 1962), pp. 84–92.

[123]Although the educational regime embodied in the so-called liberal arts had been well established by Ptolemy's day, its curricular structure had not yet been crystallized into the seven-fold one within the trivium/quadrivium envelope that is so familiar to students of the Middle Ages. Still, certain disciplines, notably the Pythagorean tetrad of arithmetic, geometry, music, and astronomy, were considered central. For a useful sketch, see David L. Wagner, "The Seven Liberal Arts and Classical Scholarship," in David L. Wagner, ed., *The Seven Liberal Arts in the Middle Ages* (Bloomington: Indiana University Press, 1983), pp. 1–31. It should be noted that, with the rise of Rome to dominance in the Mediterranean, the focus of the liberal arts education became increasingly practical so that the textual basis for teaching, particularly in technical subjects, was appropriately etiolated. The result was a proliferation of handbooks and epitomes that offered a brief, superficial outline of the given subject for students who were in a hurry to gain a smattering of technical education as a complement to their rhetorical needs; see Stahl, *Roman Science*, pp. 65–133.

[124]An architect, Vitruvius loftily informs us in the beginning of his *De architectura*, "should be a man of letters, a skilful draughtsman, a mathematician, familiar with historical studies, a diligent student of philosophy, acquainted with music; not ignorant of medicine, learned in the responses of jurisconsults, familiar with astronomy and astronomical calculations" (I.1.3, trans. Frank Granger, *Vitruvius I: De Architectura Books I–V* [Cambridge: Harvard University Press, 1933], p. 9). Altogether, Vitruvius mentions twelve disciplines necessary to the proper education of an architect.

[125]For instance, Geminus (fl. c. 70 B.C.) included optics among five other mathematical disciplines that deal with "sensibles": i.e., mechanics, astronomy, geodesy, canonics, and calculation; see Morrow, *Proclus*, p. 31. Vitruvius specifies optics as one of the requisite disciplines for the well-rounded architect; see *De architectura* I.1.4.

A fundamental grasp of linear perspective was useful in surveying, as was a basic comprehension of both linear and color-perspective in scenography. Likewise, some understanding of lighting-effects was essential in the design of buildings, and at least a passing knowledge of reflection was needed for the creation of various optical effects in those buildings.[126] Finally, if we are to take Galen as an example, no physician worth his salt was wholly ignorant of ocular anatomy and its relation to the visual process.[127] The issue, then, is not whether knowledge of optics was considered useful in late Antiquity. It is, rather, just how profound that knowledge had to be in order to satisfy the practical demands of the day. The answer, in a nutshell, seems to be "not very." A brief look at the few optical texts that survive from classical Antiquity will help clarify this point.

The most obvious characteristic of these texts is that, by comparison with Ptolemy's *Optics*, they are as superficial as they are pragmatic. For example, the account of vision offered in Euclid's *Optics* is what one might expect from a surveyor's manual; not only does it lack an articulated theoretical basis, but it ignores physical and physiological issues completely. As a result, nothing is said about color-perception, and only the barest lip service is given to binocular vision.[128] Hardly less superficial is the study of mirrors in the *Catoptrics* attributed to Euclid. Granted, its subject-matter is somewhat more complex than that of Euclid's *Optics*—particularly in the case of convex and concave mirrors— but the *Catoptrics* is inferior to the *Optics* in terms of both geometrical rigor and thematic structure.[129] Nor do we find much, if any, improvement with Hero's *Catoptrics*. Aside from a brief preliminary discussion of principles, this loosely-organized collection of theorems is devoted primarily to what might be called "funhouse"

---

[126]Vitruvius and Geminus both mention scenography as an important practical offshoot of optics; see *De architectura* I.1.4, and Morrow, *Proclus*, p. 33. On lighting-effects, see Vitruvius, *De architectura* I.1.4. Several theorems in Hero's *Catoptrics* are devoted to creating visual effects in buildings with the use of mirrors; see esp. chapters 16 and 18.

[127]See, e.g., *De usu partium* 10.14, trans. May, p. 502.

[128]Binocular vision is at issue, albeit implicitly, in propositions 25–27 of the *Optics*, where Euclid analyzes the case of a sphere seen with two eyes when the distance between the eyes is greater than, less than, or equal to the diameter of the sphere.

[129]Two cases in point are the false concluding proposition (30) about the focusing properties of a spherical concave mirror and the *ad hoc* postulate (6) describing refraction, which has no relevance to the propositional content of the work. Following Heiberg's judgment of its non-authenticity, ver Eecke characterizes the *Catoptrics* as "geometrically inferior to the *Optics*," its propositions not being demonstrated "with the same Euclidean rigor" (*Euclide: L'Optique et la Catoptrique*, p. xxix).

effects created by changing the shape or juxtaposition of various mirrors.[130]

In a sense, the very survival of these treatises into late Antiquity highlights the abject failure of Ptolemy's *Optics* to supersede them and, in the process, bury them, as, for instance, Euclid's *Elements* and the *Almagest* did to their forebears.[131] This failure, in turn, indicates that, as far as optics is concerned, late Antique scholars, even those of relatively high intellectual attainment, were content with an astonishingly low level of technical analysis. Just how low that level was is suggested by Galen's apology at the end of his account of the eye in the *De usu partium*:

I have explained nearly everything pertaining to the eyes with the exception of one point which I had intended to omit lest many of my readers be annoyed with the obscurity of the explanation and the length of the treatment. For since it necessarily involves the theory of geometry and most people pretending to some education not only are ignorant of this but also avoid those who do understand it and are annoyed with them, I thought it better to omit the matter altogther. But afterward I dreamed that I was being censured because I was unjust to the most godlike of the instruments . . . and so I felt impelled to take up again what I had omitted and add it to the end of this book.[132]

Far from obscure, in fact, the resulting geometrical account is simplicity itself, involving no more than a rudimentary understanding of the visual cone. Small wonder, then, that in choosing to lecture on optics some two centuries after Galen, Theon of Alexandria saw fit to base his commentary not on Ptolemy's *Optics* but, rather, on its inferior but more easily assimilated Euclidean counterpart.[133]

Ptolemy's *Optics* would therefore appear to have been surprisingly ill-suited to the demands of its time, in great part because it

---

[130]Hero's *Catoptrics* in its present form is incomplete, which may account in part for its rather chaotic nature; see Lejeune, *Recherches,* p. 182. The principles established at the opening of the treatise are interesting primarily for their reliance on the physical analogy between visual radiation and projectile motion.

[131]That the *Elements* and the *Almagest* so completely superseded their sources as virtually to obliterate them has long posed a problem for historians attempting to reconstruct the actual course of development in geometry and mathematical astronomy before Euclid and Ptolemy.

[132]*De usu partium* 10.12, trans. May, pp. 490–491.

[133]For the critical text of Theon's recension of Euclid's *Optics*, see J. L. Heiberg, ed., *Euclidis opera omnia,* vol. 6, pp. 123–281; for a French translation, see ver Eecke, *Euclid: L'Optique et la Catoptrique,* pp. 53–98.

overfulfilled them, particularly at the practical level. What use, after all, was the extended analysis of binocular vision in books 2 and
3 of the *Optics* to an aspiring lawyer, artist, architect or even physician? What was the practical benefit of the detailed treatment in
books 3 and 4 of image-location in convex and concave mirrors? Or,
for that matter, what purpose was served by the experimentally-
based account in book 5 of refraction through air, water, and glass?
Granted, a proper understanding of refraction might help improve
astronomical observation, but Ptolemy foreclosed on even that possibility.[134] Unlike the *Almagest*, *Tetrabiblos*, *Geography*, or *Harmonics*,
therefore, Ptolemy's *Optics* was limited from its very inception to a
single constituency, and that a very narrow one consisting of disin-
terested intellectuals in pursuit of theoretical knowledge for its own
sake. Worse yet, as time wore on and the late Roman Empire wit-
nessed serious intellectual decline, this narrow constituency was
threatened with extinction. That being the case, we should perhaps
wonder less that testimony to the existence of Ptolemy's *Optics*
during late Antiquity is so meager than that such testimony exists
at all.

The Arabic Context

Precisely how the *Optics* made its passage from Greek into Arabic is a matter of pure speculation. Lejeune conjectures that, "in all
probability," the *Optics* was rendered into Arabic during the flurry
of translation that accompanied the reign of the great Abbasid
caliph, al-Ma'mūn (813–33).[135] This conjecture, while certainly plausible, has no real evidential support beyond the fact that some of
Ptolemy's works, particularly the *Almagest*, are known to have been
translated during, or shortly before or after, that period.[136] The earliest actual sign, however tentative, of the *Optics'* diffusion within
the Arabic world comes from somewhat later, with the Arab poly-
math Abū Yūsuf Ya'qūb ibn Isḥāq al-Kindī (d. 873). It is possible,
although not certain, that he drew upon Ptolemy's work in com-
posing his own optical treatise, the *De aspectibus*. Especially significant in this regard is his discussion of visual acuity in proposition
12, which is reminiscent of Ptolemy's account in II, 20. Yet nowhere
in the *De aspectibus* does al-Kindī cite Ptolemy by name, and in cer-

---

[134]See p. 46 above.
[135]*L'Optique*, p. 29*.
[136]See Carlo Nallino, *Raccolta di scritti editi ed inediti*, vol. 5, ed. Maria Nallino (Rome: Istituto per l'Oriente, 1944), pp. 257–266 and 460–463.

tain respects, particularly in his account of visual radiation in proposition 14, al-Kindī departs significantly from him.[137]

Claiming to have been inspired in his own study of atmospheric refraction by the account in book 5 of the *Optics*, the mathematician Abū Saʿd al-ʿAlāʾ ibn Sahl (fl. after 950) provides the first undisputed testimony of which we are presently aware to the use of Ptolemy's work within the Arabic context.[138] In fact, as has recently been shown, Ibn Sahl not only surpassed Ptolemy in his analysis of refraction, but he actually anticipated the theoretical breakthroughs of Kepler, Snel, and Descartes in the early seventeenth century.[139] Nonetheless, despite such exceptions, the *Optics* seems to have had remarkably little effect upon the study of optics until the early eleventh century, when Ibn al-Haytham undertook his own optical researches, one result of which is the critique of Ptolemy that we noted earlier.[140] But the fundamental effect of the *Optics* upon Ibn al-Haytham's thinking comes through most clearly in his own optical masterpiece, the *Kitāb al-Manāẓir*, to whose very warp and weft Ptolemy's influence extends.[141] Thus, for instance, the basic analytic structure of the *Kitāb al-Manāẓir* is identical to that of the *Optics*, reflecting the same tripartite division into optics (books 1–3), catoptrics (books 4–6), and dioptrics (book 7).[142] Likewise, Ibn al-Haytham followed Ptolemy in analyzing visual perception as a

---

[137]For the critical text of al-Kindī's *De aspectibus*, see Axel Björnbo and Sebastian Vogl, eds., *Alkindi, Tideus und Pseudo-Euklid: Drei optische Werke* (Leipzig: Teubner, 1912), esp. pp. 17–19, 23–25, and 53–55. Another indication that al-Kindī may not have used Ptolemy's *Optics* is his silence in the *De aspectibus* on the issue of refraction; see A. I. Sabra, *The Optics of Ibn Al-Haytham*, vol. 2, pp. lviii–lix. Further reason to doubt that al-Kindī knew of Ptolemy's *Optics*, at least of the fifth book, comes from a short treatise of his entitled "The Size of Figures Immersed in Water," in which he explains the apparent enlarging of objects seen in water on the basis of reflection; for details, see Roshdi Rashed, "Fūthītos (?) et al-Kindī sur 'l'illusion lunaire,'" *Collection des Etudes Augustiniennes: Série Antiquité* 131 (1992), pp. 533–559.

[138]See ibid., pp. lii–liii and lix–lx.

[139]The context for Ibn Sahl's breakthrough is a treatise entitled *On Burning Instruments*, in which he deals with both burning lenses and burning mirrors. In the process of analyzing lenses, he adduces the correct law of refraction in precisely the same form as Snel is supposed to have given it. For a detailed account, see Roshdi Rashed, "A Pioneer in Anaclastics: Ibn Sahl On Burning Mirrors," *Isis* 81 (1990): 464–491.

[140]See p. 6 above. In addition to the "Doubts on Ptolemy," Ibn al-Haytham is also said to have written a summary of both Euclid's and Ptolemy's *Optics*, in which he attempted to reconstruct the missing first book of the latter. He is also credited with a "Memoir on Optics after the Method of Ptolemy." Altogether, by the count of Ibn Abi Uṣaybiʿa (d. 1270), Ibn al-Haytham wrote sixteen works on optical matters; for the list, see Sabra, *The Optics of Ibn Al-Haytham*, pp. xxxii–xxxiii.

[141]For a detailed account of this influence, see my "Alhazen's Debt to Ptolemy's *Optics*," in T. H. Levere and W. R. Shea, eds., *Nature, Experiment, and the Sciences* (Dordrecht: Kluwer, 1990), pp. 147–164. See also Sabra, *The Optics of Ibn Al-Haytham*, pp. lx–lxi.

[142]As Sabra quite rightly observes, Ibn al-Haytham, like Ptolemy, ignored both burning mirrors and the rainbow in his analysis; see idem. As far as catoptrics is concerned, Ibn al-Haytham went far beyond Ptolemy in his account of curved mirrors, adding an extensive

three-phase process that leads from physical radiation, through brute sensation, to perceptual judgment.[143] Like Ptolemy, moreover, he not only used projectile motion as a means of explaining various facets of physical radiation, particularly reflection and refraction, but he also treated the mathematical ray as a mere analytic device rather than a real entity.[144]

By no means a slavish disciple of Ptolemy, however, Ibn al-Haytham disagreed with him in certain crucial respects. Finding Ptolemy's list of visible properties woefully deficient, for example, he expanded it to include such characteristics as roughness and smoothness, or beauty and ugliness, that are not mathematically determined. The resulting catalogue of "visible intentions" adds seventeen to Ptolemy's original five. Ibn al-Haytham also disagreed with Ptolemy about the role of light in the visual process, regarding it not just as a catalyst for sight but as an actual object of vision in its own right. By far the most critical parting of the ways between the two, however, was over the issue of visible radiation and its direction. Instead of locating the source of radiation in the centerpoint of the eye, Ibn al-Haytham transposed it to the visible object itself. Accordingly, every point on the object's surface was presumed to radiate its form omnidirectionally through the surrounding transparent medium. On this basis, Ibn al-Haytham ultimately transformed Ptolemy's cone of visual rays emanating outward from the center of the eye into a cone of light- and color-rays passing inward toward that centerpoint. In carrying out this transformation Ibn al-Haytham effectively dissociated the physical cause of vision from its subjective consequences, thus making it possible to treat light-radiation as a purely objective phenomenon.[145] In short, Ibn al-Haytham paved the way for the modern science of physical optics.

---

analysis of both cylindrical and conical mirrors to that of plane and spherical mirrors. At a technical level, too, he went far beyond Ptolemy in his study of mirrors. The clearest example of his technical superiority can be found in his treatment of "Alhazen's Problem" in the fifth book of the *Kitāb al-Manāzir*. The problem itself is to determine the point of reflection on any spherical mirror, whether concave or convex, when the object-point and the center of sight are given. In deriving a general solution to this problem, of course, Ibn al-Haytham improved immeasurably upon Ptolemy's case-by-case approach in Theorems IV.1–21, IV.24–25, and IV.27–29. For details on Ibn al-Haytham's solution, see A. I. Sabra, "Ibn al-Haytham's Lemmas for Solving 'Alhazen's Problem,'" *Archive for History of Exact Sciences* 26 (1982): 299–324.

[143]Unlike Ptolemy, however, Ibn al-Haytham supplied considerable detail about ocular anatomy and physiology, all of it based on Galenic principles, including the imputation of first visual sensitivity to the crystalline lens rather than to the cornea.

[144]See my "Extremal Principles," esp. pp. 134–139.

[145]See Gérard Simon, "L'*Optique* d'Ibn al-Haytham et la tradition Ptoléméenne," *Arabic Sciences and Philosophy* 2 (1992): 203–235.

The Latin Context

It is tempting to suppose that the obscurity into which Ptolemy's *Optics* fell within the Arabic ambit after Ibn al-Haytham's death was the result of its being so completely overshadowed by the *Kitāb al-Manāẓir*. Indeed, we know of only three citations of the *Optics* from the mid-eleventh century on, one by Abū al-Ḥasan 'Alī ibn Riḍwān (d. 1068), a second by Sa'id al-Andalusī (d. 1071), and a third by Joseph ben Judah (d. 1226).[146] Yet, as the more or less concurrent translation and dissemination of both works in the Latin West reveals, the marked superiority of Ibn al-Haytham's treatise did not necessarily cause interested readers, even those of intelligence, to reject the *Optics* out of hand. On the contrary, it was precisely among the most avid disciples of Ibn al-Haytham in the West—the so-called Perspectivists, whose members included Roger Bacon (fl. c. 1265), Witelo (fl. c. 1275), and John Pecham (fl. c. 1280)—that Ptolemy's *Optics* found its most eager audience.[147]

Surely the most eager member of this audience was the exuberantly eclectic English friar, Roger Bacon, who cites the *Optics* at numerous reprises throughout both the *De multiplicatione specierum* and the *Opus maius*, particularly the fifth part, which commonly goes under the title *Perspectiva*. In one case he actually quotes at length from Ptolemy's account of atmospheric refraction in the fifth book of the *Optics*, and in another instance he repeats Ptolemy's claim in II, 86 that people with deep-set eyes see farther than normal because the visual flux is compressed as it exits their eyes.[148] Bacon's freewheeling acceptance of both the *Optics* and the *Kitāb al-Manāẓir* may appear grossly inconsistent at first glance; after all, Ibn al-Haytham's theory is intromissionist whereas Ptolemy's is extramissionist. But in fact Bacon regarded the two accounts as complementary rather than con-

---

[146]See Sabra, *The Optics of Ibn al-Haytham*, n. 88, p. lx; see also Lejeune, *L'Optique*, pp. 29*–30*.

[147]For details on the Perspectivists, see Lindberg, *Theories*, pp. 104–146. For a discussion of Perspectivist theory, see my "Getting the Big Picture in Perspectivist Optics," *Isis* 72 (1981): 568–589.

[148]The quotation from book 5 of the *Optics* is found in *De multiplicatione specierum* 2.4; for the text, see David C. Lindberg, ed. and trans., *Roger Bacon's Philosophy of Nature* (Oxford: Clarendon Press, 1983), p. 120; see also Lejeune, *L'Optique*, pp. 31*–32*. Additional references to Ptolemy's *Optics* in this work can be found in 1.1, p. 8; 1.2, pp. 32 and 34; 2.1, p. 95; 2.2, p. 98; 2.3, pp. 106 and 110; 2.4, pp. 120, 122, 126, and 128; 2.5, p. 130; 2.6, pp. 136 and 183; and 3.3, p. 196. Bacon's repetition of Ptolemy's account of farsightedness is found in *Opus maius* 5.2.1.1; for the text, see John H. Bridges, ed., *The Opus Majus of Roger Bacon*.

tradictory.[149] As a consequence, he tended to regard Ptolemy more as a corroborating authority than as a primary source, often perverting Ptolemy's theoretical intent in order to cast him in that role.[150]

Considerably more restrained in his eclecticism than Bacon, Witelo was far more singleminded in his dependence upon Ibn al-Haytham as an optical source. Indeed, so closely is his *Perspectiva* modeled after the *Kitāb al-Manāẓir* in both theoretical framework and analytic structure that Witelo earned Giambattista della Porta's derision as "Alhazen's Ape" in the sixteenth century.[151] Nevertheless, Witelo's evident allegiance to Ibn al-Haytham did not prevent him from drawing upon Ptolemy's *Optics* on occasion. Unfortunately, the full extent of his debt to Ptolemy is impossible to determine, because, unlike Bacon, he was chary of citing his sources. Some instances of direct appropriation are clear, however. One such can be found in the fifth book of the *Perspectiva*, where at least two propositions—62 and 63—were lifted directly from the *Optics*.[152] Another, even clearer instance of appropriation occurs in the tenth book of the *Perspectiva*, where Witelo caps his analysis of refraction by repeating almost verbatim Ptolemy's tabulations for refraction from air to water, air to glass, and water to glass.[153] These figures he could only have gotten from the *Optics*, because Alhazen chose not to include them in his own account. Such direct borrowings aside,

---

[149]See, e.g., *Opus maius* 5.1.7.2–4. Bacon makes it clear than the visual process entails intromission as well as extramission but that neither of the two radiations involves a passage of matter. On the contrary, both radiations—outward as well as inward—entail a passage of form or quality that he dubs "multiplication of species." Every active entity in the universe multiplies its species as a token of its activity, and the eye is no exception, multiplying its species outward in order to establish its intentionality in the visual act.

[150]For instance, in *Opus maius* 5.1.10.2, Bacon completely misconstrues—or misrepresents—Ptolemy's account of how light and color manifest their visibility, imputing to Ptolemy the theory that they radiate their species to the eye.

[151]See Lindberg, *Theories*, p. 118. That this appellation was not entirely unjust is corroborated by Sabetai Unguru, the recent editor of books 2 and 3 of the *Perspectiva*; see *Witelonis* Perspectivae *liber secundus et liber tertius: Books II and III of Witelo's* Perspectiva. *A Critical Latin edition and English Translation with Introduction, Notes and Commentaries, Studia Copernicana* 28 (Wroclaw: Ossolineum Press, 1991), p. 31.

[152]The source-propositions in the *Optics* are Theorems IV.40 and IV.41 (IV, 175–182). For Witelo's text, see my *Witelonis* Perspectivae *liber quintus: Book V of Witelo's* Perspectiva. *An English Translation with Introduction, Commentary, and Latin Edition of the First Catoptrical Book of Witelo's* Perspectiva, *Studia Copernicana* 23 (Wroclaw: Ossolineum Press, 1983).

[153]*Perspectiva* X, prop. 8; for the text, see Friedrich Risner, ed., *Opticae thesaurus: Alhazeni libri septem . . . item Vitellonis thuringopoloni libri X* (Basel, 1572; rpt. New York: Johnson Reprint, 1972). Unaccountably, Witelo changes the first value of *r* for refraction from air to water from 7° 30′ to 7° 45′. That Witelo did not attempt to confirm these tabulations experimentally is indicated by his inclusion of grossly incorrect reciprocal values (i.e., for refraction from water to air, etc.), which could only have been derived on the basis of faulty reasoning.

though, Witelo seems to have used the *Optics* more sparingly than Bacon, and this trend continues with John Pecham, whose *Perspectiva communis* reveals no unequivocal borrowings and only a handful of possible ones.[154]

With the continued dissemination of Ibn al-Haytham's treatise and the proliferation of Perspectivist works during the late thirteenth and early fourteenth centuries, Ptolemy's *Optics* was bound to lose its status as a legitimate source in optics. To confirm this point, we need only look at the textual histories of these works during the Middle Ages and Renaissance. As to the *Optics*, we noted earlier that thirteen manuscript copies are known to exist and that, of these thirteen, the lion's share (eight) date from the sixteenth century. This clumping, we also noted, probably indicates not an awakening of scientific interest in the treatise but the enthusiasm of Renaissance scholars eager to revive classical Antiquity and all its works. Now, by contrast, take the Latin version of Ibn al-Haytham's *Kitāb al-Manāẓir*, which currently exists in twenty-two manuscript copies. Add to this the single extant manuscript of a fourteenth-century Italian translation for a total of twenty-three.[155] The difference between this total and that of the *Optics* may not seem particularly large, but bear in mind that Ibn al-Haytham's treatise is not only considerably longer, but also far more technically demanding than the *Optics*. Equally demanding technically, and more than half again as long as the Latin version of the *Kitāb al-Manāẓir*, is Witelo's *Perspectiva*, of which at least twenty-nine manuscript copies survive.[156] Meanwhile, Bacon's *Perspectiva*, which is much shorter and less technical than the previous two works, survives in no fewer than thirty-nine manuscripts, either separately or as part of the *Opus majus*; and his *De multiplicatione specierum* survives in twenty-six.[157] Of John Pecham's even shorter epitome, the *Perspectiva communis*, we presently have no fewer than sixty-two manuscript copies.[158] Unlike the *Optics*, moreover, all of these treatises except the *De multiplicatione specierum* saw publication during the

---

[154]Perhaps the likeliest instance of borrowing occurs in *Perspectiva communis* III, prop. 12, for which Pecham may have drawn from Ptolemy's account of atmospheric refraction in V, 24–31; but even in this case the evidential support is slim. See David C. Lindberg, ed. and trans., *John Pecham and the Science of Optics: Perspectiva Communis Edited with an Introduction, English Translation, and Critical Notes* (Madison: University of Wisconsin Press, 1970), pp. 222–225.

[155]See Lindberg, *Catalogue*, pp. 17–19. In addition to the twenty-one Latin mss that Lindberg lists, another has turned up recently in the Bibliothèque Nationale.

[156]See Unguru, *Witelonis Perspectivae liber secundus et liber tertius*, p. 32.

[157]See Lindberg, *Catalogue*, pp. 40–42, and *Roger Bacon's Philosophy of Nature*, p. lxxv.

[158]See Lindberg, *Pecham and the Science of Optics*, pp. 52–56.

Renaissance, some more than once.[159] For instance, the Latin text of Pecham's epitome appeared in print eleven times between 1482 and 1627. It was also published in Italian translation in 1496. Of Witelo's *Perspectiva*, the first printed edition appeared in 1535, to be followed by a reprint in 1551. Alhazen's *De aspectibus* saw its *editio princeps* at the hands of Friedrich Risner, who published it in tandem with a new recension of Witelo's *Perspectiva* in the monumental *Opticae thesaurus* of 1572. And some four decades later, in 1614, Bacon's *Perspectiva* appeared for the first time in print.

Obviously, then, given both the superiority and ready availability of these Perspectivist sources, no serious scholar of the later Middle Ages or Renaissance was going to waste much, if any, time on the *Optics*. And even those who might have found it handy as a basic introduction to ray-theory had a better alternative in Pecham's *Perspectiva communis*, which was technically superior yet no more difficult to assimilate. There is thus little cause to wonder that the *Optics* slipped back into relative oblivion during this period, an oblivion so deep that, by the later sixteenth century Friedrich Risner, the remarkably erudite editor of both Ibn al-Haytham and Witelo, had evidently never laid eyes on it.[160] All in all, therefore, the period of effective influence for the *Optics* was surprisingly brief, covering no more than the three-and-a-half centuries separating Ibn Sahl and the later Perspectivists. Yet, despite its blemished record, the *Optics* played a crucial role in the history of optics. For without it Ibn al-Haytham might never have been inspired to undertake his own optical studies, and without them, the foundations of modern optics might have been laid quite differently, if at all.

---

[159]For details, see the relevant sections of Lindberg's *Catalogue*.

[160]This conclusion is based on the fact that, although he cites other ancient optical works as "sources" at appropriate points in Witelo's *Perspectiva*, he fails to cite Ptolemy's *Optics* where it is obviously relevant.

ENGLISH TRANSLATION

ADMIRAL EUGENE'S PREFACE

[1] Since I am convinced that Ptolemy's *Optics* is indispensable for those who love science as well as those who investigate the natures of things, I have not hesitated to undertake the burdensome task of translating it into Latin in the present volume. We find, however, that every kind of language has its own idiom and that translation from one to another is not easy, particularly for the faithful translator. This is especially so for one who wants to render Arabic into Greek or Latin, which is all the more difficult as the difference between them is considerable, as much for the [forms of the] verbs and nouns as for the composition of the letters. Hence, because certain points in this work do not perhaps seem clear, I thought it worthwhile to summarize the author's intent, which is more clearly understood in the Arabic version, in order to ease the way for readers.

[2] Although it has not been found, the first book—as is explained in the opening part of the second book—contains [a discussion of] how visual flux[1] and light interact and assimilate to one another, how they differ in their powers and operations,[2] and what their differences and accidents are.[3]

[3] The second book explains what the visible properties[4] are and what characterizes each of them.[5] It also explains that none of those

---

[1]Latin = *visus*; as used by Ptolemy (or his translator), this term has a wide variety of meanings, among which are included: sight, center of sight, viewpoint, eye, visual flux, visual cone, visual faculty, and visual ray. See Lejeune, *Euclide et Ptolémée*, pp. 18–21, for a discussion of the probable Greek correlatives.

[2]Latin = *motus*; in this case, "motus" clearly refers to the effects of light and visual flux in terms of the potency/act polarities of Aristotelian philosophy.

[3]Distinction according to "differences" and "accidents" is fundamentally Aristotelian in nature, harking back to his discussion of the levels of predication in *Categoriae* 4 and 5, *De interpretatione* 11, and *Analytica posteriora* 1.4–6. In particular, however, it was with Porphyry's *Isagoge* (c. 300 A.D.) that Aristotle's scheme was systematized and canonized. According to that scheme, "differences" are essential, definitive attributes (such as "rational" and "mortal," which, added to "animal," define Man), whereas accidents are incidental attributes (such as "white" or "bald") that are inessential to the thing's nature. Or, to put it another way, differences truly specify, whereas accidents merely describe.

[4]Latin = *res videnda*; taken in its general sense, this term means "visible object," but it also carries the specific meaning of "visible property." Unfortunately, context is often the only guide for which meaning is intended at any given point in the *Optics*.

[5]Latin = *qualis habitus sit in unaquaque earum*; technically the term *habitus* (Greek = *echein*) refers to a particular type of predicate within the Aristotelian scheme—that of "having" or "affection," or by extension, "being in a certain state or disposition" (Aristotle's examples are

properties is perceived by sight without some illumination and something to block penetration [by the visual rays], and it informs us that among the visible properties themselves, some are intrinsically visible, some primarily visible, and some secondarily visible.[6] It also tells us that, among the remaining senses, touch alone shares with sight the fact that it perceives [all of] the aforementioned visible properties, except color, which is perceived only by sight. In addition, it describes things that appear more or less [clearly],[7] and it explains that the properties that are intrinsically as well as primarily visible are seen by means of a passion[8] occurring in the visual flux (one such passion being "coloring," another "breaking," and another "diplopia")[9], whereas the properties that are secondarily visible are seen by means of accidents conveyed through that passion.

[4] The same book explains how things appear upward and downward, right and left, and near and far. It also informs us that what is seen with one eye appears at a single location and, likewise, that what is viewed with both eyes appears at a single location as long as it is viewed simultaneously by rays that correspond in order—that is, [rays] in each of the [two] visual cones[10] that have a corresponding position with respect to their own axis. This happens when the axes of both cones fall on a given object, as is by nature customary in sight.[11] But if the visual faculty is forced to exceed its customary limit in any way and is directed to another object so that the rays from the eyes strike the previous object in noncorresponding order, then that single object will appear in disparate locations. Two objects will even appear in three or four [different]

being shod or armed); see *Categoriae* 9. In this context, as in most others throughout the *Optics*, the term is used as a sort of catch-all to denote a characteristic condition that is subject to change or qualification.

[6]Latin = *que videntur vere; que videntur primo; que videntur sequenter*. See II, 4–7 below for Ptolemy's account of these categories of visibility.

[7]That is, the conditions affecting visual clarity.

[8]Latin = *passio*; see II, 22, n. 31 below.

[9]Latin = *revolutio*, connoting the turning of the visual axes away from their wonted focal accommodation; I have followed Lejeune, *L'Optique*, p. 65, n. 85, in translating this as "diplopia" for lack of a more specific modern term. All of these "passions" affect the visual flux specifically, so that, e.g., the "breaking" involves either complete or partial bending of the visual flux, in the form of visual rays, at reflecting or refracting surfaces. In fact, contrary to the sense of the passage, "breaking" affects visual perception of the *secondarily* visible, not the intrinsically or primarily visible properties.

[10]Latin = *piramis visibilis; piramis* rather than *conus* is the standard term for "cone" in medieval Latin, where *conus* can actually be used to denote "vertex"; see, e.g., Witelo, *Perspectiva* V, prop. 22, in my *Witelonis Perspectivae Liber Quintus* (Warsaw: Ossolineum, 1983), p. 210. Eugene does, in fact, use the term "conus" for "cone" at one point, in II, 105 below.

[11]In other words, in binocular vision, the object appears single as long as the two visual axes touch the same point on the object's surface; that way, both cones share the same base (see II, 27 and 28 below). Otherwise, in the case of diplopia, the object appears double.

locations, as is shown by means of [an apparatus consisting of] a ruler and cylindrical pegs that Ptolemy shows us how to set up.

[5] Furthermore, [the second] book discusses how [apparent] size varies with [visual] angles, as well as with distance and orientation.[12] It also tells us how straight and circular lines, as well as plane, [spherical] convex[13] and [spherical] concave surfaces are perceived. And it describes not only the kinds of effects and illusions that are due to the visual faculty, or to the mind, or to the visible properties themselves, but also the misperceptions[14] and errors that occur in the visual faculty with regard to the visible properties.

[6] The third book deals with appearances due to reflection in plane and convex mirrors, starting out with an experiment using a bronze plaque by means of which it is demonstrated that every reflection occurring in the three [basic] types of mirror—i.e., plane, [spherical] convex, and [spherical] concave—takes place at equal angles.[15] After that, there is an experiment with a colored tablet, in which it is demonstrated that a single object is seen in [two] different locations and two objects in one location. By means of this apparatus, moreover, those actual locations are specified.

[7] The fourth book deals with things that appear in concave mirrors and compound mirrors as well as in [combinations of] two or more mirrors.

[8] In the fifth book, although it is incomplete, Ptolemy talks about the refraction[16] of visual rays, which always occurs at unequal angles; and he speaks of the things that appear in such a case, when two different media, one denser than the other, lie between the viewer and visible objects. What is seen [along a line of sight directed] from a rarer to a denser medium (e.g., what is seen [along a line of sight directed] from air to water) always appears larger than it actually is. And the deeper the denser medium is, the larger the

---

[12]Latin = *positio*, which can be taken as "disposition" or, by inference, slant or orientation.

[13]Latin = *curvus*; with very few exceptions, *curvus* denotes "convex" rather than merely "curved."

[14]Latin = *fallacia* = "illusion" or "misjudgment"; technically, this term pertains not to all visual illusions but to visual illusions that are due to some visual misjudgment. Thus, as in its specific logical context, so in this one, "fallacy" involves reaching a false conclusion—a false *perceptual* conclusion in this case; see II, 22, n. 32 below.

[15]That is, in reflection the angle of incidence is always equal to the angle of reflection. Evidently, Ptolemy thought that the basic form of curved mirror was spherical rather than circular; but the vast majority of his theorems about curved mirrors involve circular sections only, and it seems likely that his experimental evidence for appearances in such curved mirrors was derived from cylindrical rather than spherical sections.

[16]Latin = *flexio*. Ptolemy (or at least his Arabic or Latin translator) has no consistent term for "refraction" to differentiate it from reflection. He does use the terms *fractio* and *refractio*, but they can denote both reflection and refraction. Unlike reflection, of course, refraction is marked by the inequality of the angles of incidence and refraction, as Eugene says at the end of the clause.

object appears to be. On the other hand, what is seen [along a line of sight directed] from a denser to a rarer medium appears smaller than it actually is; and the deeper the rarer medium is, the smaller the object appears to be. He demonstrates all these facts by means of diverse experiments, one of which involves a vessel called a *fostir*;[17] another entails a bronze plaque with a glass semicylinder attached to it; and [yet another is carried out] by means of a cube, a cylinder, and a cube [with one side] concave, all made of glass.

[9] However, concerning the aforementioned things that are seen by means of refraction, it must be understood (although Ptolemy does not explain it in what has been discovered of the fifth book) that they ought to be viewed straight on, not at an angle. For an object that is completely immersed in water and looked at obliquely from the air does not appear larger at all but, in fact, smaller than it actually is.[18]

---

[17]While Eugene uses the term *fostir* here in his own précis, in the actual translation he uses the term *baptistir*, which is probably closer to the actual Greek form used by Ptolemy. For a technical discussion of the possible implications of this difference, see Lejeune, *L'Optique*, n. 4, pp. 8–9.

[18]Here we have a strong indication that Eugene not only translated the *Optics* but also carried out at least some of the experiments detailed by Ptolemy.

BOOK 2
THE BASIC ANALYSIS OF VISION

Topical Resume

Introductory Section [1–3]: [1] summary of book 1 and statement of goal for book 2, [2] list of visible properties, [3] the two modes of existence for these properties

The Objective Mode [4–21]

The Three Categories of Visible Properties [4–12]: [4] luminous compactness as intrinsically visible, [5] illuminated color as primarily visible, [6] the remaining properties as secondarily visible, [7–12] the remaining visible properties as dependent on color

Color [13–17]: [13] color as the proper object of sight, [14–15] color as an inherent, objective property of physical objects, [16–17] relationship between color and light

Visual Clarity [18–20]: [18] visual clarity as a function of the intensity of the illumination or of the visual flux striking the visible object, [19] visual clarity as a function of physical and spatial conditions, [20] visual clarity as a function of position within the visual cone

Darkness [21]: darkness as the absence of visibility

The Subjective Mode [22–82]

General Principles [22]: vision of the intrinsic and primary visible properties stems from a passion created in the visual flux; vision of the secondary visible properties is accidentally conveyed by that passion; visual perception involves an interplay between objective conditions and subjective interpretation

Color-Perception [23–25]: [23] relationship among color, light, and visual flux, [24] color is not a mere affect of the visual flux, [25] bodies are seen as such by virtue only of their being colored and opaque

The Perception of Place [26–46]

Monocular Vision: [26] perception of place depends upon relationship of visible object to rays within the visual cone with reference, ultimately, to the visual axis and the vertex-point

Binocular Vision: [27–29] in binocular vision an object is seen properly (i.e., singly) as long as the two visual axes converge at one point on it; otherwise, the object will appear doubled, [30–32] description of what happens when the focal point is displaced outward from the eyes, [33–37] **Experiments II.1 and II.2**, involving outward and inward displacement from the eyes, [38–44] **Experiments II.3–II.5**, involving displacement from side to side parallel with the eyes, [45–46] concluding remarks about binocular vision

Size-Perception [47–63]: [47] size-perception is ultimately dependent upon the visual angle subtended by the object, [48–51] size-perception is not a function of the number of visual rays striking the object, nor is failure to see an object at a great distance due to its falling between discrete visual rays; in fact, the visual flux is perfectly continuous, [52–55] analysis of size-perception as a function of three factors taken together: visual angle, spatial orientation (i.e., slant) of the visible object, and distance from the eye, [56–62] **Examples II.1–II.6**, illustrating size-perception according to these three variables, [63] concluding remarks on size-perception

Shape-Perception [64–73]: [64] shape-perception is a function of the shape of the base of the visual cone picked out by the visual rays striking the surface of the visible object, [65–68] perception of straightness and curvature, planes and curved surfaces, convexity and concavity, and depth of curvature, [69–71] perception of curvature and its depth depends on distance of object from viewer, [72–73] the role of spatial orientation in perception of shape and curvature

Perception of Change [74–82]: [74–75] discussion of change in general and its visual apprehension, [76–77] locomotion and its visual apprehension, [78–81] **Examples II.7 and II.8** illustrating the visual grasp of motion, [82] motions that are too fast or too slow are not perceived

Visual Illusions [83–142]

Illusions Due to Physical Circumstances [83–101]

General Observations: [83] transitional paragraph, [84] basic differentiation of illusions according to level of perception,

Illusions Due to Perceptual Judgment [122–142]

> Distance: [122] transition, [123] some of these illusions are due to the viewer's not having adequate criteria for judging properly, [124] brighter objects appear closer; color-perspective in painting, [125] misperception of distances for viewer looking from high altitudes

> Size: [126] darker or more vaguely colored objects appear larger than brighter, more vividly colored objects of the same size

> Shape: [127–128] illusion of convexity or concavity by shading, [129] reversal of apparent relief in cut glass

> Motion: [130] moving objects that disappear in the middle of motion appear to move faster than they really do, [131–132] misperception of relative movement; perception of passing scenery as actually moving, [133] the eyes of a portrait seem to follow the viewer

> Intellectual Intervention: [134–136] general summary of criteria of judgment and their shortcomings as cause of such misperceptions, [137] the intellect's role, [138] interpreting image-reversal in reflection, [139] misjudging celestial distances, [140] misjudging parallels carried to a distance, [141] misjudging relative speeds of chariot-wheels and horses pulling the chariot, [142] brief statement of conclusion

The Second Book of Ptolemy's *Optics*

(previously translated from the Greek language into Arabic, and now from Arabic into Latin by the Admiral Eugene of Sicily on the basis of two versions, the more recent one, from which the present translation was drawn, being the more accurate. The first book has not yet been found.)

[1] In the preceding book we have explained everything that one can gather about what enables light and visual flux to interact, how they assimilate to one another, how they differ in their powers and operations, what kind of essential difference characterizes each of them, and what sort of effect they undergo. In this book, and in those that follow, however, we shall assemble the facts that pertain to the sensible action of the visual faculty according, as is fitting, to an order and sequence of propositions, first discussing in particular what the visible properties are and what characterizes each of them, along with their first and last distinctions, so that how the sense [of sight] operates may become more obvious.

[2] Accordingly, we say that the visual faculty apprehends corporeity, size, color, shape, place, activity, and rest.[1] Yet it apprehends none of these without some illumination and something [opaque] to block the passage [of the visual flux]. We need say no more on this score but must instead specify what characterizes each of the visible properties.

[3] We contend, therefore, that these visible properties exist in two ways, one of which depends upon the disposition of the visible property [itself] and the other upon the action of the visual faculty.[2] So let us speak separately of each, and let us start with the way that is determined by the disposition of the visible properties, some of which are intrinsically visible, some primarily visible, and some secondarily visible.

[4] Now luminous compactness[3] is what is intrinsically visible, for objects that are subject to vision must somehow be luminous, either in and of themselves or from elsewhere, since that is essential to [the functioning of] the visual sense; visible objects must also be compact in substance in order to impede the visual flux, so that its power may enter into them rather than pass through without incident effect.[4] Thus, it is impossible for anything to be seen without these two conditions' being met, nor [can anything be seen] when one of them is met without the other.

[5] On the other hand, colors are primarily visible, because nothing, besides light, that does not have color is seen. Still, colors are not intrinsically visible, since colors are somehow contingent on the compactness of bodies and are not visible per se without light. Indeed, colors are never seen in darkness, except for [the color of] an object that shines from inherent whiteness or that is exceedingly

---

[1]Latin = *corpus, magnitudo, color, figura, situs, motus et quies.* I have rendered *corpus* as "corporeity" to indicate that it is not so much body as the fact that something is a body that is apprehended by sight. This list, with *corpus* and *color* excepted, is clearly reminiscent of Aristotle's list of common sensibles: see, e.g., *De anima* 2.6.418ª17, where Aristotle enumerates magnitude, shape, motion, rest, and number; it is probable, given other instances, that Aristotle meant this list to be exemplary rather than comprehensive. For further discussion, see Smith, "Psychology," pp. 201–202.

[2]Ptolemy is drawing a crucial, albeit rough, distinction between the objective grounds of visibility (according to the disposition of the visible properties themselves) and its subjective grounds in perceptual interpretation (according to the action of the visual faculty). The implication here, which is elaborated upon in II, 22 as well as succeeding paragraphs, is that the viewer must act intentionally toward the visible object if it is to be visually perceived.

[3]Latin = *lucida spissa* = "things that are dense and luminous"; as will become clear in due course, Ptolemy's use of density as an optical variable is problematic.

[4]In other words, it is luminous compactness that enables vision to infer corporeity immediately and, thus, to be alerted that there is something to be visually perceived.

polished, for each of these is a case of brightness, and brightness is a kind of luminosity.[5]

[6] All the rest of the aforementioned visible properties are secondarily visible, because the visual faculty apprehends things as bodies by means of their [inherent] colors and characteristics, whereas objects that have no compactness, but are exceedingly tenuous and have no color, are neither sensed nor perceived as bodies by the visual faculty. Furthermore, size, place, and shape are perceived only through the mediation of bodies' surfaces, which coincide with the colors upon which external light falls. Activity and rest, as well, are apprehended by means of an alteration, or lack thereof, in any of the aforementioned visible properties.[6]

[7] While the faculty of sight apprehends illuminated colors immediately, it apprehends the rest of the visible properties by means of such illuminated colors, not insofar as they are colors but only insofar as they have boundaries. For the visual faculty apprehends shapes and dimensions by means of the boundaries of the colored object, while place is apprehended by means of its location. The visual faculty also apprehends the motion or rest of these same colors by means of their change or lack thereof. And the motion or rest of shapes, dimensions, and location is perceived by means of the motion or rest of the boundaries or places of the colored object.[7] For instance, an object that appears white certainly does not appear round, small, near, or stationary on account of its whiteness, just as the characteristics of a resisting object that the hand feels are not grasped through its hardness. On the contrary, these properties are discerned by means of either the boundary of whiteness or the boundary of hardness—the boundary in this case being an essential characteristic [of that hardness] or its size according to its extended nature.

[8] And that is why the hand apprehends differences among the resisting objects it feels and why the eye apprehends differences

---

[5]See II, 19 below. Notice, then, that light is not visible per se; it is a mere catalyst for visibility and becomes visible itself only insofar as it has color—brightness being a sort of inherent whiteness. Color is therefore the sole proper object of sight, as Aristotle contends in *De anima* 2.6.418ª11–14. As such, it is seen immediately, even though its visibility is ultimately contingent on luminous compactness. Or, to put it another way, color cannot exist as a visible quality without being embodied and illumined. See Beare, *Theories*, p. 64.

[6]Notice, then, that the secondarily visible properties coincide perfectly with Aristotle's common sensibles insofar as 1) they are inferred mediately from the proper sensible, and 2) they can be inferred by more than one sense. Such properties, being accessible through inference, are thus not so much sensible (*sensibilia*) as perceptible (*perceptibilia*); see, e.g., *De anima* 2.6. 418ª8–11.

[7]All of this is to say that, taken objectively, the secondary visibles are merely intentional entities within the context of visibility. See n. 13 below on Aristotle's conception of "place."

among colors in and of themselves without something else's intervening. However, what characterizes the remaining [visible properties] is only perceived by means of these two [primary qualities], since those [remaining] properties are not apprehended except through something that is contingent on something else. For the apperception of those contingent properties depends on a perception of sensibles that pertains specifically to the sense.[8] Indeed, because they are inseparable from the bodies [in which they inhere], shapes and sizes are perceived by means of boundaries; but place, activity, and rest are perceived because they are separable [from] and conveyed [by such basic properties].[9]

[9] This is why, when we are surrounded by clear air and open our eyes, we do not perceive the shape of the nearby air enveloped by the visual cone, because the air's superadded color is extremely subtle and does not have enough intensity to render it sensible. But when the nearby air enveloped by the visual cone is condensed, then we perceive its color and shape, because its color extends far into its interior and becomes more substantial and evident, as is the case with water.

[10] Furthermore, if we take a liquid of some color and add to it another [medium of] identical color for the purpose of using that identical color to draw images and shapes in the liquid, then the colors of the drawings will be seen together with the surrounding liquid of the same color, but their shapes will not be seen, nor their distances, nor the relative differences in size among them.

Now, everything pertaining to colors appears in some way or another by means of the adjoining boundaries of dissimilar[ly colored] objects, just as the location of defining outlines appears by means solely of the lines separating them;[10] and that [apparent location] is created either by contrasts among all the colors or by the correct imitation [of separation] that appears through [linear] drawings.

[11] Likewise, the outlining of shapes [within the liquid] causes a change in the uniformity of the representations, but we do not on that account say that the representations are actually seen. As we have said, first, since the continuous color of the defining lines, the color of the edges, and the colors of all the drawings in no way

---

[8]In short, the mediate sensibles (i.e., the *perceptibilia*) are perceived only by means of the proper sensibles (i.e., the *sensibilia*).

[9]Notice the division of the secondary visibles into properties that are intrinsic to their subjects and those that are extrinsic to them. Thus, while shape and size are absolutely determined, place, activity, and rest are relatively determined by comparison of the given object with other objects or with itself over time.

[10]See II, 7 above.

differ [among each other] through any change in the quality of their color, they happen to lack [distinction] by virtue of sameness. Also, as far as the observation of their shape is concerned, since no defining line appears in any of these figures, and because of the absolute uniformity [of color] that obtains among these different shapes, some color is needed to disclose their outlines. Finally, nothing about them will appear clearly without color. On the contrary, according to differences among colors alone, the shape of dissimilar colors, being intrinsic to them, must appear along with them. The same condition holds in each of the remaining objects that are seen along with a corresponding one—all of which involves a false perceptual inference.[11]

[12] So too, since light and visual flux strike the surfaces of bodies together, it is quite appropriate that the first thing to be sensed in all visible objects is a characteristic of their surfaces. And color is more properly attributed to the surface than to the interior of things. For this reason, the ancients used to equate surface and color, because color is a certain property affixed to the substance of an illuminated thing, and the genus "surface" is like that; and so it is an apt designation for it.[12] As far as the remaining visible properties are concerned, corporeity is not surface, because surface is its boundary, yet all the remaining visible properties depend upon something in bodies having to do with surface. For instance, size is the boundary of a surface's quantity, whereas shape is a qualitative arrangement of surface, and place is the boundary of the location of a surface.[13] Activity, though, depends upon surface insofar as it is attributed to any of those properties [i.e., size, shape, and place]—for instance, the activity of alteration, or of growth, or of diminution, or of locomotion.[14]

[13] A [sole] proper sensible can be found that is appropriate to each of the senses; e.g., the quality of "resisting the hand" for touch, savors for taste, sounds for hearing, and odors for smell.[15] But

---

[11]The basic point is simple enough: adjacent objects will be seen only if they are of contrasting color or if their separation is delineated by a boundary-line of some different color. The concluding phrase (Latin = *quod fit secundum fallaciam* = "all of which involves a false perceptual inference") might refer to the way we interpret line-drawings as if they represented objective reality when, in fact, they do not; such representation is, of course, illusionistic rather than real.

[12]According to Aristotle, *De sensu* 3.438ᵃ30–33, it was the Pythagoreans who equated surface and color; see Beare, *Theories*, pp. 59–60.

[13]This is clearly reminiscent of Aristotle's definition of place (in *Physics* 4.4.212ᵃ4–6) as, in essence, the innermost boundary of what contains the outermost boundary of what is contained. Consequently, the place of any moving body shifts with the body itself.

[14]This list of change-types is reminiscent of Aristotle's list in *Categories* 14.15ᵃ13–14; see also *Physics* 3.1.201ᵃ10–15.

[15]In this list of proper sensibles Ptolemy is simply following Aristotle; see, e.g., *De anima* 2.6.418a11–14.

among the things that are common to the senses according to the origin of nervous activity,[16] sight and touch share in all except color, for color is perceived by no sense but sight. Thus, color must be the proper sensible for sight, and that is why color is taken to be what is primarily visible after light.[17]

[14] In view of this fact [that color requires light to render it actually visible], it seems that color is not really a proper sensible, as certain people have supposed, claiming that color is something accidental to the visual flux and to light and that it has no real subsistence,[18] because none of the [proper] sensibles needs anything extrinsic [to make it sensible], whereas colors need light.[19] Hence, it seems that this missing subsistence is provided by the visual flux, not by the visible objects. For objects that are seen and that are affected by the visual flux are not of such a nature as to appear to the visual faculty without light.

[15] However, [if we grant] that the subsistence of different colors, as well as their generation, depends solely upon a change in light and in the visual flux, it follows that [all] bodies that maintain precisely the same location with respect to a given luminous source and a given viewpoint should appear to be of the same color. Yet we find the majority of such bodies to be of various colors. We find this to be so not only in the case of several objects, but also in the case of a single object that remains perfectly fixed in relation to both the luminous source and the viewpoint. [Take], for example, the animal called a chameleon, or the redness that arises in certain people from blushing, or the pallor that overtakes others from fear. Furthermore, what happens in these cases occurs without any noticeable change appearing in the things themselves or in the external conditions, except for a change in color.

---

[16]Latin = *secundum principium nervosum*; this origin-point, where the *virtus regitiva* resides, apparently corresponds to the Stoic *hegemonikon*—the "Governing Faculty" in the brain that interprets all sense-perception. That Ptolemy may also have thought of this origin-point in terms of Aristotle's common sense (*De anima* 3.1.425ª27–29) is quite possible; see II, 22–23 and 76 below for further references to the Governing Faculty. For an account of the Stoic conception of the *hegemonikon*, see Josiah B. Gould, *The Philosophy of Chrysippus* (Albany: SUNY Press, 1970), pp. 126–137.

[17]Ptolemy is quite clearly following Aristotle in making color the proper sensible or object of sight; see n. 5 above. Aristotle, *De sensu* 4.442b5–8 remarks on the closeness of touch and sight in the perception of common sensibles.

[18]Latin = *subsistentia*; this is a technical term to denote the state of actually existing.

[19]Here Ptolemy is referring to those, such as the atomists, who would make color a sort of byproduct of light (see, e.g., Plato, *Timaeus* 67c–68a) or those who would make it a subjective effect of some more fundamental physical interaction (see, e.g., Theophrastus' account of Democritus in *De sensibus* 63–78). Ptolemy is quite clearly following Aristotle in according color true objective existence in physical reality. See David E. Hahm, "Early Hellenistic Theories of Vision and the Perception of Color," in Peter Machamer and Robert Turnbull, eds., *Studies in Perception* (Columbus, Ohio: Ohio State Press, 1978), pp. 60–95.

[16] It is therefore obvious from what we have said that color truly inheres in these objects and belongs to them by nature, and it is seen only when light and visual rays combine to make it effective. And that is what has prompted the claim that nothing proper to any color is ever seen, since no color may be seen without light somehow shining [upon it]; either something that is self-luminous or something that is rendered luminous in some similar way should highlight colors by mixing with them at the surface, because [light] is generically related to none of the visible properties but color. It relates to itself, however, as [if providing] "form" to the "matter" of color.[20]

[17] It is for this reason that objects are seen more or less clearly, not only because of variation in the condition of the visual flux, but also because of internal variation and, even more important, because of variation in the condition of the objects that illuminate the visual field. In each of the two types [of radiation], moreover, this variation is a function sometimes of the intensity of their powers and sometimes of the quality of their operations.[21]

[18] As far as variations in intensity are concerned, an object is seen more clearly when more visual flux[22] impinges upon it or when more light shines upon it—e.g., what is looked at directly is seen more clearly than what is looked at by means of reflection or refraction.[23] Even so, an aggregation of [uninterrupted] visual rays is weakened when they are extended out to a great distance. Also, what is looked at with both eyes is seen more clearly than what is looked at with either eye alone. And what is self-luminous is seen more clearly than what is illuminated by something else. Also, the larger or more numerous the luminous sources, the more clearly the object upon which their light shines is seen.

[19] On the other hand, among objects whose appearance depends upon the quality of [radiative] effects, those that lie directly in front of, and at right angles to the rays are seen more clearly

---

[20]Latin = *Communicat autem sibi ipsi, ut forme coloribus quoque ut yle*. In Aristotelian natural philosophy, of course, "yle" (= i.e., *hyle*) designates pure matter. What Ptolemy means here is that, like unformed matter, unillumined colors have only potential existence in relation to any viewing subject; to become visually actualized (i.e., "informed"), they must have superadded light.

[21]The two types of radiation at issue here are that of visual flux and that of light. As is clear from the succeeding three paragraphs, "intensity of powers" has to do with the concentration of flux or light, whereas "quality of operations" has to do with the angle of incidence.

[22]Latin = *claritas visus*; Lejeune, *L'Optique*, p. 19, n. 15, suggests quite plausibly that the use of *claritas* (= "brightness") here implies an essential connection between light and visual flux; after all, as Ptolemy himself says in II, 5 above, "brightness is a kind of luminosity."

[23]According to Ptolemy's account, reflection and refraction are impact-phenomena. They therefore tend to impede the radiation of visual flux, thus causing a weakening of visual activity; see, e.g., III, 21 and 22, as well as V, 1 and 2 below.

than those that do not. For everything that falls orthogonally strikes its subjects more intensely than whatever falls obliquely.[24] Also, what is polished is seen more clearly than what is rough, because there is disorder in a rough object resulting from the fact that its parts are not arranged in a regular way. But the parts of a polished object have a certain regularity, and [so] brightness is inherent to it. Dense objects, as well, are more clearly seen than rare ones, because rare bodies give way to the impinging [ray], whereas dense bodies resist it. Objects that radiate [light] by themselves are also seen more clearly than those that are lit by something else, and so, for example, is an object that is seen in illuminated, rarefied air.[25] Furthermore, objects that lie a moderate distance from the viewpoint appear more clearly [than those that do not], because objects that are [too] near the eye are enveloped by visual cones that fall within the internal humor of the eye and that impede the visual flux.[26] On the other hand, objects that lie far away [from the eye] appear less clearly, since the visual rays, as they stream outward, take on some of the blackness of the air through which they pass.[27] Thus, distant objects appear nebulous, as if seen through a veil.

[20] And since [each] visual ray terminates at its own unique point, what is seen by the central ray—i.e., the one that lies upon the axis [of the visual cone]—should be seen more clearly than what is viewed to the sides [of the visual axis] by lateral rays. The reason is that those rays lie nearer to [the edge of the visual cone where there is an increasing] absence [of rays], whereas those rays that approach

---

[24]Here Ptolemy is drawing on a specific dynamic model—that of physical projection—to explain variations in intensity of illumination. Hero of Alexandria, in *Catoptrics*, chapter 2, has recourse to much the same dynamic model in accounting for the rectilinear propagation of flux.

[25]Latin = *in aere subtili claro*; this implies that visibility depends not just on the luminousness of the visible object but also on illumination of the air through which it is seen. Perhaps Ptolemy has in mind Aristotle's distinction between potential and actual transparency (see *De anima* 2.7.418b4–19), or perhaps he is adverting to the Stoic idea that light provides a necessary pneumatic alteration of the air to make it perspicuous; see Samuel Sambursky, *Physics of the Stoics* (Princeton: Princeton University Press, 1959), pp. 27–28.

[26]The Latin for "internal humor" is *humiditas interioris*. Perhaps, as Lejeune, *L'Optique*, p. 20, n. 18, suggests, Ptolemy discussed ocular anatomy in the lost first book. Lacking such an account, we can only speculate about his grasp of that anatomy from a few remarks, such as this one, scattered throughout the *Optics*. By Galenic standards, of course, this "internal humor" would be the vitreous humor. Presumably, it is its tendency to impede (Latin = *repellere*) the visual flux that explains why we cannot see something that is placed directly upon or, worse, within the eye itself.

[27]Ptolemy seems to envision two different modes of coloring for the visual flux. The first, and primary, mode involves the kind of essential coloring that occurs when the flux takes on the color of visible objects by direct contact. The second mode involves a sort of accidental coloring of the flux by incidental contact as it passes through various tinged media; see, e.g., II, 107 and 108 below.

the [visual axis] lie farther from [such an area of] absence.[28] The same holds for objects that lie toward the middle of spherical sections whose centerpoint is the apex of the visual cone, because the generating point of the sphere itself and powers that approach their generating sources are more effective. The farther such powers extend from their sources, then, the weaker they become—as, e.g., [the power of] projection [in relation to] the thrower, or of heat in relation to the heater, or of illumination in relation to the light-source.[29] Therefore, since the visual ray within the cone has two primary referents, one being the centerpoint of the [ocular] sphere where the vertex of the visual cone lies and the other being the straight line that originates at this point and extends the whole length [of the cone] to form its axis, it necessarily follows that the visual perception of what lies far from the vertex of the cone is carried out by a more weakly-acting ray than the visual perception of something lying at a moderate distance. The same holds for objects that lie far from the visual axis in comparison to those that lie near it.[30]

[21] We have therefore accounted for the properties that are intrinsically visible, those that are primarily visible, and those that are secondarily visible. We have also discussed how objects are seen more or less clearly. Darkness, however, is never seen; instead, we apprehend it through a deprivation that occurs in the visual faculty, just as we in no way hear silence but apprehend it by means of the absence of sound. And, generally speaking, we recognize various types of [sensible] deprivation through an absence of sensation. We apprehend the extent of those [missing sensibles], though, by means of the [contrasting] limits of the actual [sensible qualities

---

[28]The rationale here is unclear. On the one hand, Ptolemy may simply mean that, since the visual radiation stops completely at the edge of the visual cone, rays that lie closer to that edge necessarily lie closer to the absence of radiation it demarcates. He could also be considering the radiation of visual flux from a central point in the eye to be ideally spherical in nature. Accordingly, the visual cone marks out a section in that sphere. Assuming that the radiation is uniform throughout the sphere, and assuming that the base of the visual cone (in external objects) is essentially planar, then the closer to the axis of the cone, the more concentrated the radiation, whereas the farther from that axis, the more dispersed (i.e., the closer to "absence") it is.

[29]Note Ptolemy's recourse to dynamics in order to explain changes in radiative intensity; see notes 21 and 23 above. Ptolemy's approach to the weakening of visual acuity with distance is in sharp contrast to that of Euclid, who traces it to the dispersal of discrete rays as they pass farther from the center of radiation; see Euclid, *Optics*, props. 2 and 3.

[30]Overall, the point Ptolemy wants to make here is that objects are seen more clearly the closer they are (within certain limits) to the eye and the more closely the line of sight lies toward the axis of the visual cone. Note how vague Ptolemy's formulation of the relationship between distance and intensity in light-radiation is; the implication, in fact, is that the relationship is a simple inverse one. It was only with Kepler that the proper inverse-square relationship was finally specified (for light rather than visual flux) in the early seventeenth century (see *Ad Vitellionem paralipomena* 1, prop. 9). See Damianos, *Hypotheses*, paragraphs 9 and 10.

bracketing them], just as we apprehend the size of a dark spot by means of the boundaries of light surrounding it, and just as we apprehend the duration of silence by means of the end and beginning of sound.

[22] We said before that there are two basic modes [determining] how visible properties are seen, and now that we have sufficiently explained the first of these modes, which depends upon the disposition of the visible properties themselves, we must explain the second mode, which depends upon how the sense of sight itself functions. And we should say, first, that all the intrinsically or primarily visible properties are in fact seen by means of a passion[31] that arises in the visual flux, whereas the secondarily visible properties are only seen through accidents conveyed by that passion. In every visual perception, neither that passion nor whatever accident it conveys owes its existence to the Governing Faculty [alone], nor in fact does it stem from the visible objects [alone]. Rather, it depends upon a relationship or rational interchange[32] that is established between those objects and what pertains to and arises from the Governing Faculty.[33] Nor do we apprehend any of these passions and accidents according to such relationships alone, but also according to a moderate sensibility that the [perceptual] source has of objects under its scrutiny. That point will be demonstrated when we have analyzed it according to particular visible properties.

[23] According to what we have presupposed, we see any luminosity or color by means of a passion arising in the visual flux, while we see the secondarily visible properties that remain through the accidents that this passion conveys. Indeed, the passion arising in the visual flux is [called] "illumination" or "coloring." Illumination by itself, however, is a sort of excess-condition[34] in luminous objects, so it hurts and offends the [visual] sense.[35] Illumination is also

---

[31]At a superficial level, "passion" (*passio*) is merely the natural correlative of "action" or "operation" (*actio/actus*). At a somewhat deeper level, however, *passio* connotes "suffering" and, taken in that sense, implies that the act of seeing entails pain of some sort; see, e.g., n. 35 below. While this may not reduce sight to a species of touch, it certainly links the two senses closely; see n. 17 above and n. 84 below.

[32]Latin = *ratiocinatio*; evidently, then, perceptual interpretation has a sort of logic which, when traduced, leads to "fallacy"; see Preface, n. 14 and II, 11 above.

[33]Latin = *virtus regitiva*; see n. 16 above. What Ptolemy means to say here is that visual perception involves more than a merely passive reception of objective visible qualities; it is also a function of subjective intentionality that finds its ultimate source in the Governing Faculty.

[34]Latin = *quedam de superhabundantiis habitudinum*.

[35]This notion that excessive brightness can harm the eye is a commonplace in classical thought (see, e.g., Aristotle, *De anima* 3.2.426^{b}1 and 3.13.435^{b}7–19). As is clear from Aristotle's account, however, such a notion implies a sort of physical effect, perhaps something akin to stamping or etching, on the eye by an external agent. This indicates a conceptual inconsistency on Ptolemy's part: on the one hand, his basic theory is clearly extramittive; yet, on the other, the physical model of stamping or etching by an external agent is clearly intromittive.

created along with coloring in objects that are struck by light from outside. Light and color are also transformed into one another by a transition of one into the species of the other, since luminosity provides the genus for both. And if light falls upon it, color becomes luminous, while light, if it is colored, is obviously altered.[36] The visual flux, on the other hand, provides nothing qualitative to either of them, for it is necessary that the sense [of sight, which is] perspicuous, should have no qualification but should be pure and should suffer the qualification [passed to it] by light and color, because it shares their genus. Nevertheless, it does undergo a straightforward qualitative alteration from all colors and light.[37] And this [alteration] is not always sensible, except when the intensity of the passion [it arouses in the visual faculty] is adequate to the perceptual capacity of the Governing Faculty. So too, differences among colors struck by the visual flux are not apprehended if the colors of the objects in question are merely separate while the rest of their [qualifications] are identical; but [they will be apprehended] if there is a sensible degree of difference among those colors.[38] Moreover, objects whose colors differ by some [small] amount appear distinct from one another [when viewed] from nearby, but not from afar, because vision becomes weak in distinguishing among remote objects.

[24] From what we have said, then, it is clear that the visual flux apprehends color by the accident of coloring. For instance, it apprehends whiteness because it is whitened, whereas it recognizes blackness because it is blackened, and the same holds for each of the intermediate colors.[39] Some have thought that whiteness is perceived through a spreading out of the visual rays, while blackness is perceived through a constriction of them, but that is not so.[40] In

Perhaps Ptolemy had something akin to Democritus' theory of *emphasis* (= "reflection" or "imprinting") in mind. As described by Theophrastus, in *De sensibus* 50–54, that theory was predicated on the notion that visible images were imprinted in the air and, thence, on the surface of the eye. Ptolemy's reference to the cornea in III, 16 and IV, 4 below as *aspiciens* (= "viewer") implies that he understood it as visually sentient, which lends credence to this interpretation; see Kurt von Fritz, " Democritus' Theory of Vision," in E. A. Underwood, ed., *Science, Medicine and History*, vol. 1 (London: Oxford U. Press, 1953), pp. 83–99 (esp. 93–95). See also Beare, *Theories*, 25–29.

[36]Ptolemy adverts to this generic relationship in II, 16 above.

[37]In short, while light and color can affect one another qualitatively, as they can also affect visual flux, visual flux for its part has no qualitative effect whatever on light or color.

[38]Evidently what Ptolemy means here is that, while two colors may be physically distinct— e.g., by inhering in perfectly contiguous objects—they will not be perceived as distinct unless they are set off by something beyond mere physical distinction, such as degree of intensity. See Ptolemy's previous discussion in II, 10 and 11 above.

[39]Ptolemy's basic conception of color seems to be consistent with Aristotle's theory that all colors are intermediate between the extremes of white and black; see, e.g., *De sensu*, 3. Actually, this notion of black and white as polar primaries is held by many classical thinkers other than Aristotle; see Hahm, "Early Hellenistic Theories" for a general account. See also Beare, *Theories*, pp. 68–74.

[40]This explanation is propounded, for instance, by Plato in *Timaeus* 67e.

the first place, we find no necessary reason why the rays should be spread out by the one and constricted by the other, nor [do we know] how the intermediate colors—i.e., red, rose, and blood-red—that are composed of these [extremes] might be apprehended.[41] Besides, the larger of identically white objects ought to appear larger yet, because it is apprehended by many [more] rays, not only on account of a [greater] aggregation of [incident] rays, but also because the angle [at the vertex] of the visual cone does not remain constant but dilates if the objects are white, given that such objects spread the rays out. And if these objects are black, that [visual angle] is reduced on account of the constriction [of the rays]. The visible portion of the sky ought therefore to appear larger in daytime than at night, but nothing of this sort seems to happen.[42]

[25] Accordingly, the visual flux apprehends these colors by means of the quality of the passion that is aroused in it, and it apprehends bodies by means of the passion aroused in it of their simply being colored and impeding the passage [of visual rays]. Moreover, it recognizes them in a general way insofar as surface is an intrinsic property of bodies. Thus, bodies that do not arouse a passion of this kind in the visual faculty (such as those that we said are not compact but subtle) are in no way seen. For they are not perceived by the visual faculty, much less perceived as bodies by it. Yet whatever has a more concentrated color appears to be denser in substance, even though it may not actually be, as [in the case of] milk in comparison to glass.[43]

[26] The visual faculty also discerns the place of bodies and apprehends it by reference to the location of its own source-points[44] [i.e., the vertices of the visual cones], which we have already discussed, as well as by the arrangements of the visual rays falling from the eye upon those bodies. That is, longitudinal distance [is determined] by how far the rays extend outward from the vertex of

---

[41]It is not clear from this passage whether Ptolemy is citing the three intermediate colors of red, rose, and blood-red as primary or merely exemplary. If the former, then he seems to be at least modifying Aristotle's account of the three primary colors (red, green, violet) of the rainbow as explained in *Meteorology* 3.4. If he means only to give exemplary intermediates, then Ptolemy, in citing these particular colors, is well within the classical tradition; see, e.g., Aristotle, *De sensu* 3 and Pseudo-Aristotle, *De coloribus*. For insights into contemporary artistic views of color, and their connection to philosophical theory, see Vincent Bruno, *Form and Color in Greek Painting* (N.Y.: Norton, 1977), pp. 53–102.

[42]Lejeune, *Euclide et Ptolémée*, pp. 44–48, explains how Ptolemy might have determined the extent of the visual field (= 180 degrees) observationally on the basis of a particular sighting-instrument used later in his analysis of reflection and refraction in books III and V.

[43]This is an instance where the relationship between visibility and density proves to be problematic (see n. 3 above).

[44]Latin = *principium*, which is the vertex of the visual cone; note that, although connected to the general sensory source (the Governing Faculty; see n. 16 above), the visual flux has its own distinct source in the center of the eye; see Lejeune, *Euclide et Ptolémée*, pp. 51–57.

the cone, whereas breadth and height [are determined] by the symmetrical displacement of the rays away from the visual axis. That is
how differences in location are determined, for whatever is seen
with a longer ray appears farther away, as long as the increase in
[the ray's] length is sensible.[45] (We should bear this point in mind in
the case of all types of illusion, for a certain kind of illusion is created on the basis of sensible differences other than those [objective
ones] arising from the actual visible properties—but we will deal
with these later.)[46] Objects that appear higher are seen with rays that
incline more toward the tops of our heads, whereas objects that appear lower are seen with rays that are lower and more inclined toward the feet. So too, whatever is seen with right-hand rays appears
to the right, and whatever is seen with left-hand rays appears to the
left. Therefore, objects are more clearly perceived through the ordered disposition of rays, because [those rays] are not directed
everywhere [indiscriminately] toward all the particular points that
are seen from the vertex [of the visual cone]. Otherwise, in fact, it
would happen that up and down or right and left would appear the
same, and the position of a thing would never appear definite, or
else every position would appear undifferentiated.[47]

[27] Furthermore, every body appears at a single location to the
[left or right] sides of anyone who looks with one eye alone, and it
is seen at a single location by anyone looking with both eyes if he
apprehends it with rays that are correspondingly arranged—that is,
rays that have an identical and equal position within both visual
cones in relation to their own axis. This happens when the axes of
both cones converge at a given [spot on a] visible object, as is the
case when we fix on visible objects with the normal glance that, by
nature, lets us scrutinize them carefully.[48]

---

[45]The notion that the distance of objects is somehow sensed by means of ray-lengths marks
a major difference between Ptolemy's and Euclid's treatment of spatial perception.

[46]This parenthetical phrase makes little or no sense in the context, perhaps because it has
been moved from its original location over the course of text-transmission. This possibility is
heightened because the phrase is followed immediately by eight lines of text that are clearly
misplaced in the original Latin version and that Lejeune repositioned to II, 47, where they fit
the context far better.

[47]The concept of directional privilege for the visual rays is enunciated by Euclid, albeit
without explicit reference to the axis of the visual cone; see *Optics*, def. 5 (perception of
up and down) and def. 6 (perception of right and left). Anna de Pace, "Elementi Aristotelici," pp. 249–51, emphasizes the importance of this notion of directional privilege as
an indication of Ptolemy's Aristotelian commitment to *physical* rather than mere geometrical explanation.

[48]See Preface, n. 11 above. By modern lights, of course, binocular vision is a matter of pro-
jecting properly corresponding (but not absolutely identical) images on properly corresponding sections of the retina; see Lejeune, *Euclide et Ptolémée*, pp. 124–129, for a useful
summary of the modern account.

[28] It seems, moreover, that nature has doubled our eyes so that we may see more clearly and so that our vision may be regular and definite. We are naturally disposed to turn our raised eyes unconsciously in various directions with a remarkable and accurate motion, until both axes converge on the middle of a visible object, and both cones form a single base upon the visible object they touch; and [that base] is composed of all the correspondingly arranged rays [within the separate visual cones].[49]

[29] But if we somehow force our sight from its accustomed focus and shift it to an object other than the one we wanted to see, and if the [new] object toward which our sight is directed is somewhat narrower than the distance between our eyes, and if the visual rays [that] fall together from our eyes on that [new] object are not correspondingly arranged, then that same object will be seen at two places. But when we close or cover either of our eyes, then the image in one of the two locations will immediately disappear, while the other will persist, [and the image that persists is] sometimes the one directly in front of the covered eye and sometimes the one directly in front of the other eye. This point will be easily understood if we try to explain it in the following way:

[30] Let a short ruler be set up, let two long, thin cylindrical pegs be stood vertically upon it, and let the distance between the two pegs themselves and between the pegs and the edges of the ruler be moderate. Let either edge of the ruler be placed between the eyes so that the pegs lie in a straight line at right angles to the line connecting the eyes.[50]

---

[49]Pace, "Elementi Aristotelici," p. 253, defines corresponding rays as " those that are positioned on the same side (right or left) [of the proper axis] and that are removed from the proper axis an equal distance," the phrase "equal distance" being her interpretation of the Latin, *positio equalis*. What might "equal distance" mean in this case? On the one hand, it cannot reasonably be construed in terms of angular displacement from the proper axis, as it is by Lejeune (*Euclide et Ptolémée*, pp. 141–142 and *L'Optique*, p. 27, n. 34): for instance, in figure 3, let E_1 and E_2 represent the two eyes, let AB represent the visible object, and let the two axes E_1B and E_2B focus on point B. Thus, E_1A and E_2A are corresponding rays insofar as they converge on the same point and lie on the same side as their respective proper axes. Yet they do not necessarily form equal angles with respect to their proper axes (i.e., angle AE_1B ≠ angle AE_2B). On the other hand, the "equal distance" at issue could be measured (as it is by Pace, "Elementi Aristotelici," p. 256) by the separation AB between both rays and their respective axes, but this is a pointless specification, since their convergence already dictates this equality. I think it best to follow Gérard Simon in construing *positio equalis* as "equivalent position"; see *Le regard*, pp. 144–145.

[50]In pointing to the similarity between the apparatus described by Ptolemy and one described earlier by Archimedes in his account of the dioptra, Lejeune, *L'Optique*, p. 28, n. 37, suggests that this similarity may not be coincidental. For Archimedes' description see Lejeune, "La dioptre d' Archimède," *Annales de la Société Scientifique de Bruxelles*, series 1, fascicule 1 (April, 1947), p. 31.

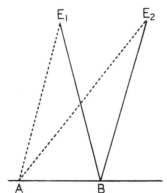

Figure 3

[31] Accordingly, if we focus our eyes on the nearer of the pegs, we will see it as one, whereas we will see the other, which is farther away, doubled. And if we close either eye, the peg that appears directly in front of that same eye and that forms one of the doubled images will disappear. On the other hand, if we focus our eyes on the farther of the pegs, we will see it as one and will see the nearer one doubled. And if we close either eye, the one of the doubled pegs that appears to be opposite to this same eye and that forms one of the two images will disappear.

[32] That each of these phenomena occurs and is a logical consequence of what we have proposed will be seen from the following diagram:

[33] *[EXPERIMENT II.1]* Let points **A** and **B** [in figure II.1] be the vertices of the visual cones, and let **B** lie at the right eye and **A** at the left. Let two pegs, **G** and **D**, be erected vertically upon line [**GD**, which is] perpendicular to **AB**, and from each vertex of the two visual cones let rays **GA**, **GB**, **DA**, and **DB** be extended to the two pegs. Then let us first focus on **G**, which is nearer.

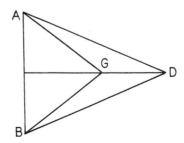

Figure II.1

[34] **AG** and **BG** will therefore lie upon the axes themselves. Of the remaining two rays, however, **AD** will be one of the left-hand rays [in the cone whose vertex is at **A**], and it is obvious that **BD** is one of the right-hand rays [in the cone whose vertex is at **B**]. It necessarily follows, then, that **G** is seen at one location, insofar as each of the axes corresponds with the other. On the other hand, **D** must be seen at two locations, since **AD** is one of the left-hand rays of the left eye, while ray **BD** is one of the right-hand rays of the right eye. Thus, when we cover the left eye, the left-hand [member of the doubled image] will disappear, and when we cover the right eye, the right-hand [member] will disappear.

[35] Now if we focus on **D**, the opposite will happen. This is demonstrated from the fact that **AD** and **BD** will lie on the [visual] axes. Thus, **D** will be seen as one, and **G** will be seen double, since **AG** happens to be one of the right-hand rays of the left eye, while **BG** is one of the left-hand rays of the right eye. So the opposite of what happened before will occur: that is, if we cover the left eye, the image seen on the right-hand side by ray **AG** will disappear, and if we cover the right eye, the image seen on the left-hand side by ray **BG** will disappear.[51]

[36] *[EXPERIMENT II.2]* However, if our focus is such that both axes do not converge on a visible object but are to all appearances parallel, like **AG** and **BD** [in figure II.2], then each of the pegs will appear double according to the principles we have outlined.[52]

[37] For this to be manifest and demonstrable, the nearer peg, which lies at **L**, should be white, while the farther peg, which lies at **M**, should be black. Points **L** and **M** will thus [each] be seen at two

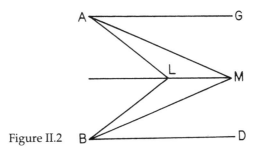

Figure II.2

---

[51]For a simple graphic description of experiment II.1 and its results, see Lejeune, *Euclide et Ptolémée*, pp. 132–134, or Simon, *Le regard*, pp. 134–135.

[52]The expression "to all appearances parallel" (Latin = *sensu equidistantes*) seems to indicate not that the axes must be actually parallel but that they must be unfocused with respect to the two cylinders.

locations flanking the [actual] locations of **L** and **M**. Accordingly, if we cover the left eye, [both] pegs that [appear to] lie on the right-hand side will disappear, whereas, if we cover the right eye, [both] pegs that [appear to] lie on the left-hand side will disappear. The reason is that, when rays **AL**, **AM**, **BL**, and **BM** are extended, **AL** will certainly lie farther to the right [of axis **AG**] than **AM**, and **BL** will lie farther to the left [of axis **BD**] than **BM**. As a result, the right-hand [images of the] pegs will be seen by the left eye and the left-hand [images of the] cylinders by the right eye.[53]

[38] *[EXPERIMENT II.3]* Again, if we suppose both [visual] axes to be parallel to each other and set up the ruler [lengthwise and] parallel to the line connecting the eyes so that the white peg faces the left eye and the black one the right eye, and if the distance between both pegs is equal to that between the eyes, then the two pegs will appear as three.[54]

[Let points **A** and **B** (in figure II.3) be the vertices of the visual cones, let **B** lie at the right eye and **A** at the left, and let **BD** and **AG** be their respective axes. Let the two pegs **M** and **L** be erected vertically upon the ruler so that **M**, the black peg, lies on axis **BD**

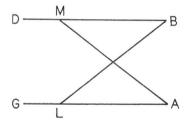

Figure II.3

[53]See Lejeune, *Euclide et Ptolémée*, pp. 134–135, or Simon, *Le regard*, pp. 135–136.

[54]As Lejeune notes in *L'Optique*, p. 32, n. 38, the demonstration that follows this enunciation in II, 39 and 40 not only makes no sense with reference to figure II.3, which is actually provided in the text, but is subject to internal inconsistencies in terms both of theoretical principles and of the phenomena described. In this context, Lejeune further notes that two other figures (II.3a and II.3b), which are interpolated in the text, seem tailored to fit the conditions. For these reasons, he is led to conclude that the only original part of this section was the enunciation and figure II.3; the remaining text, along with its figures, represents a botched attempt, perhaps by the Arabic translator, to fill a lacuna that had already developed in the Greek version. Although I find Lejeune's reasoning plausible, I think it at least equally plausible to assume that II, 39 and 40 were originally preceded by a now-lost paragraph in which the text is linked specifically to figure II.3. Accordingly, the paragraph in brackets between II.38 and 39 represents my reconstruction of the missing text as it might have been in the original Ptolemaic version. Notice that the reasoning used in this reconstruction is the reverse of that used in II.38 and 39, which indicates that the logic in those two paragraphs was badly mangled in the course of text-transmission. Cf., however, Pace, "Elementi Aristotelici," p. 259, for a counter-argument based on the assumption that the theorem as it stands is original and, moreover, in strict conformance with Ptolemy's theoretical principles for binocular vision.

and **L**, the white peg, lies on axis **AG**. Accordingly, each peg will be seen singly by the non-corresponding rays **AM** and **BL**, the white one appearing to the left along **BL** and the black one to the right along **AM**. But a third peg, composed of both colors, will be seen between the two single images by axial rays **BD** and **AG**, which are corresponding.[55] Therefore, if we cover the right eye, the left-hand peg of the two pegs flanking the middle one—i.e., the white one—will disappear along with the black component of the middle peg composed of both colors. And if we cover the left eye, the black peg on the left will disappear along with the white component of the middle peg composed of both colors.][56]

[39] [Assume, now, that the two pegs **M** and **L** do not lie upon the visual axes **BD** and **AG** but, rather, to one side or another of them (as in figures II.3a and II.3b).][57] Each of these pegs will be seen singly by corresponding visual rays, even though neither of the [visual] axes falls upon them, because the pegs are placed to the sides.[58] But a third peg, composed of both colors, will be seen between [the two single images] by rays **AM** and **BL**, which are not corresponding.[59]

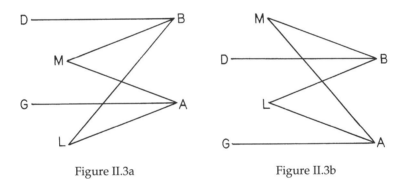

Figure II.3a   Figure II.3b

---

[55]In short, though they are directed at two different objects, such corresponding rays will, by their nature, fuse the respective images into one, just as they will when they are all focused to the same single spot.

[56]For a simple graphic description, see Simon, *Le regard*, pp. 136–137; Simon mistakenly treats this reconstruction as if it were Ptolemy's own account.

[57]Here, then, we have the second of two cases, in which, instead of actually lying on the visual axes (as in the case above), the two pegs lie to the left or right of those axes. In both cases, however, the line connecting the two pegs must be parallel to the line connecting the vertices of the visual cones.

[58]Note that here, the "proof" has the corresponding (i.e., parallel) rays produce two separate images rather than one fused one.

[59]As in the previous case, so in this one, the causal analysis is reversed: the fused image should be seen by the corresponding rays (i.e., parallel rays **BM** and **AL**), while the disparate images should be formed by the non-corresponding rays **AM** and **BL**. Note that in this case—which in essence represents the limiting case—where the visual axes are parallel, the corresponding rays actually do form equal angles with respect to their proper axes; cf. n. 49 above. For an overall analysis of this experiment, see Lejeune, *Euclide et Ptolémée*, pp. 137–141.

Therefore, if we cover the right eye, the right-hand peg of the two pegs flanking the middle one—i.e., the black one—will disappear along with the white component of the middle peg composed of both colors. And if we cover the left eye, the white peg on the left will disappear along with the black component of the [middle] peg composed of both colors.[60]

[40] Indeed, if we connect the visual rays [to pegs **M** and **L**], then both pegs will be seen at a single location by rays **AM** and **BL**, which are not corresponding, [the resulting single peg being] the one in which the two colors coincide. However, of the other two visual rays falling upon **L** and **M**, the black peg will be seen by the right-hand one, while the white peg will be seen by the left-hand one. Therefore, if we cover the right eye, the black peg to the right will disappear along with the white portion of the composite [middle] peg. And if we cover the left eye, the white peg to the left will disappear along with the black portion of the composite peg. This is demonstrated by the following figure.[61]

[41] *[EXPERIMENT II.4]* If the distance between the pegs is not equal to that between the eyes, the two pegs will be seen in four places.

[42] Accordingly, if the distance between **L** and **M** [in figure II.4] is greater than that between **A** and **B**, and if the pegs lie to the outside of the [visual] axes, then the black one will of course be seen in two locations on the right, because [rays] **AM** and **BM** both lie to the right [of their respective axes]. The white peg, on the other hand, will be seen in two locations on the left, because [rays] **AL** and **BL** lie to the left [of their respective axes]. Thus, if we close the left eye, the more displaced of the two black pegs—i.e., the one seen by ray **AM**—will disappear, because that ray lies more to the right [of its axis **AG**] than **BM** does [of its axis **BD**]; of the two white pegs, the one seen by ray **AL** toward the middle [of the visual field] will also disappear, for that ray lies less to the left [of its axis **AG**] than **BL** does [of its axis **BD**]. On the other hand, if we cover the right eye, the more displaced of the two white pegs, which is seen by ray **BL**, as well as the black peg that is seen toward the middle [of the visual field] by ray **BM**, will disappear.[62]

---

[60]Of course the very opposite will occur when the respective eyes are covered. It is in view particularly of this observational evidence, of which Ptolemy must have been aware, that Pace's attempt to rationalize the preceding "demonstrations" seems badly misguided; see n. 54 above.

[61]Here, too, the opposite will result when the respective eyes are covered. Presumably, the "following figure" is represented by figures II.3a and II.3b. Lejeune opines that it was at this point that someone, recognizing the inconsistency between the demonstration and figure II.3, attempted to rectify matters by interpolating figures II.3a and II.3b.

[62]See Lejeune, *Euclide et Ptolémée*, pp. 135–36, or Simon, *Le regard*, p. 137.

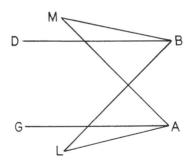

Figure II.4

[43] *[EXPERIMENT II.5]* If the distance between **L** and **M** [in figure II.5] is smaller than that between **A** and **B**, and if the visual axes lie to the outside of **L** and **M**:

[44] The black peg will be seen on the right-hand side by ray **AM**, because that ray lies more to the right [of axis **AG**] than **AL** does; then the white peg seen by ray **AL** appears next in [right-to-left sequence]. And if we cover the left eye, these two images will disappear. After this comes **BM**, which is the one of the middle rays [**BM** and **BL**] that is inclined to the left, and the black peg will again be seen by this ray, because it lies less to the left [of axis **DB**] than ray **BL**. Finally, the white peg seen by ray **BL** will appear farthest to the [left] side. These [two] images will likewise also disappear if we cover the right eye.[63]

[45] These phenomena that we have discussed arise and appear only because of the lateral separation [between the eyes], because both eyes are evenly disposed as far as height and depth [i.e., up and down] are concerned, whereas they are unevenly disposed as far as right and left are concerned. For in this type of disposition, both axes of the [visual] cones turn in the horizontal direction until their bases coincide on [the surface of] the visible object. It is also possible for the eyes to turn in a direction other than right-to-left, but it is impossible for them to shift up and down separately, be-

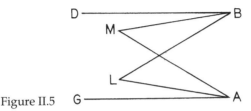

Figure II.5

---

[63]See Lejeune, *Euclide et Ptolémée*, pp. 136–147, or Simon, *Le regard*, pp. 137–138.

cause neither eye is set higher than the other, nor has nature re-
quired this, since they are not connected along the vertical but are
evenly disposed [according to that direction]; and [so] their rays
converge by turning to the sides.

[46] In order not to interrupt what we have set out to show [in this
section], we will [postpone] discussing problems involving the phenomena that arise in these cases, and [we will also postpone] any
detailed analysis of each particular appearance until the proper
time and place.[64]

[47] The visual faculty apprehends size when the cross-sections
of the base [of the visual cone] lying on the visible object are sensibly proportionate[65] to the distance between us and that object,
which is the case when the rays subtended by it form a perceptible
angle at the vertex of the cone. In fact, not every object that has dimension has a perceptible subtended [visual] angle, for when the
projections around the edge of an object are [seen from] extremely
far away, their visual angles dwindle so much that they cannot be
perceived.[66] However, that a magnitude somehow apprehended
under larger visual angles becomes imperceptible at vast distances
ought generally to be understood as a result of the weakness of the
visual flux. If anything having dimension lies exceedingly far from
the viewer—as do the bodies seen in the heavens—then, when the
distance from the viewer is not excessive, the smallness of the visual
angles themselves causes their dimensions to be disproportionate to
their distances.[67] And that is why many magnitudes that are seen
when they are nearby do not appear from afar, because the visual
rays meet [the object] at remote distances and form a subtended angle of imperceptible size.

[48] Although it is the case that a larger angle within a single visual cone contains more visual rays, the visual sense does not perceive that an object is larger because of the abundance of its rays
striking the object, as some have supposed, attributing the cause to
distance. In other words, [they claim that] when a given magnitude
is seen from afar, it looks smaller because of the paucity of rays
striking it, while it does not appear at all when none of the rays

---

[64]The study of binocular vision is resumed in III, 25–62 above.

[65]Latin = *habent sensibilem proportionem.*

[66]Cf. Euclid, *Optics*, prop. 9. In the original text, this sentence is placed at II, 69, where it
constitutes a clear *non sequitur.* For that reason, I have transposed it to this location, where it
seems by logic to fit best. As noted earlier (n. 46 above), the next two sentences were transposed by Lejeune to this location from II, 26.

[67]As Lejeune explains it, the main point of this rather cryptic account is that bodies can lie
so far away that, like the stars, they appear virtually dimensionless (i.e., as points) because of
the smallness of the visual angle under which they are viewed.

reaches it, because it falls within the tiny gap between rays.[68] But neither of these opinions is true, because nothing appears larger or smaller on account of the abundance or paucity of rays alone, since the quantity of the rays does not vary with the size of the angle but with accumulation and concentration. [On this basis], then, a thing does not appear larger or smaller, just as an object struck by light does not appear larger or smaller, on account of the abundance or paucity of incident rays but rather because of the largeness or smallness of its cross-section.

[49] Each of these points will become clear from the observable phenomena themselves. For if we look directly with both eyes at the same magnitude under the same conditions, we will see it more clearly than when we look at it with one eye. Moreover, when we look at it without anything in the way, it will appear clearer than it will when subtle media, which resist the passage of the visual flux somewhat, are interposed between us and it. Yet in none of these cases is there any [way] that [the object] might appear larger. For if someone places something subtle enough for the visual rays to penetrate between a luminous object and the objects struck by its light, then less light will strike them than before, but their cross-sections will not be larger than before.[69]

[50] Also, it is not because it falls within the gap between visual rays that an extremely small visible object is not seen. On the contrary, it must be understood that, as far as visual sensation is concerned, the nature of visual radiation is perforce continuous rather than discrete. But if we set up mathematical demonstrations and treat the visual rays as if they were straight lines, [it follows that] large magnitudes lying the same distance away as small ones that are invisible because of their smallness will still be clearly seen. This would not happen if the visual rays at that distance were diminished [in number] and spread out; instead, a large and a small object would look the same at that location. For if all the radiation that falls on their cross-sections along the entire base [of the visual cone] is composed of discrete rays, each of which apprehends a [single] point [on the given object], then, assuming that there is some real, spatial separation between those points, whatever lies far away ought not to be seen because the visual flux does not fall on those [interpunctal gaps]. Not even the points will be seen, because

---

[68]This theory underlies Euclid's account of size-perception in *Optics*, def. 4 and prop. 3.

[69]Later, in II, 126, Ptolemy claims that bright objects actually appear *smaller* than dimmer objects of the same size seen at the same distance; this apparent size-disparity involves a perceptual misjudgment (i.e., a *fallacia*) rather than an alteration in the size of the visual angles.

they have no size and do not subtend any angle. Hence, every such object [seen in this point-by-point way] will be invisible.[70]

[51] But if one claims that some of the visual rays will be discrete and some continuous at the same distance, then, since it has no necessary reason, this claim leads its proponent into ambiguity and error about things that are actually seen. For, according to that claim, every large object ought to appear fragmented [like a mosaic] rather than continuous, and small objects lying the same distance from the eye ought to appear and disappear by turns as they are moved to the sides.

[52] Furthermore, it follows from what we have proposed that differences in the size of objects must be determined and perceived according to differences [in the size of] corresponding visual angles. This is so, provided, as we have said, that those angles are perceptible, that they vary in no other way than in [terms of brute] sensation, and that there follows the perception[71] that the sizes differ among themselves not according to their actual extent, but because they subtend different angles.[72]

[53] All things being equal and equally disposed in everything except distance, orientation, and angles, then, as far as distance is concerned, whatever is closer gives the visual impression of being larger, while, as far as orientation is concerned, whatever faces us more directly appears larger. For in both of these cases, the visual angles become larger. And when we say " directly facing," we mean that the visual ray that falls on the center of the visible object is perpendicular to it, whereas we say that an object is slanted when none of the visual rays is perpendicular to it or when a ray other than the one falling on its center is perpendicular to it. We also say that an object is nearer when the ray that falls on its center is shorter, whereas we say it is farther away when that same ray is longer.[73]

---

[70]Clearly, then, the visual ray, as a discrete, mathematical line, is an analytic fiction for Ptolemy. The problem is how to justify the use of ray-analysis in accounting for intensity-variation; how, for instance, can we talk about "aggregation" and "concentration" of something that is continuous? As with Ptolemy, so with subsequent theorists, the response to this contradiction was to treat radiation (whether visual or luminous) as virtually, not really discrete, with the flux or light being treated as particulate and the ray being reduced to a trajectory of sorts; see, e.g., my "Extremal Principles," pp. 123–127 and 134–140.

[71]Latin = *ymaginatio*; I have translated it as "perception" in order to highlight the fact that it is the result of a subjective scrutiny and judgment of the initial visual data.

[72]In other words, at the most primitive level of visual perception, the apprehension of size depends solely upon the size of the visual angle, not on the actual spatial extent of the given magnitude.

[73]As Lejeune remarks in *L'Optique*, p. 38, n. 46, Ptolemy's inclusion of orientation and distance as determinants in size-perception is in response to Euclid's simplistic postulate (*Optics*, def. 4) that the perception of size is due solely to the visual angle. Pace, "Elementi Aristotelici," pp. 261–266, observes that, unlike Euclid's purely ideal, geometrical approach

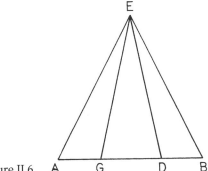

Figure II.6   A   G   D   B

[54] Now the aforementioned two variables [i.e., distance and orientation] by which the apparent size of objects is determined make no difference in the sensible impression, but the remaining one does, for, if there is a difference in that third variable—i.e., in the visual angles—then the object will appear larger when the angle subtended by it is larger.[74] For instance, if there are two magnitudes, such as **AB** and **GD** [in figure II.6], and if they lie the same distance away at the same orientation but subtend unequal angles, then **AB**, which subtends the larger angle at point **E**, will appear larger.

[55] And if there is a difference in [either] of the two remaining features alone, then the object will never appear larger, no matter whether it faces us more directly or whether it is closer. Thus, it will appear either smaller or equal, and in each case the apparent size will depend on relative differences in actual size.[75]

[56] *[EXAMPLE II.1]* For instance, if two magnitudes, **AB** and **GD** [in figure II.7], have the same orientation and subtend the same angle at **E**, then, since **AB** does not lie the same distance as **GD** [from point **E**] but is closer to it, **AB** will never appear larger than **GD**, as seems appropriate from its proximity [to **E**]. Instead, it will either appear smaller (which happens when the distance of one from the

---

to visual analysis, Ptolemy's approach is self-consciously concrete and psychological. Thus, Ptolemy conceives of "physico-perceptual reality as a complex synthesis" whose very complexity and concreteness make it irreducible to mere geometrical abstraction.

[74]The point here, which Ptolemy elaborates upon in II, 63, is that the initial, primitive apprehension of size-differences depends upon the size of the subtended visual angles; distance and obliquity are ancillary variables that allow us to refine the data provided by those angles (see also II, 58 below).

[75]See Lejeune, *Euclide et Ptolémée*, pp. 95–99, for a brief outline of the analysis of size-perception that follows in paragraphs 56–62 below.

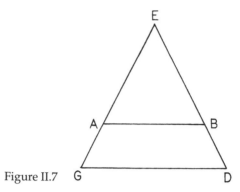

Figure II.7

other is perceptible), or it will appear equal (which happens when the difference in [relative] distance is imperceptible).

[57] *[EXAMPLE II.2]* Likewise, if there are two magnitudes, such as **AB** and **GD** [in figure II.8], that subtend the same angle at **E** and lie the same distance from it, and if their orientation is different, [so that] one of them, **AB**, faces **E** directly and the other obliquely, then **AB** will never appear larger than **GD** on account of its facing orientation. Instead, it will either appear smaller than **GD** (which happens when the difference in orientation between the two magnitudes is perceptible), or it will appear equal to it when the difference in orientation is imperceptible.

[58] It seems, moreover, that size-comparison among these objects ordinarily springs from judgment rather than from the actual nature of the orientation or distance. For, if a given sensible impression of the angles is aroused, if certain objects appear oblique or remote, and if one judges any of them to be smaller—even though it may be the only one—still, while one object may not be vi-

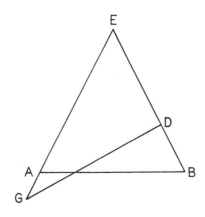

Figure II.8

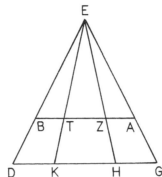

Figure II.9

sually sensed as smaller than another, but as equal [to it], we judge one of them to be larger. If the difference in obliquity or distance is imperceptible, then the two objects appear equal, but if that difference is perceptible, then they do not.

[59] *[EXAMPLE II.3]* For example, in the [earlier] figure [i.e., II.7], where the orientation was the same, if we draw the small angle formed by lines **HZE** and **ETK** [in figure II.9], then magnitude **GD** will always appear larger than **ZT**, because it is farther away, and the angle it subtends is larger. But **HK** will never appear larger than **AB**, since the judgment based on the angle is not outweighed by a judgment based on distance alone.[76] However, **HK** will appear smaller than **AB** if the distances and the angles differ perceptibly. But when they differ imperceptibly, the magnitudes will appear equal, just as in the first case.

[60] *[EXAMPLE II.4]* On the other hand, in the [previous] figure [i.e., II.8], where the distance was the same but the orientation was different, let us construct the small angle formed by lines **KTE** and **EZH** [in figure II.10]. Accordingly, magnitude **GD** will always appear larger than **ZK**, for the size of [its subtended] angle and the [amount of its] obliquity conspire to make it appear larger. Yet **HT** will never appear larger than **AB**, because the judgment based on the angle is not outweighed by the judgment based on orientation alone.[77] **HT** will, however, appear smaller [than **AB**] if the obliquity

---

[76]According to Ptolemy, then, even if **HK** were longer than **AB**, it would never appear to be longer to the viewer at **E**, because the judgment based on visual angle takes precedence over that based on distance alone. What this means, of course, is that size-perception cannot really be explained on geometrical grounds.

[77]Here we are faced with basically the same inconsistency as adverted to above.

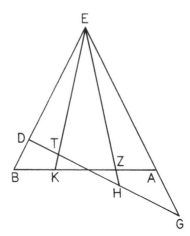

Figure II.10

and angles differ perceptibly, whereas, if they differ imperceptibly, it will appear equal to it.

[61] *[EXAMPLE II.5]* If we leave both magnitudes oriented as they are in the preceding figure and connect line **GB** [in figure II.11], then **GB** will always appear larger than **AB**, since it is farther away and more oblique, and since there is no difference in the angles. Moreover, **GB** will appear larger than **GD** when the judgment based on distance outweighs that based on obliquity. But it will appear smaller than **GD** when the judgment based on distance is outweighed by that based on obliquity. Finally, they will appear equal when the judgments based on both variables are equal to the sense.

[62] *[EXAMPLE II.6]* If, however, both magnitudes differ in all three previously cited respects, as happens when we construct the

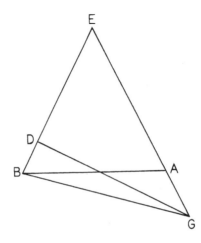

Figure II.11

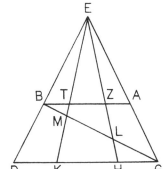

Figure II.12

figure [II.12] with both magnitudes similarly oriented and draw line **GLMB**, then **GB** will always appear larger than **ZT**, because all three variables (i.e., size of angle, amount of obliquity, and distance) conspire to make it appear larger. It also appears larger than **HK**, because **GB** has two of the variables (i.e., size of angle and obliquity) that make an object [look] larger. On the other hand, **LM** always [appears] smaller than **GD**, since two variables make it do so—i.e., smallness of angle and proximity [to **E**]. The only variable belonging to **GD** to make it appear smaller is its facing orientation. Nevertheless, **LM** does not appear smaller than **AB** in all cases, for **AB** has nothing but the size of [its] angle to make it appear larger, whereas **LM** has greater distance and obliquity in its favor. Still, **LM** appears larger when the difference in these two features together outweighs the difference in angles. If the difference in these two features [together] is outweighed [by the difference in angles], though, **LM** will appear smaller than **AB**, and if the differences balance out, the magnitudes will appear equal.

[63] That, then, is why we should not consider as adequate [the explanation of] those who have claimed that [in size-judgment] only distances should be taken into account along with angles and their resulting impression, whereas, they have claimed, any difference in orientation should be nullified in cases where the distances are equal. For, although there may be no difference in distance, a difference in orientation can still frequently occur. From this comes a perception of size different from the one that is due to the angle, as long, again, as we suppose this difference to be perceptible. Generally, therefore, since it seems that the size of visible objects is perceived by means of the size of the visual angles, that [angular] variable should be more fundamental and apposite [than the others in size-judgment]; and the same holds if it appears that differences

among sensibles are perceived by means of apposite and proportional differences [among all the visual rays]. It is in fact natural for the sense-impression of objects to be produced by the genus shared by their accidents. Now the genus common to magnitudes and angles is the size of the very continuous objects that are being sensed. In addition, the genus common to distances is the amount of distance in each of those objects. And the genus common to the orientation of [those] objects is their spatial disposition. On these bases, then, the visual faculty perceives distances by means of ray-lengths, and it perceives orientations by means of how the bases [of the visual cone] are spatially disposed. So it follows that the size of visible objects is apprehended primarily and properly by means of the size of the base, which is referred back to the vertex [of the visual cone], whence it is determined by means of the size of the subtended angle.

[64] The visual faculty perceives shapes by means of the shapes of the bases upon which the visual rays fall. And it perceives the outlines of the shapes by distinguishing the lines encompassing those bases as straight or circular,[78] these being the two principles that differentiate shapes. But it perceives [the shape of] the entire defining surface [of the object]—in terms of whether it is plane or spherical—by means of the surface of the [visual cone's] whole base.

[65] In addition, the visual faculty perceives [that] a line [is] straight when the length of the line that lies between the endpoints of [two subtended] visual rays is as tightly stretched as it can be; and that is why the straight line stands alone as the shortest of all the lines that have endpoints, and this is the natural definition of a straight line.[79] Thus, we gauge the straightness of objects, first, by means of bodies whose parts are uniform, such as hairs or woolen threads, which we stretch as tightly as we can;[80] and we do the same thing using rulers, dioptras, and the plumb-line.[81] The visual faculty perceives curvature, however, when all the visual rays that fall on the curve form equal angles at the vertex of the visual cone

---

[78]Latin = *curvus*; although the term can be taken generally to mean "curved" (or "convex"), later context makes it clear that it is intended here to denote "circular."

[79]See Archimedes, *On the Sphere and Cylinder* I, postulate 1.

[80]Cf. Hero's definition of a straight line, as quoted by Heath, *The Elements of Euclid*, vol. 1, p. 168.

[81]Latin = *id quod est erectum penes orizontem*; I am following Lejeune in translating this as "plumb-line." Taken generically, the dioptra was a straightforward sighting-device used in surveying and astronomy.

with the one, particular line that extends [from the vertex] to the center of the circle.[82]

[66] Now a surface is perceived [as] plane when straight lines fit everywhere throughout the entire base upon which the visual rays fall, whereas a surface is perceived [as] spherical[ly curved] when circular arcs fit everywhere on all portions of the base.

[67] A surface or line appears concave to us when the ratio between the visual ray orthogonal to it and the oblique rays is greater than the ratio between the corresponding [perpendicular and oblique] rays falling on a plane surface or on [its constituent] straight lines. Such [surfaces or lines] appear convex to us when the [former] ratio is smaller [than the latter one].[83] These two shapes [i.e., concave and convex] are alike, though, [insofar as] they are proportional to one another and belong to the category of things, such as ascent and descent, that are deemed "reciprocal." Each of them is also congruent with the other, the convex with the concave and the concave with the convex, for what encompasses the outside is concave, whereas what is encompassed inside is convex. So, too, objects are apprehended [as] concave by means of the surfaces of convex bases [defined by impinging visual rays], whereas objects are apprehended [as convex] by means of the surfaces of concave bases, just as such objects are perceived by touch, convex ones being apprehended through the concavity of the encircling hand, and concave ones being apprehended through the convexity of the encircled hand.[84]

[68] In the same way, all shapes composed of the primary shapes will be [visually apprehended]. But, as far as what is set forth about these shapes is concerned, it is well worth repeating ourselves to say that the differences in their [perceived] sizes must be a direct func-

[82]In other words, we perceive circular curvature when the circle directly faces the eye so that the axial ray is orthogonal to the circle's plane and cuts it at the centerpoint. Notice, first, that, according to this condition, we cannot recognize circular curvature with both eyes, since both axes cannot strike the circle's centerpoint orthogonally. Notice, second, that, according to this condition, we cannot recognize circular curvature when the circle is viewed obliquely. Cf. Euclid, *Optics*, props. 34 and 35.

[83]Let **A** in figure 4 represent the vertex of the visual cone; let **CF**, **DG**, and **EH** represent sections of concave, plane, and convex surfaces, respectively; let **AB** represent the orthogonal (i.e., axial) ray; and let **ACDE** represent an oblique ray. According to Ptolemy's definition, then, **AB : AC > AB : AD > AB : AE**, and these proportionalities hold no matter how closely **ACDE** approaches **AB**.

[84]Here Ptolemy reduces vision to something very much like, if not identical to, touch or feeling, the visual rays representing a sort of percipient extension through the air of the *principium nervosum* adverted to in n. 16 above. This notion links Ptolemy far more tightly at this point to the Stoics than to Aristotle, who explicitly rejects the model of touch for sight; see, e.g., *De anima* 2.11.423^{b}17–23 and *De sensu* 3.442^{a}30–442^{b}14.

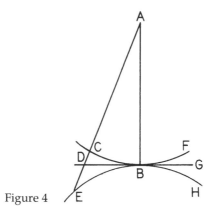

Figure 4

tion of the sizes of the angles in [equivalent] figures consisting of straight lines.[85]

[69] Now differences in depth [of curvature] among spherical shapes must be large enough to bear a perceptible ratio to [i.e., to be perceptibly different from] the distance that lies between them and the eye, because if they do not have a perceptible ratio, objects that are composed of straight lines will appear rounded.[86] If the afore-mentioned ratio [or difference] is imperceptible, convex or concave lines and surfaces will appear straight or plane. And the difference [i.e., the depth and type of curvature] in any of these shapes will not be apparent on account of the distance.

[70] Hence, we see spheres that are near us as convex,[87] but we see the shape of the sun and moon as disk-like from all quarters, because a circle is formed by the ends of the visual rays that de-lineate the outlines of each of them, whereas the central ray is per-pendicular (which happens in the case of spheres no matter how they are disposed but does not happen in the case of a disk-like figure unless it faces us directly).[88] A perceptible difference arises when the object's convexity blocks the impinging perpendicular [ray] long before it reaches the [object's] center. Indeed, when the radius of the sphere has no perceptible ratio to the perpendicular ray on account of the great distance, there will be no perceptible

---

[85]In other words, when we judge the relative sizes of concave or convex objects, we are judging on the basis of their planar cross-sections.

[86]The gist of Ptolemy's argument here seems to be that, on the one hand, if we could per-ceive depth of curvature when the requisite difference was imperceptible, then we might well also perceive planes as curved, since the same difference is imperceptible. Note that I have transposed the next four lines of text in the Latin edition from here to II, 47 above.

[87]Latin = *curvus*.

[88]Cf. Pseudo-Aristotle, *Problemata* XV, 8.

difference between the perpendicular [ray] that falls on the surface and the line [continuous with this ray] that extends to the center, nor between the circumferential rays and the ones falling on a plane surface containing a line that encompasses the visible object. For this reason, then, the convexity is not perceived except as a plane surface.[89]

[71] On this account, we do not see the breadth of disk-shaped objects when they are situated in such a way that [the planes] of their surfaces pass through the vertex of the visual cone; instead, they appear to us as straight lines. So, too, all [plane] surfaces will appear as [straight] lines—and a straight line [will appear as] a point—when they are extended to pass through the vertex of the cone. Again, when a disk-shaped figure lies [sideways] near the viewer, it will appear convex, as has been said; but when it is far away, it will appear as a straight line for the very reason that we have just given.[90]

[72] Yet, when [such] surfaces do not face the eye directly and [when their planes] do not pass through the vertex of the visual cone, they will appear as surfaces, to be sure, but their shapes will not appear the same as when they faced the eye directly. Hence, squares and circles will appear oblong, because, among equal sides and diameters, those at right angles to the axial ray of the eye subtend a greater [visual] angle than those inclined [to that ray].[91] Also, whatever lies nearer us subtends a greater [visual] angle than whatever lies farther from us, and those objects that are enlarged by perceptible increments and that lie at moderate distances appear larger.[92]

[73] Distortion in visual perceptions of this kind depends on the orientation of the figures and on the displacement of the viewers, but it is impossible for one, single case of such distortion to represent every one. If, however, one is disposed to examine many such cases together, the more carefully [he does so, the more] miraculous the natural capacity of the visual flux to arrange visual information turns out to be, both in its outward reach, which occurs in perfect order as it propagates, and in the sensitivity it exhibits in seeing and discerning differences among the objects it strikes, no matter how

---

[89]Lejeune's citation in *L'Optique*, p. 49, n. 54, of Euclid, *Optics*, props. 23–27, is inapposite in this context.

[90]Cf. Euclid, *Optics*, prop. 22.

[91]Cf. ibid., props. 35, 36, and 58.

[92]Apparently, Ptolemy is taking perspective into account here by noting that, when a square is viewed obliquely but straight on, its back side is shorter than its front side; or, to put it another way, its lateral sides seem to slant inward.

they are placed. Moreover, it does this swiftly, without delay or interruption, and it carries out a careful scrutiny with a marvelous, nearly incredible power, and it does this unconsciously because of its speed. This is possible for us to understand, particularly in the case of transparent media, when breaking occurs in any incident light-ray[93] and causes numerous distortions among the shapes of bodies that lie near those [transparent] objects. We immediately perceive this because of the alteration of shapes and the shifting of their locations in transparent media. The visual flux also acts the same way in shadow and in light; and it does so regularly and properly according to its position and the direction toward which it is inclined.

[74] So let what we have said to this point suffice, for the chain of reasoning has brought us to the verge of explaining the natural operation of vision that involves no illusion or misperception. Let us therefore discuss change and how the visual faculty apprehends each of its subsets, such as the one involving magnitude that is called "growth" and "diminution," or the one involving shapes and colors that is an alteration proper to them, or the one involving bodies that is [called] "locomotion." And after this follows [a discussion] of the nature of misapprehensions that arise in the visual faculty concerning the particular visible properties.

[75] First, we contend that, except for locomotion, the category of change is simple for everything. It is of course by means of a passion that arises in it, as well as by means of an accident conveyed by that passion, that the visual faculty apprehends shapes, colors, and bodies. It then perceives them and recognizes any change in them when they are noticeably altered in [no less than] the least perceptible time; for example, it perceives growth when the [visual] angle subtended by a given magnitude that remains spatially fixed becomes larger than it was before, whereas [it perceives] diminution when [that angle] becomes smaller [than before]. Also, it perceives a change in shape when the shape of the visible object alters the surface at the base of the visual rays falling on a given [area of that] object. And it perceives a change in color when the color-effect of a given body on the visual faculty is altered, and the visual faculty [thereby] suffers a different passion from the passions of colors arising in it [earlier] from the object.

---

[93]Latin = *radius luminis*. Presumably the "light" to which Ptolemy is referring in this case is that which emanates from the eye and passes through the transparent medium to the visible object.

[76] The particular way in which locomotion is perceived, however, requires a special discussion if we are to explain how it happens and make sense of the sort of change it represents. Visual
perception of the phenomena associated with locomotion depends
primarily on the visual faculty itself [rather than on the object un-
dergoing the change].[94] A particular object appears stationary either
when the vertex of the visual cone undergoes no perceptible motion
over a given continuous span of time, and the endpoint of a given
ray apprehends sensibly one and the same spot on the object, or
when it does undergo some motion, but it appears to the viewer that
the displacement of the viewpoint and the displacement of the rays
by which the object is visually grasped are equal during equal spans
of time.[95] The motion and rest of the visual flux, as well as their degree,[96] is apprehended not by the visual faculty but by the sense of
touch that extends to the Governing Faculty, in the same way that
we do not discern the motion of our hands by sight when our eyes
are closed but by means of a continuous [sense-link] that reaches
to the Governing Faculty.[97] We do, however, see a given visible object move when the viewpoint [at the vertex] remains immobile
over some minimal perceptible span of time while a noticeable
change somehow occurs in the visual flux or when the displacement
of the moving viewpoint is not equal to the displacement of the
visible object vis-à-vis succeeding rays during equal spans of time.

[77] We can understand the points we have made concerning the
movement and rest of visual flux more clearly now by means of the
following account.[98]

[78] *[EXAMPLE II.7]* Take, first, the case of visible objects that
move directly toward or away from the eye. Let **AB** [in figure II.13]

---

[94]As Pace rightly observes in "Elementi Aristotelici," pp. 266–269, Ptolemy's conception of
motion, as visually perceptible, is relative rather than absolute because it is ultimately dependent on a changing relationship between external objects and the vertex of the visual
cone. A moving visual cone can therefore perceive a stable object as moving, and vice-versa—
which is to say that the judgment of motion is as much psychologically as physically or mathematically determined; see, e.g., II, 131 and 132 below.

[95]In other words, if the amount by which the rays themselves are moved by the moving eye
is matched by the amount by which the object appears to be left behind as the rays sweep
over it, then the object is perceived as stationary; see II, 80 below.

[96]Latin = *quantitas*; presumably, the "degree" of rest would be its duration, whereas the degree of motion would be its speed.

[97]Pace, "Elementi Aristotelici," p. 267, cites this intrinsic "sense of touch" that extends to
the Governing Faculty as the ultimate safeguard against what she appropriately terms "absolute relativism" in Ptolemy's analysis of the perception of motion.

[98]See Lejeune, *Euclide et Ptolémée*, pp. 110–114, for a summary of the analysis that follows
in paragraphs 78–82 below.

Figure II.13   A   G   D   B

be a given visual ray, and let **A** be the vertex of the visual cone. Then, if the eye is stationary and does not move during some continuous span of time, and if the visual ray apprehends the visible object at point **B**, it will indeed appear stationary. If the object is apprehended elsewhere than at point **B**, it will of course appear farther away when it moves outward from point **B**, whereas it will appear nearer when it moves inward from point **B**.

[79] But if the eye moves directly toward or away from the object, e.g., from **A** to **G** along line **AB**, and if the visible object **D** moves to **B**, then, if the place at which the visible object [**D**] is [finally] encountered is where point **B** previously lay, it will appear stationary, and ray **AB** will appear to have moved the same distance, **AG**, as the eye. If the visible object is encountered in some place other than **B**—if it is in fact at point **D**—then it will appear to move toward the eye at the same speed [as the eye is moving]. If, however, it is encountered at some point closer in than **D**, it will appear to move toward the eye but more quickly. And if it is encountered at some point between **B** and **D**, it will [appear to] move in the same direction but more slowly. Finally, if it is encountered anywhere beyond **B**, it will [appear to] move in the opposite direction.

[80] *[EXAMPLE II.8]* Now let us take another example involving lateral motion. From point **A** [in figure II.14], which is the vertex of the visual cone, let several rays, such as **AB**, **AG**, and **AD**, be drawn, and let them remain fixed, so they do not move. When the location of the visible object is the same—e.g., when it remains in line with **AG**—then it will appear stationary. But if its location is [subsequently] different—e.g., if it is [subsequently] in line with **AD** or **AB**—it will appear to move.[99]

[81] On the other hand, if the visual cone pivots around its vertex **A**, and if we suppose that the distances between the given rays are equal throughout and that the motion occurs in the same [counterclockwise] direction, such that **B** [moves] to **G**, while **G** [moves] to **D**, then any [fixed] object that used to be seen along ray **AG** will be seen along ray **AB**, which is the next ray in line, and again that object will appear stationary, because the position of **AB** will be precisely the same as **AG**'s [used to be]. But if the object

---

[99]Cf. Euclid, *Optics*, prop. 51.

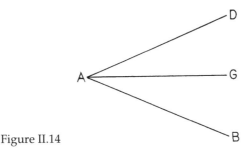

Figure II.14

continues on to any of the rays other than this one, it will appear to move. For if it [continues] to be viewed along [moving] ray **AG**, its [new] position will be in line with **AD**, and it will appear to move in the same direction and at the same [angular] velocity as the eye. If it is seen by any of the rays ahead of **AG** [which will have moved to **AD**], it will appear to move in the same direction [as the eye] but faster. And if it is seen by any of the rays that lie between **AB** and **AG** [which will have moved to **AD** and **AG**, respectively] it will appear to move in the same direction [as the eye] but more slowly. Finally, if it is seen by any of the rays behind [**A**]**B** [which will have moved to **AG**] it will appear to move in the opposite direction.[100]

[82] Therefore, since a motion that occurs in the smallest perceptible span of time must be noticeable [in order to be perceived], we contend that if either of these [conditions] is not met, it is impossible for there to be noticeable motion. For, if the distance over which the motion occurs is moderate and the time within which it occurs is imperceptible (in fact, this often happens in circular motions, as is evident in the case of a potter's wheel or the wheels of horse-drawn chariots when they are violently turning), the motion is not apparent, because the [moving objects] return to their original positions in less than the minimum perceptible time-span while the visual flux remains fixed for what strikes the sense as a continuous time-span on the places that the visible object occupies—and that is what characterizes [the visual perception of] objects that do not move. On the other hand, if an unnoticeable motion occurs during a brief perceptible time-span (as frequently happens in the case of objects that move when there is a great distance between them and

---

[100]Lejeune, *Euclide et Ptolémée*, p. 112, observes that, as in the perception of size, so in that of motion, a "corrective judgment" is necessary if the perceiver is to untangle his own motion from that of the observed object. Like Pace, he points to the intrinsic "sense of touch" as the requisite vehicle insofar as it provides us with a sort of physical self-awareness.

viewers—e.g., in the case of the bodies that revolve in the heavens or objects that [move] far off at sea), then [such objects] will not appear to move. And since every magnitude and every place is perceived by means of the distances that the visual flux senses, it is generally the case that, if the place to which a moving object makes its passage is imperceptible, neither the object nor its motion will be seen. And if this happens during a tiny perceptible [period of] time, we perceive neither when it occurs nor how long it took in terms of the [amount of] time we call "now." Under these circumstances, then, nothing will appear [to happen] during a continuous span of time, but such objects will appear to move only if, by ignoring the sense-data, we plot their motion in another way on the basis of their original location.[101]

[83] We have thus demonstrated how each of the visible properties is apprehended through passions arising in the visual faculty, as well as through the accidents that are conveyed by those passions, and we are satisfied with what has been said on that score. However, it follows that we should differentiate among the illusions that pertain to these [visible properties] so that, on this basis, we can resolve the issues that arise from the scientific investigation of optics.

[84] Of the illusions that involve the properties themselves, some involve causes that are common to all the senses, and some involve things that are specific to the visual faculty in the actual visible properties; some, moreover, arise in the sight, and some in the mind. It is therefore worthwhile to differentiate among them and to explain which illusions arise in the sense itself and which ones arise in the judgment of the mind.[102]

[85] Now in the illusions that are common to all the senses, misperception should not be ascribed to the eye, for in these cases one of the variables that naturally bear on every such phenomenon shows up. Moreover, we find that such illusions stem from a variety of causes [such as]: variation in the intensity of the visual power or differences among sensible objects themselves, when they are [mentally] compared to one another, or when one is [physically] placed beside the other. This happens in the science of optics with

---

[101]Ptolemy is obviously referring here to the tracking of planetary motion over long periods of time.

[102]In short, visual illusions can be categorized according to the level at which they occur: a) at the level of the brute sense or b) at the level of perceptual interpretation (presumably by the Governing Faculty).

regard to the comparison of one [visual] power to another, as, for example, [when we address] the issue of why some people do and some do not see the same objects at the same distance, and the same holds for objects that are nearer or farther away. In all these cases, it is a matter of greater or less [intensity of power].

[86] It is because of an abundance of visual power that objects are seen at a distance. Hence, older people always look at an object up close, because, along with the rest of their faculties, the visual power is produced more weakly in them.[103] On the other hand, those who have deep-set eyes see farther than those who do not have such eyes; and the reason for this is that their visual power is compressed, for when it emanates from narrow places, the visual flux is stretched and elongated.[104]

[87] The cause of farsightedness, however, is the eye's [humoral] moisture, part of which is taken away along with the visual flux. Thus, when there is little moisture, the visual flux leaves the accompanying moisture behind at once as it radiates outward and [so] close things are seen clearly. But when there is a great deal of moisture, an object will be seen at a great distance, so that one who wishes to see with certainty must look from far off.[105]

[88] What happens in vision, moreover, is like what happens in touch. For a given object feels softer to one whose body is harder than to one whose body is softer. Furthermore, it feels cooler to one whose body is hotter than to one whose body is cooler.[106]

[89] In addition, since the whole of a given magnitude appears larger, while its parts appear smaller, when the sense grasps the magnitude as a whole, it apprehends the quantity of the whole better [than it does that of its parts], but when it grasps one part at a time, it apprehends the whole less [well] than the part. And when each of the objects that are apprehended is smaller than the whole that is composed of them, then the magnitude, if it is continuous,

---

[103]As Lejeune points out in *L'Optique*, p. 56, n. 69, this reasoning seems to contravene the evidence: it is generally the case that the older one gets, the more farsighted, not nearsighted, one gets; cf., e.g., Pseudo-Aristotle, *Problemata* XXI, 25.

[104]See Pseudo-Aristotle, *Problemata* XXXI, 8 and 16, where squinting in myopia is explained in terms of concentrating "vision" (*opsis*) for a clearer view; cf. Beare, *Theories*, p. 81.

[105]Presumably, then, farsightedness is a function of youth. Within this context, Ptolemy's claim that older people are nearsighted finds justification in Aristotle's theory that the process of aging involves desiccation. Accordingly, the more desiccated the eye becomes, the less hampered the flux is at short distances by the intraocular moisture; see *On the Length and Brevity of Life* 5; cf., however, *Meteorology* III, 4, 374a21–23.

[106]By analogy, then, to someone with an abundance of visual flux, dimly illuminated objects appear brighter than they do to someone with a deficiency of such flux.

will have a certain disposition with respect to its particular elemental parts.[107]

[90] So too, nearly the same as what we have described arises from the comparative difference of objects according to "greater" or "less"—e.g., what happens when objects are seen next to something larger and are judged to be smaller [than they really are], or when objects are seen along with more brightly colored ones and are judged to be duller [than they really are]. This of course stems from a defect in judgment when one [sense-impression] is overshadowed by another according to the perception proper to objects that are overshadowed by others and the [perceptual comparison] is not made proportionately.

[91] In fact, this affects all the senses, for an impression sometimes diminishes and sometimes augments not just in vision, but also in smell, and in taste as well as in sound. For at the same time some things are sensed more [intensely] than others of the same kind. Not only should accidents of this kind be segregated from those about which vision errs, but also accidents that depend upon a change in the visible properties that becomes so extreme as to cause an object to appear other than it would if the visual faculty and mind were not then somehow disturbed and their state were not changed.

[92] Now when particular colors are not seen according to their proper nature but are mixed together, no one would say that this [mixed appearance] is due to an illusion occurring in the visual faculty either on account of their multiplicity or on account of the place they occupy, because the images of colors do not appear other than they were on account of the way light falls on them. Instead, their appearance varies in terms of more or less for the same reason that what is not illuminated does not appear.[108]

[93] The moon, however, has its own color, which appears during eclipse, when light is absent, but does not appear at other times. This problem is solved insofar as, during the time of eclipse, the moon lies in a sort of shadow. The earth, by means of which the blocking [of sunlight] occurs, is quite far from the moon at that time. At other times, however, the moon lies in darkness, since the por-

---

[107]As Lejeune suggests in *L'Optique*, p. 57, n. 72, this passage is reminiscent of Pseudo-Aristotle's *Problemata* XVI, 7. The point here seems to be that we can look at a composite in two ways: in terms of its totality rather than of its parts, or vice-versa. Still, even when viewed in terms of its constituent parts, the whole emerges as a distinct, supervening entity. All of this presumably has to do with some variation in perceptual " sensitivity" toward composite objects (i.e., vis-à-vis the constituents or vis-à-vis the composite whole).

[108]In short, variations in light-intensity cause variations in the intensity of the colors upon which they shine (i.e., from bright to dull).

tion that creates the blockage—namely, half the lunar sphere—is contiguous with the portion that is thereby blocked. But more light falls on what lies in shadow than on what lies in darkness.[109]

[94] Likewise, we ought not to assume that our sight is deceived when the image of colors changes in some way, as happens in the case of light that is colored by certain flowers or other coloring agents and then strikes visible objects. For everything that appears to be the same color by dint of the color of something else shining on it should not be deemed to have changed except through an effect created in the visible objects themselves from a mixture stemming from both [agent and patient] that produces a common state in the eye, in the visible object itself, and in the sensation arising from them.

[95] Under the same classification are grouped those objects whose color appears homogeneous not because of a mixture of another color with its own but because of the various colors that it possesses. They appear this way on account of distance or quickness of motion, because in either of those cases the visual power is weakened in its capacity for seeing and discerning individual constituents. For, if the distance of the visible objects is such that the [visual] angle subtended by the entire object is sufficiently large while the constituent angles subtended by different [constituent] colors are imperceptible, then, on the basis of an apprehension of the undiscerned constituent parts, when the overall sense-impression is assembled, the color of the object will appear uniform rather than composed of individual constituents.

[96] The same happens in the case of extremely quick motion, for instance, the motion of a potter's wheel daubed with several colors. For a given visual ray does not stay fixed on one particular color, since that color leaves it behind on account of the speed of rotation.

---

[109]The reasoning here seems to be this: shadow and darkness are different entities, the former involving a relative absence of light, and the latter a complete absence of it; during eclipse, the moon falls into the earth's shadow, so there is still enough ambient light striking it to render its inherent color visible; at other times, the unlighted portion of the moon is in complete darkness, so none of its inherent color is visible. Unfortunately, Ptolemy never specifies what the source of the ambient light in full lunar eclipse is. By Ptolemy's time, the problem of "moonglow" and its source during eclipses was long recognized. Two basic explanations are cited by the first-century thinker, Plutarch, in his dialogue *Concerning the Face which Appears in the Orb of the Moon* 21, 933F–935C. The first of these, which was held by the Stoics and which persisted to the time of Galileo, was that the moon is endowed with its own faint luminousness that becomes visible in the absence of occluding light. The second theory, which is ascribed by Plutarch to the "mathematicians" and which has a suspiciously modern ring, was that the glow is caused by a modicum of light that somehow shines around the earth's shadow to strike the moon's surface. Ptolemy's account is clearly more consonant with the latter (Mathematical) than with the former (Stoic) explanation.

And so, in falling on all the [constituent] colors, the same ray cannot distinguish between first and last nor among those that [now] occupy different locations. In fact, all the colors appear simultaneously throughout the disc as a single, uniform color that is the same as the color that would actually be formed from a mixture of the constituent colors. By the same token, if spots of a color different from that of the disc are marked on it (provided they are not on its very axis), they will appear to form circles of the same color [as the given spot] when the disc is rapidly spun. On the other hand, if [differently colored] lines are drawn on the disc's surface through the axis, then the entire surface of the disc will appear to be of a uniform color when it is spun. For, since the color spins through a perceptible distance in equal perceptible time-intervals, it is adjudged to touch all the locations through which it passes. The visual impression that is created in the first revolution is invariably followed by repeated instances that subsequently produce an identical impression. This also happens in the case of shooting stars, whose light seems distended on account of their speed of motion, all according to the amount of perceptible distance it passes along with the sensible impression that arises in the visual faculty.

[97] In addition, the illusions that involve size and shapes are created in much the same way as we have already said, [for instance,] when the same object appears smaller and seems to be more rounded [than it actually is] at a great distance, because the subtended angles diminish and become tiny.[110] Also, when an object similar to a disc revolves swiftly about one of its diameters, it appears egg-shaped to anyone viewing it from the side rather than along the axis of its rotation. The reason is that during each revolution its surface is sometimes perpendicular to the visual ray falling on its center, at which time it appears circular; but sometimes that same surface will be aligned [lengthwise] along the line whose extension intersects the vertex of the visual cone, in which case it appears as a straight line. More often than not, however, it is oblique[ly disposed toward the viewpoint] and thus appears distended. Since the longer shape of the object predominates over the others in continuous motion, then, the shape of the entire object is noticeably distended throughout because of the speed of rotation. And so it appears to the observer to have the shape of an egg.

[98] An illusion can also arise regarding the motions of visible objects without any deception of the visual sense. But this is due to the

---

Differentiating Faculty,[111] when something like what we said happens in the case of the rapid motion of potter's wheels occurs by means of the aforementioned [visual] effects. When [such wheels] revolve, it is assumed that their motion goes unperceived because of the brevity of time within which they rotate. This happens as well in the case of moving objects that are seen from afar, for they are adjudged to be stationary on account of the immobility of the visual ray, since the rays do not in this case move perceptible distances in a [sufficiently] short time.[112]

[99] If, indeed, we move some given distance that does not bear a sensible proportion to the distance of a visible object, then the visible object will seem to travel along with us in the same direction and at the same speed. This often happens with the moon and all the other celestial bodies on account of their brightness and considerable distance [from us]. And so, when we move and direct our eyes toward such a visible object, our sight does give us the impression of a perceptible displacement, because the distance we move is as a point to the distance of the object away from us. On the other hand, objects that a given eye apprehends as moving move with it. It necessarily follows that, as long as they travel through a perceptible distance that is detectable to the sense, and as long as the distance [between eye and object and the distance of the object's travel] are not [sensibly] proportional, those objects will be adjudged to move with the same speed and in the same direction as we do.

[100] Likewise, too, of objects that have the same speed, those that are closer appear to move faster.[113] Among those objects whose cross-sections are equal, the ones that are nearer to the viewpoint subtend a greater visual angle.[114] Those things, moreover, that mark out greater arcs in equal times appear to travel more swiftly.

[101] . . . even though they are smaller, [certain objects], such as small boats and small arrows, that are moving at the same rate as larger ones in fact appear to move more rapidly.[115] Indeed, putting it another way, objects that traverse [incremental] spaces longer than themselves in several stages during equal times are adjudged

---

[111]Latin = *virtus discernitiva*; this is evidently another designation for Governing Faculty, perhaps to specify one of its particular functions?

[112]See II, 82 above.

[113]Cf. Euclid, *Optics*, prop. 54.

[114]Cf. ibid., prop. 5; this is, of course, a non-sequitur within the context.

[115]As Lejeune notes (*L'Optique*, p. 63), the grammatical context of this first sentence-fragment indicates that there is a lacuna of some sort between the end of paragraph II, 100 and the beginning of II, 101 in the Latin text as it presently stands.

to move faster than those that traverse [incremental] spaces equal to their own length, since a small object measures a given distance several times more than a large one. It therefore follows that, when objects of unequal size travel at the same speed over given distances, the smaller ones appear to move in more stages. Accordingly, then, since the object passes over the distance apprehended by the visual flux in equal times in more stages than [does the other object passing] over the same distance, it appears to move faster.[116]

[102] As we have said, these difficulties and their ilk must be imputed to the passions and natural accidents that befall the visual sense, for in those features through which the reality [of visible objects] is perceived, there arises a misperception based on variations that are due to the visible objects themselves. There are, however, cases where this is not the cause. In such cases, where the properties of the visible objects are unaltered and there arises a certain perception, the sense does not grasp the visible properties according to its custom and nature. There is, instead, another passion or accident in them that creates an illusion, in which case the cause lies in the visual faculty itself. We must present this case in the present chapter and discuss it, distinguishing first among the things responsible for the illusion, some of which affect the visual faculty from the visible objects themselves, others of which arise from elsewhere.

[103] Now a visual illusion springing from the visible object arises in the case of colors only by the action of something besides the visible object, whereas for all the remaining visible properties an illusion stems from the various defining characteristics that pertain to those objects themselves.[117] For these visible properties are perceived mediately, while colors are perceived immediately and by themselves. As far as characteristics that seem to pertain to the visible object are concerned, the cause of the visual illusion itself is that the visual sense does not grasp the features that provide the intrinsic, defining characteristics for [visual] perception. Rather, it grasps features of something else that has a mediate status [with respect to those primary features]. But the reason for illusion in the case of those properties that create visual passions apart from the visible object is that visible objects cannot be sensed without an accompanying sensation of neighboring bodies. This does not happen be-

---

[116]Ptolemy is referring here to the impression we have of the apparent rapidity with which small animals, such as mice, scurry, not so much because of their actual velocity as because of their proportional velocity.

[117]In other words, altering the secondary properties (i.e., those that are secondarily visible) of visible objects will not alter their color, whereas such alteration can change other secondary properties. Thus, for instance, change in shape will not affect color, but it can affect size.

cause their action is such in all rare and subtle bodies but because what does the first sensing must to some extent not be sensible, just as the prime mover must by no means itself be mobile; so we ought not to impute a passion to everything and [to suppose] that nothing remains according to its own nature.[118] For things that are assumed to be of such character are found to be of a different kind, and thus the first effect will be destroyed, since the end in these cases would lead to infinity.

[104] Since, therefore, we have already discussed visual illusion and categorized it according to a) the kind arising in the passion that we have said is proper to the visual faculty itself and b) the kind that arises during interpretation, let us explain each of them separately according to the particular visible properties that are subject to them. But let us speak first of those things that arise in the passion; and let us assert that, among the passions that arise in the visual flux, one is called "coloring," another "breaking," and another "diplopia."

[105] It is natural and customary for visual flux to fall in a pure and direct fashion upon visible objects while maintaining a corresponding arrangement of the two visual cones.[119] On the one hand, coloring involves the privation of purity, while, on the other, breaking involves the privation of a direct line of sight to an object. Diplopia, finally, involves a lack of corresponding arrangement [of the two visual cones]. Each of these passions is of two kinds. Coloring, in fact, can be either anterior or posterior. Anterior [coloring] is the kind that occurs in front of the visible object, whereas posterior [coloring] is the kind that occurs behind the visible object. Breaking, for its part, involves either refraction or reflection. Refraction is the sort of breaking that happens [during the visual ray's passage] through a resisting medium, whereas reflection entails a [complete] breaking [and rebound of the ray] from [the surface of] the resistant medium. Diplopia, finally, is either anterior or posterior. Anterior diplopia involves focusing of the [visual] axes in front of the visible object, whereas posterior diplopia involves focusing of the [visual] axes behind the visible object.

[106] An illusion arises in the case of colors by the visual flux's being colored by something other [than the visible object itself]. Ante-

---

[118]The surface-sense of this passage seems to be that there must be an impassive final (or first) cause in visual perception that arbitrates without itself being subject to arbitration—hence, the avoidance of infinite causal regress, as indicated in the final sentence of the paragraph. What Ptolemy intends within this context, however, is far from clear; perhaps we are dealing here with another transposition of text.

[119]Here we have the single instance in which Ptolemy uses *conus* instead of *piramis* (see Preface, n. 10, above).

rior coloring is generated by itself and also by the two types of breaking. Posterior coloring, on the other hand, is generated only by the two types of breaking.

[107] Anterior coloring is generated by itself when we have looked for a long time at some very bright color and then look away to something else. For in that case what is last looked at seems to possess something of the color of what was first observed because the impression of bright colors lasts a long time in the visual faculty. And so, after we have looked at such [bright colors], we see neither clearly nor without some impairment.[120] Anterior coloring also happens when we look at something through thin, threadbare cloths of a red or purple hue. The visual flux passes through the weft and warp of the cloths without breaking and, in the process, takes on something of the color of the threads it brushes by. Thus the visible object appears to be tinged with the color of media traversed by the visual flux.

[108] In addition, anterior coloring arises in reflection when the mirror is immobile, dense, and colored. For the colors of objects that are seen by reflection from such mirrors appear to be commingled with the color of the mirrors.[121] Anterior coloring also arises in refraction when the transparent media are not particularly rare but, rather, cause significant refraction yet are neither very weakly nor very intensely colored. For if they are intensely colored, they do not allow the colors of the objects seen by means of refraction to mingle with their own after the visual flux has passed through; in effect, they predominate and color the visual flux by the force of their own coloring. And if the transparent media are only tenuously colored, the visual flux does not undergo coloring [from them]. If, however, the mingling is proportionate, then some sort of coloring occurs,

---

[120]Latin = *lesio* = "impairment" or "injury"; in short, excessively bright visual impressions can actually harm the visual faculty; see II, 23, n. 35 above. Ptolemy is of course referring to after-images here; understandably enough, he fails to realize that, depending on certain conditions, such images can be either positive (i.e., the same color as the original stimulus) or negative (i.e., the complementary color of the original stimulus); see, e.g., M. Luckiesh, *Visual Illusions: Their Causes, Characteristics and Applications* (1920; rpt. New York: Dover, 1965), pp. 128–130.

[121]The most common material for mirrors in Ptolemy's day was bronze, although other materials, such as tin, lead, gold, silver, polished stone (e. g., obsidian), and even dark glass were used; indeed, in the notorious reflection-experiments described in III, 8–12 below, Ptolemy specifies the use of iron for the three mirrors involved. According to Forbes, the earliest mention of a glass mirror with metal backing comes from the early third century A.D., and it was not until the very early sixteenth century that the modern glass mirror, backed with a mercury-tin amalgam, was developed. Thus, the mirrors of Antiquity all lent varying degrees of color to the images reflected in them. See R. J. Forbes, *Studies in Ancient Technology*, vol. 5 (Leiden: Brill, 1966), pp. 187–189.

and the visual flux takes on incidental color, as happens, for instance, [when it passes] through a thin cloud that is neither white nor excessively tinged, or through sheets of unblemished horn, or through the subtle vapors above the earth's surface, or through glass fragments, or through other lightly colored objects.[122] In such cases, the colors of visible objects will appear to be commingled with the color of the transparent medium through which the visual flux passes, [the resulting color] being a composite of both colors.

[109] Posterior coloring, as well, arises in the case of refraction, for instance, when transparent media take on something of the colors of the objects that lie within them or that appear behind them. The same holds for reflection as it occurs from mirrors. For, if mirrors are so disposed that the images seen in them lie on the surfaces of the mirrors themselves, then they are deemed to have the color of the object that the visual faculty apprehends after reflection.[123] This sort of thing also happens in another sort of disposition, when the reflection takes place at extremely acute angles to the surface of the mirrors. For the place where the object['s image] appears is determined by where the ray emanating from the eye to the mirror's surface meets the orthogonal dropped from the visible object to that surface [i.e., the cathetus of reflection], given that each of these two lines lies in the same plane, which is normal to the mirror's surface.[124] Therefore, when the angles at which the reflection takes place from the surface are small, the cathetus of reflection will be short, and so the location of the image will nearly coincide with the surface itself.

[110] These facts, as well their logical implications, will be demonstrated in due time in a later discussion that will include appropriate details, so that, in attempting throughout to demonstrate specific points about individual instances, we do not create confusion in our account. Let this suffice for now, therefore, and let us

---

[122]Thin plates of horn, as well as of marble and other less durable materials (e.g., parchment), were often used in the place of glass in Antiquity; see Forbes, *Studies*, vol. 5, pp. 184–187 and *Studies*, vol. 6, pp. 167–168. Depending on where it was manufactured and whether (and what) coloring agents were used, glass of that era tended to take on a variety of shadings from light yellow to dark blue.

[123]Ptolemy describes this supposed blending of the image and the mirror's surface in his later discussion of concave mirrors in book 4 (see IV, 25 below). Such blending occurs, Ptolemy claims, either when the image-location is indefinite or when it coincides with the center of sight.

[124]Here, for the first time in the *Optics*, Ptolemy articulates the principle of image-location in mirrors, as well as the principle that the plane of reflection is normal to the surface of reflection (or to the plane tangent to that surface at the point of reflection); see III, 3 below for a recapitulation of these principles.

limit ourselves to discussing only what is pertinent here: i.e., to account for the [type of] illusion that is due to the visual faculty.[125]

[111] According to the preceding account, if a fire or light lies slightly above the horizon and there is a pond or pool of water near the observers, and if that water is somewhat roiled, then, to the visual flux that is reflected from the water's surface to the fire or light, the image of the light in that water's surface will appear distended; and [that image] will seem to move in the same direction as the observers so as to lie in a direct line between them and the visible object. For that reason, there appears to be light where in fact no light exists, but this appearance is due to reflection. For the light of the luminous object is considerable and is diffused in various directions, but the portion of it that is seen in the water represents a mere fraction that is projected only longitudinally. Moreover, the luminous body itself does not shift with the movement of those viewing it but lies in the same place for all who look at it from different directions. The image in the water does shift with the movement of those viewing it, though, and when there are several observers, each will see it in a different location: i.e., where it appears [to lie] in a direct line with the generating object.

[112] These, then, are the characteristics of objects that appear by themselves and of those that are seen by means of images cast in other objects. As long as it is moderate, any unevenness in the water's surface causes light to appear elongated and dispersed. When the surface is smooth and absolutely flat, then the image seen in it is identical in shape to the actual objects [generating it], because in that case reflection can take place at only one spot on the plane surface, the angles being equal when there is reflection from one place to another. And when the [reflecting] surface is uneven and unpolished, it is possible for a visible object to appear at several spots on the composite surface defined by the [waves and troughs of the] surfaces under discussion. That is how the visible object's image becomes distended, since the image will lie at those spots from which reflection at equal angles occurs, and those spots vary according to the juxtaposition of convex and concave surface-segments. Moreover, the location of the [composite] image that appears at common spots will coincide with the [reflecting] surface, since the visible object will lie outside the center [of concave curvature]. On the other hand, each of the images that appear below the surface will appear only slightly

---

[125]Ptolemy is referring to the segregation in II, 84 above of illusions according to whether they are due to the visual faculty or whether they are due to the mind and its subsequent evaluation of the visual data.

removed from it on account of the obliquity of the visual ray.[126] But when the water is violently roiled, the [composite] image is broken up and dispersed according to the crests and troughs, whose unevenness is considerable. This is why each particular image produced by such surfaces can be seen clearly, some of them appearing higher than others and some appearing lower. When, however, the surface is only slightly agitated, the image appears more continuous because of the smallness of the reflecting surface-segments that are contiguous to one another. And when the differences in height between such contiguous surface-segments is undetectable, the entire [composite] image will appear as a single, elongated object and will seem to lie at the same level as far as the sense is concerned.

[113] From what we have said, then, it is obvious that if some portion of the sun overlooks a relatively calm sea at dawn or dusk, there will appear on the water an image that is not only continuous with, but also of the same size as the portion of the sun appearing above the horizon. If, however, the water is somewhat roiled, that image will appear longer [than the portion of the sun generating it].

[114] So any illusion involving colors that is due to a passion arising in the visual faculty is created at most according to the types discussed earlier. However, the illusion involving location that is due to breaking and diplopia is created in either of the following ways.

[115] In the case of diplopia, the same object appears in two places, or two objects appear in one place, just as we showed earlier.[127] In the case of reflection and refraction, though, the object appears [to lie] directly in line with the [incident] visual ray, even though this is not actually the case insofar as the ray is broken. Moreover, among objects that are seen in this way, some appear to lie closer than they actually are—the true measure being the distance between the objects and the viewers in combination with the distance between the objects and the surface at which the breaking occurs. Other objects, however, appear to lie at the same distance as the true one, and others yet appear to lie at a greater distance [than the true one].[128] In addition, certain objects are seen by those visual

---

[126]Ptolemy is attempting to explain the distension of the image in this case according to two basic principles articulated in II, 109 above. On the one hand, the images cast by the troughs will coalesce with the reflecting surface since the reflected light-source lies well beyond the center of curvature (i.e., the image cast is a real one); on the other hand, the images cast from around the top of the crests will coalesce with the reflecting surface on account of the acuteness of the angles of incidence and reflection. As a result, the composite image (crests + troughs) will tend to coalesce lengthwise along the reflecting surface.

[127]See II, 27–44 above.

[128]In other words, the true distance of the object from the observer in the case of refraction and reflection is measured by the sum of the incident and reflected or refracted ray, whereas

rays that lie on the same side with them, whereas others are seen by different rays, depending on the surface-shapes of the bodies that cause the breaking; this is the case, for example, when we see left-hand objects with right-hand rays or higher objects with lower rays,[129] and vice-versa.

[116] It is for this reason that elevated islands appear lower than they actually are when seen at sea, because the rays that strike the sea below them are raised higher by the reflection while, meantime, the air that is perceived directly by the visual flux is rendered higher than the islands. This point becomes even clearer when a red cloud lies above the island, for in that case an identical red cloud appears below the island.[130]

[117] So too, it is possible in reflection for the same object to be seen in several places, as happens in the case of concave mirrors and in mirrors that display several images. They do so because they are set up in such a way that certain of their constituent parts reflect rays to the visible object while others reflect them elsewhere. Accordingly, several disparate rays show the object according to their particular direction, and the number of locations where the visible object appears is equal to the number of [such disparate] rays.[131]

[118] Therefore, in the case of all visible objects, an illusion arises from the passion that affects the visual flux. In the case of sizes and shapes, this is due to either kind of breaking, whereas in the case of motion it is due to radial sweep[132] [of the visual flux].

[119] In reflection and refraction, which define the kinds of breaking, an illusion comes about when the surfaces of reflection or refraction are not plane. In fact, this is due to the convexity and concavity of the [given] surface, because, if the visual angles subtended by the images are larger or smaller than the angles subtended by the [generating] objects as seen [directly] and properly, and if the distance [between eye and image] is the same as that [be-

the apparent distance, being measured along the imaginary prolongation of the incident ray, can vary from this in either direction. The apparent distance of images is, of course, subject to various distortions in convex and, more particularly, concave mirrors.

[129]For instance, up and down are reversed when we view objects reflected in water, and the viewer's own image often appears upside-down in concave mirrors.

[130]Ptolemy seems to be referring to mirages in this rather confusing account. The logic seems to be that, when we view an island from afar, we are seeing its reflected image as if it were an actual part of the island. Meantime, so the logic seems to go, the visual rays that pass unreflected because they are somewhat higher take on the color of the intervening air (misty?) which presumably blocks direct perception of the island. As a result, we see the image, as if it were the island itself, overtopped by the blocking mist.

[131]Ptolemy is describing composite mirrors made up of several plane mirrors; presumably, then, a concave mirror can be thought of as composed of innumerable tiny plane segments.

[132]Latin = revolutio.

tween eye and] actual object, then the image is rendered larger or smaller than the object itself as seen directly.

[120] Furthermore, when individual rays distributed over the image of the object are longer or shorter than the rays that fall upon the actual object itself when it is directly seen, then, for the previously-given reason, the shape of the image appears different from that of its generating [and directly viewed] object. For that reason, too, straight lines that appear behind a transparent object whose surface is not plane do not appear straight, because, in that case, the broken visual ray does not maintain its proper [spatial] arrangement, so that it will not strike the visible object from the point directly facing it but, rather, from a point located elsewhere. Thus, even though the line is actually straight, its image does not appear so. On the other hand, if the [surface of the] transparent body is plane and the object seen [immersed] in it is straight, then, when part of that immersed object lies outside [the medium] while the other lies within it, as happens in the case of oars, the resulting perception is that the object itself is bent.[133] In fact, the part outside [the medium] is seen in its true location, whereas the part lying within the transparent medium appears closer to the refracting surface [than it really is]. Therefore, the two aforementioned segments of the visible object will not appear to lie in a straight line with one another; instead, the object will look broken.

[121] A particular kind of continuous sweep of the visual ray gives the impression that the visible object is moving. This sort of sweep sometimes occurs at the very source of the visual flux, as happens, for instance, in the case of vertigo and fainting, whose effect reaches to the eye itself. For, while striking the visible object, the ray continues to follow the motion of its source, which has changed the direction of its focus, but the sense does not detect this sort of [internal] motion.[134] Consequently, the perceiver assumes that, in passing over the object seriatim, the visual ray is moving.[135] This illusion is created at times because of visual contact with transparent

---

[133]The apparent breaking of an oar in water was a source of fascination for many in Antiquity; see Schöne, *Damianos*, p. 29, n. 9, for various references.

[134]Evidently, what Ptolemy has in mind here is the so-called oculogyral illusion, which is created when the viewer is rotated rapidly and, after stopping, senses the room spinning about in the opposite direction for awhile, after which the direction of apparent motion reverses for a short time; see, e.g., J. O. Robinson, *The Psychology of Visual Illusion* (London: Hutchison University Library, 1972), pp. 211–212. Not surprisingly, given Ptolemy's explanatory framework for this illusion, he fails to recognize, or at least to acknowledge, the second phase of this illusion.

[135]This seems to contradict the preceding point; since we do not sense the interior motion, then the sweep of the ray over the object ought to arouse the impression of the *object's*, not the ray's, motion.

media, as happens in the case of an object seen in moving water. It actually seems to move with the flow of the water, because, as the water's surface moves, [its constituent parts continually] take up different locations. Therefore, several visual rays view the visible object [seriatim], and, being seen at several [successive] locations [in terms of the moving portions of water], the object seems to move.[136]

[122] This, then, is all we have to say concerning the types of illusion that are due to the passions aroused in the visual faculty from the visible properties taken generally. We are accordingly obliged to turn our attention to the phenomena that arise from the perceptual scrutiny of those properties.

[123] But let us begin with illusions that have to do with location. Now illusions that involve colors come about solely through a passion of the visual sense. But the illusions we just mentioned that involve location and all [the rest of] the visible properties appear in two ways at most. On the one hand, such an illusion arises from the colors inherent in a variety of visible objects, those colors being misapprehended by the visual faculty when it is unable to carry out its proper perceptual scrutiny and is led to apprehend the object by means of its initial sense-impression. On the other hand, such an illusion arises from the actual arrangement of the visual rays, in which case what is discerned about the visible properties is not perceptually scrutinized as it should be in terms of variations [in the proper arrangement of the visual rays] but in some easier way.

[124] Of objects that lie in the same region, those that are brighter seem closer. But in these sorts of illusion, visual perception of [how far away] the dim object [lies] depends not on the length of the visual ray when the distance is considerable, but, rather, on the difference [in brightness] of colors. Likewise, the locations of bright objects, such as the sun and the moon, are judged to be nearer, whereas dim objects appear to be remoter, even if they are actually nearer. Thus, mural-painters use weak and tenuous colors to render things that they want to represent as distant.[137]

[125] Furthermore, when we look in some given direction from elevated places without [also] looking downward, we judge that the terrain that is far away from us lies below our level. This illusion arises from the fact that the basis upon which the judgment should be made is the length of the rays. In that case [i.e., of properly based

---

[136]In short, if we fix on the flowing water itself, it will seem to stand still while the image in it appears to fall behind; see also II, 131 below.

[137]This technique of rendering so-called atmospheric perspective by the use of weaker colors was well established among painters of Ptolemy's time.

judgment], though, the [central rays of the] visual flux fall on the actual ground, so the illusion does not arise. However, since the object [i.e., the distant terrain] has been sensed by the lower of the visual rays or by a ray that is similarly oblique, and since things that are seen by such rays are wont to appear below our feet, and since the viewpoint rarely happens to be in an elevated place, the object is judged to lie below us.[138]

[126] The same sort of illusion arises in the case of size according to both of the ways that have been discussed. For example, when objects subtend equal [visual] angles and lie at equal distances [from the eye], the one that is less brightly colored appears to be larger.[139] But if the distances are unequal, the more remote one seems larger than it did at equal distance insofar as [judgment of] distance now enters into the account. This sort of illusion arises because, instead of judging [the sizes of] these things by means of the size of [subtended] angles, one judges in this case on the basis of a perception of the increased length of the visual ray. According to that criterion, objects that are the same size appear smaller when they move away from the viewpoint. From this it necessarily follows that, among those objects that should be [judged] equal in size according to the subtended visual angles, the ones that lie at the greater distance should appear larger.[140] The same illusion also stems from differences in colors, for a object whose color is dimmer seems farther away and is therefore immediately assumed to be larger, just as happens with objects that actually are—i.e., when objects are seen under equal angles while some of them lie at a greater distance.[141]

[127] A similar sort of illusion occurs in the case of shapes when the shape of an object is perceived and recognized not by means of the shape of the [base of the] visual cone as defined by the visible object with which it is in contact, but by means of one of the afore-

[138]Lejeune's citation in *L'Optique*, p. 75, n. 105, of Euclid, *Optics*, props. 10 and 13, is not wholly apposite here, since the phenomenon described by Ptolemy, unlike that treated by Euclid, involves an abnormal rather than a normal perceptual operation.

[139]In fact, the very opposite is the case, as we know from the so-called Helmholz irradiation illusion: i.e., the brighter object will appear larger than the dimmer one; see, e.g., Luckiesh, *Visual Illusions*, pp. 114–123. Perhaps Ptolemy's misapprehension here stems from his understanding of the so-called moon illusion, which is adverted to in III, 59 below.

[140]In short, when two objects at different distances subtend the same visual angle, the more remote one is quite properly judged to be larger, even though at the level of mere sensation it does not actually look larger.

[141]This follows from the claim made at the beginning of this paragraph that dimmer objects will appear larger than brighter objects at same distance from the viewpoint; accordingly, when both are perceived to be equal in absolute cross-section, the dimmer object will appear to lie farther away.

mentioned characteristics. For, according to the colors applied to them, surfaces sometimes appear convex and sometimes concave. Thus, a painter who wishes to represent these two shapes by means of colors paints the part he wants to appear higher a bright color, whereas the part he wants to appear concave he paints with a weaker and darker color.[142]

[128] This is why we judge a concave veil to be convex when we view it from afar. The reason is not that the wind disposes it in such a way that sunlight and the visual flux reach the area of concavity [blown inward by the wind]. Rather, the reason is that the [relatively] orthogonal rays strike the middle of the veil so that it shows forth vividly, whereas at its outer edges either no ray at all or a somewhat oblique one strikes it, which is why it appears dark [toward the edge]. Accordingly, then, the edges of the veil appear depressed while the middle appears elevated, and this is how something that is actually convex appears [to the viewer].[143]

[129] In the case of transparent objects, such as a glass [plate] etched on one of its faces, the surface itself does not appear flat to us when we view it from the unmarked face.[144] The section [of the plane face] that lies over a raised portion of the etching will appear indented, and the section [of the plane face] that lies over an indented portion of the etching on the opposite face will appear raised. For in this case the sense does not interpret the shape of the first surface it encounters by means of the shape of the base of the visual cone that strikes the surface. Instead it does so by means of the shape [of the base] formed by the flux as it passes out of the transparent body. When the visual flux passes through a raised portion of the etching, its shape [at the base of the visual cone] will be raised, whereas when it passes through an indented portion of the etching, its shape [at the base of the visual cone] will be indented. But since the sense of sight concludes that the shapes of such objects are contrary to the shape at the base of the visual cone (indeed, it detects concavity by the convexity of the base of the visual cone and

---

[142]Well before Ptolemy's time, painters in the Greco-Roman world had perfected techniques of chiaroscuro using both "linear" and "painterly" techniques to give the illusion of relief in their depictions; see Bruno, *Form and Color*, pp. 24–52 for an account of the early development of such techniques.

[143]Evidently, Ptolemy assumes that the relative weakness of the visual rays that are oblique to the axial ray causes the resulting visual impression to be weaker and thus darker than it should be; see II, 20 above.

[144]By Ptolemy's time, the art of creating cut-glass relief, coupled with the ability to manufacture fairly clear, colorless glass, was relatively new but nonetheless well-established, particularly at Alexandria, which was an important center for glassmaking; see Forbes, *Studies*, vol. 5, pp. 171–180.

convexity by the concavity of the base of the visual cone), it arrives at the same conclusion in this case too. For, since the base of the visual cone striking a raised portion of the etched side of the transparent body is raised, it concludes that the bodies themselves are indented. On the other hand, it concludes that an indented portion is raised when the base of the visual cone is indented at that point.[145]

[130] Something like this also happens in certain types of visual impressions involving motion. We can understand this on the basis of what we will [now] show: namely, that objects that do not move swiftly but nonetheless disappear quickly from sight are judged by the viewer to travel swiftly. Examples are to be found in a fire that moves for a brief time, a spark, and [luminous] bodies that pass by windows and narrow openings. This illusion [of speed] comes about because the sense does not judge how much the visual flux has been affected [by the passing object] in terms of the time in which the motion occurs but rather in terms of what appears and is perceived by means of color. For, when any object passes over the entire visual field in a brief, perceptible period of time, it appears to move swiftly and disappears at the edge of that field. But, in the case of moving objects, if any of them happens to disintegrate or vanish for some other reason—e.g., if it disintegrates before leaving [the visual field] or vanishes before reaching either extremity of that field—then such objects are judged to move swiftly because, in that case, the visual faculty grasps the speed in terms solely of the speed with which the object is lost to sight or vanishes.

[131] Furthermore, when a boat stands still in a calm, waveless river that flows swiftly, anyone in the boat who does not look at the shoreline beyond [but focuses on the river] judges that the boat is moving swiftly upriver while the water is standing still. The reason for this illusion is that the motion of the water sensed by the visual flux, being opposite to that with which the boat is assumed to move, is manifested by the contrast between the color of the boat and the color of the water. Now the contrast created by the motion of the parts of the water's surface alone is not clear to the sense because of the uniformity of the parts of the [water's] surface and the similar-

---

[145]This illusion depends on the direction of lighting. We automatically assume that a textured object is raised when the brighter side faces the light-source. Now if the etching is thought of as an intaglio, then the light will pass through the side of the carving that lies toward it and will reflect from the opposite side, which thus presents an illuminated face to us. As a consequence, if the light-source is not seen, the intaglio is perceived as if it were in relief with the light shining directly on the illuminated side; see, e. g., Luckiesh, *Visual Illusions*, pp. 144–152. Ptolemy's explanation depends on the assumption elaborated upon in II, 67, above, that the flux senses concavity and convexity by means of the convexity or concavity in the base of the visual cone in contact with the concave or convex object.

ity [throughout] of its color. Yet, according to the motion of the visual flux upon the parts of the visible object's surface, it is necessary that either the water or the boat appear to move.[146] Thus, since the water will appear calm, the motion must appear to belong to the boat. On the other hand, if we look at the water, the shoreline, and the boat all at the same time, and if we take cognizance of the fact that the shoreline is stationary, then we will see that the boat is stationary, since the boat is seen by the same rays that see the shoreline. We will also see the water moving since we will have realized that the boat and the shoreline are stationary.

[132] Likewise, if we sail in a boat along the shore during twilight, or if we move in something other than a boat, and if we do not sense the motion of the thing carrying us, then we judge the trees and topographical features of the shoreline to be moving. This illusion stems from the fact that, when the visual rays are displaced [laterally], we infer that the visible objects are moving because of the displacement of the visual ray. Although the visible objects are stationary, then, it is assumed that the apparent motion belongs to them.

[133] It is also assumed that the image of a face painted on panels follows the gaze of [moving] viewers to some extent even though there is no motion in the image itself, and the reason is that the true direction of the painted face's gaze is perceived by means only of the stationary disposition of the visual cone that strikes the painted face. The visual faculty does not recognize this, but the gaze remains fixed solely along the visual axis, because the parts themselves of the face are seen by means of corresponding visual rays. Thus, as the observer moves away, he supposes that the image's gaze follows his.[147]

[134] It must be borne in mind, though, that in all the visual perceptions that arise inferentially from sensation and perceptual scrutiny, several visible features in one and the same object are involved. And those features that are not properly judged in and of themselves cause a false perception of those that are properly

---

[146]According to Pace, "Elementi Aristotelici," p. 267, this example can be traced back at least to the Academic Skeptic, Carneades of Cyrene (213–129 B.C.).

[147]One possible interpretation of Ptolemy's argument here is the following. Assume that, when we view a portrait, we fix our gaze on the eyes so that we view them along the axes of the visual cone, focusing upon them one at a time in alternation. Thus, as we move with respect to the portrait, our line of sight along the axes toward the eyes remains relatively fixed while the rest of the flanking rays do not. For that reason, the portrait's eyes are perceived to be immobile along a reciprocal line of sight to ours while the rest of the face is perceived to move in relation to that line.

judged in and of themselves.[148] The kind of misperception formed in this way need not always arise in the cases of illusion that we have applied to the particular visible properties. It may arise, instead, when the apprehension of the characteristics to be inferred is not clear but is derived through differences among characteristics that are not supposed to be inferred and that are irrelevant to those that are. This is the case with the variations in the visual flux by which location, motion, size, or shape are judged, since these properties cannot be apprehended in and of themselves by sense [alone], whereas those characteristics by which the colors of the objects themselves are judged are more manifest.[149]

[135] When visible objects of the same size and equidistant [from the viewer] are seen from nearby, the visual sense does not judge the brighter of these objects to be nearer or smaller, nor does it judge that some flat objects are convex or that others are concave on account of differences in color. Nor does it judge that smooth, flowing water is static or that the depiction of a face follows the gaze of the viewer. The reason for this is that, on the basis of its appropriate passion, the visual faculty is able to discern differences in color at a greater distance. However, because they are apprehended by means of accidents that are conveyed incidentally by that passion, differences among the remaining visible properties are apprehended when they are close, but how [they are actually disposed] is not [apprehended]. For in each of these visible properties the accident that occurs at the base of the visual cone relates to [radial] length, which represents the distance [between viewpoint and visible object].[150]

[136] Yet, if that relation is beyond the sense's capacity to grasp, it induces an imperfect perceptual apprehension. Therefore, since it cannot see the visible object in the way properly suited to it, the sense apprehends it through the evidence of other differences. Accordingly, the visible object sometimes appears to it properly and sometimes through a false perception—one that is false, like the one arising from the illusion in the previously-discussed cases, yet true

---

[148]That is, visual impressions of characteristics that are not immediately perceptible (such as shape and size) can lead to misperception of characteristics that are.

[149]In other words, our perception of the secondarily or mediately visible characteristics depends on inferences made from such variables in the visual flux as ray-length, size of visual angle, disposition of the rays vis-à-vis the axis (i.e., laterally as well as up and down), etc., whereas our apprehension of color is immediate.

[150]In other words, our perception of the secondary visibles is always in relative, not absolute terms, because it depends on the spatial relationship between eye and object that is mediated by visual rays.

insofar as it is of a smaller, more oblique object.[151] For visual perception should naturally take place at a fairly short distance, and when the object is properly sensed because the eye is near and the visual flux strikes that object in the direction suited to it, then in such circumstances perceptual inference is not led astray in the visual faculty. And all the characteristics of objects that are perceived by means of angles will be no less evident than the others.[152]

[137] At this juncture, though, we ought to point out that everything we have said about illusion applies not only to an illusion due to the sense of sight itself but also to the perception that arises from it. And since we are deceived in several cases when the visual sense impinges on visible objects according to its nature and habit, while the mind, remaining in continuous apperceptual touch with such objects, judges them to be abnormal, we must reiterate that this is an illusion involving mental inference.

[138] This is what happens in [the perception] of position when we look into plane mirrors and the visible object [i.e., the viewer himself] faces the mirror directly. In that case our sight shows us our [right-hand and left-hand] sides in the way that is natural for it to show objects viewed directly: i.e., what is seen by right-hand rays appears to the right, while what is seen by left-hand rays appears to the left. Our mind, however, shows us right as left and left as right, because objects that actually face us are so disposed that their right is opposite to our left, while their left is opposite to our right. And this is why, when we move one of our hands [in front of a mirror] our sight tells us that the hand that moves [in the mirror] is the one facing it [i.e., right to right or left to left], while our mind tells us the opposite.

[139] An inferential error about distances and their amount also arises, as happens in daylight when we look through the air that surrounds us. Since this air is in fact thicker and more colored than that which is higher up because it lies at [the surface] where plentiful vapor arises from land and water, it is more apt to take on the light that infuses it and to absorb the visual flux. We therefore suppose that we see the sky to be of the color that is common to both the vapors and the sky. The same holds generally for all extended bodies that are rare and humid and that lie at a fairly remote

---

[151]Thus, in comparing a closer object that faces us directly to a more remote one that is slanted, we may err in judging the relative sizes or distances of those objects while yet being correct in our general perception that the one object is both more remote and oblique.

[152]In short, when the visual angles and angles of radial incidence are moderate (i.e., properly perceptible), they provide us with a true perceptual gauge for the secondary visibles.

distance from the air that surrounds us; the exceeding rarity of such bodies precludes the visual flux entirely from seeing them, even though the light does not impede their apprehension. This is what happens when the viewer stands in the dark and looks at the stars but does not see what surrounds them, in spite of the fact that light impinges upon it.[153] On the other hand, when the viewer stands in a lighted place, he does not see the stars, because the light intervening between him and them weakens the outgoing visual flux.[154] The air that is seen in daylight is judged to be more remote than everything else, insofar as nothing else appears more distant than it, but the sun and moon are judged to be nearer [than its outer reaches] because of their brightness. However, in seeing the sky as higher than anything else, the mind arrives at a false conclusion and supposes that the mere visual impression is correct and represents what it perceives; and [so] it judges that what seems very remote to it is larger than something else that is naturally and truly more remote and larger than everything else.[155]

[140] Something like this also happens in the case of shapes: namely, that in buildings whose walls are parallel along the sides, the upper edges appear wider, as do high door-lintels. This is due to a perception created in the sense, even though the upper parts of those buildings do not actually have a narrower interval or lie nearer [one another] than the lower parts.[156] Indeed, men have been accustomed to building in such a way that the result is well disposed and solid. Therefore, because it normally seems to the mind that they are wider, even though they are not, it judges them to be truly so, since it assumes of such edifices that they have parallel sides. Now, because these buildings are erected vertically, they stand perpendicular to the horizon, but things that are perpendicular to the horizon are parallel to one another. However, since this is not how it looks, the mind judges that one of the two opposite [hor-

---

[153]Presumably, what surrounds the stars in this case is ether.

[154]This explanation is confusing at best. On the one hand, although Ptolemy admits elsewhere that a very bright object can overshadow a nearby dimmer one (see II, 90 above), it is far from clear in this case what the brighter object is. The sun is an obvious candidate, to be sure, but why should it overshadow the stars that lie far to the east when it lies to the west? On the other hand, if the air is the object in question, why does it trap light during the day (by vapors?) yet not at night (cf. II, 19 above)?

[155]Evidently, what Ptolemy means here is that, in judging the heavenly bodies to lie within the sky, we are failing to take into account that most of them (i.e., the stars) lie far beyond within an encompassing sphere that is immeasurably larger than the atmospheric sphere that surrounds the earth.

[156]Latin = *quamvis superiora illorum non habeant maiorem strictitatem et propinquitatem quam inferiora*; while I take *maiorem strictitatem et propinquitatem* to mean, literally, "a greater narrowness and nearness," Lejeune's interpretation ("a greater interval") can by no means be thrown out of court.

izontal] sides is longer than it actually is, as is the case with things that do not actually have parallel sides.[157]

[141] A false inference can also be made in the case of movements, just as happens in the case of horse-drawn chariots, when their motion is not fast. For the sense does not grasp their individual constituents [separately] but takes in the horses and the wheels at the same time. When we look at them, though, we assume that the horses are going quickly, because in things of this sort that pass over equal distances, the motion [of each constituent part] takes place in equal times, and that holds for both the horses and the wheels. As far as the wheels are concerned, because of their small size, they make several revolutions in the course of their motion, and that rotary motion is fast and continuous. On the basis of a general extrapolation, the mind imputes an equality between [linear] motion and rotation and concludes that the passage of the horses is swift.

[142] Here, then, is the end in the present book of our promised discussion of illusions and the cases in which vision is deceived.

---

[157]Given the two conflicting interpretations of *maior strictitas* above, Ptolemy's argument is subject to at least two possible constructions. On the one hand, we might assume that, in the original Greek, Ptolemy was referring throughout, not to divergence, but to convergence of the upper parts on account of perspective. Accordingly, the incoherence of the passage can be traced to some sort of mistranslation, perhaps a confusion between the two Greek terms for narrower and larger, both of them quite close in spelling (see Lejeune, *L'Optique*, p. 84, n. 120). On the other hand, Ptolemy may have been referring to the optical illusion of widening toward the sides and downward bowing toward the top that Greco-Roman architects recognized in the building of temples according to strict rectilinear elements. Thus, it was supposedly to counteract such unwanted optical effects that classical builders adjusted various elements of the building in order to restore the sense of true verticality and horizontality throughout—as is apparently the case, for instance, with the Parthenon; see Vitruvius, *De architectura* III, 3; Luckiesh, *Visual Illusions*, pp. 195–197, and J. J. Coulton, *Ancient Greek Architects at Work* (Ithaca: Cornell U. Press, 1977), pp. 108–111.

BOOK 3
REFLECTION FROM PLANE
AND CONVEX MIRRORS

Topical Resume

Introductory Section [1–2]: [1] summary of book 2 and statement of goal for book 3, [2] basic difference between reflection and refraction

The Laws of Reflection [3–23]

Preliminary Analysis [3–13]: [3] the three laws specifying image-location, [4] empirical verification of first two laws, [5] definition of the surface of reflection, [6] empirical confirmation of equal-angles law, [7–12] **Experiment III.1**, demonstrating the equal-angles law, [13] summary of image-location according to established laws

Theoretical Justification [14–23]: [14–16] reason object is seen along extension of incident visual ray, [17] reason image must appear along cathetus of reflection, [18] reason image must lie on plane of reflection, [19–20] justification for equal-angles law, [21–23] why mirror-images are less clear than objects seen directly

Binocular Vision of Extended Objects [24–62]
[24–25] preliminary observations, [26–30] **Example III.1**, showing the amount of relative displacement of point-images in diplopia when object is viewed face-on, [31–32] **Theorem III.1**, extending previous example to a line-object, [33–34], **Example III.2**, relative displacement of images of a slanted line-object under same circumstances, [35–36] common axis as ultimate referent for image-location in binocular vision, [37–40] **Example III.3**, illustrating how common axis serves as ultimate referent when object is viewed face-on, [41–42] image seen with either eye alone is always displaced with respect to common axis, [43–45] **Experiment III.2**, confirming this point, [46] summary statement, [47–54] **Theorem III.2**, demonstrating image-displacement of straight line-object viewed obliquely, [55–57] **Theorem III.3**, analyzing image-displacement of curved object viewed obliquely, [58]

summary statement about axial convergence, [59] the moon illusion, [61–62] axial accommodation according to the Apex

Image-Formation in Plane Mirrors [63–96]

[63–65] preliminary points about mirror-images and factors affecting them, [66] definition of basic terms and phrases, [67–72]
Theorem III.4, demonstrating that only a single image of a given
point-object will be seen behind a plane mirror, [73–75] **Theorem
III.5**, demonstrating that the image in plane mirrors lies the same
distance from the eye as the point-object, [76] summary statement,
[77–78] specifications for analyzing image-formation of spatially
extended objects, [79–81] **Theorem III.6**, demonstrating that
plane mirrors create no size-distortion, [82–90] **Theorem III.7**,
demonstrating that plane mirrors create no shape-distortion,
[91–96] **Theorem III.8**, demonstrating that images in plane mirrors are inverted right-to-left

Image-Formation in Convex Circular Mirrors [97–132]

[97–98] distinctions between convex and concave mirrors; preliminary statement, [99–103] **Theorem III.9**, demonstrating that
only one image of a point-object is seen behind a convex mirror,
[104–106] **Theorem III.10**, demonstrating that the cathetus of reflection and the incident ray can intersect at the surface of a convex mirror, [107–109] **Theorem III.11**, demonstrating that the two
lines can also intersect above that surface, but in that case the eye
will see the image as if it lay behind the mirror's surface,
[110–116] **Theorem III.12**, demonstrating that the image lies
nearer the eye and the mirror's surface than the object itself,
[117–120] **Theorem III.13**, demonstrating that images appear
smaller than their generating objects in convex mirrors, [121–123]
Theorem III.14, analyzing shape-distortion of straight and convex objects by convex mirrors, [124–126] **Theorem III.15**, analyzing shape-distortion of concave objects by convex mirrors,
[127–130] **Theorem III.16**, showing that the image in convex mirrors is inverted right-to-left, [131–132] summary of observations
about convex mirrors

The Third Book of Ptolemy's *Optics*

[1] In the second book of this study we explained the visible properties, how each of them is seen, and the various ways in which a visual illusion arises in the process of discerning what is true about
the visible properties. Moreover, concerning cases in which error
and doubt arise, we have shown that certain of them involve direct

vision, and we have explained these cases briefly but adequately. Certain of them, however, come about because of a breaking of the visual ray, and it is to this phenomenon that we have devoted the greater part [of this treatise]. Accordingly, we have only to examine this latter phenomenon on the basis of scientifically exhaustive demonstrations and to distinguish the particular illusions according to each of its two types.

[2] Now one of these types involves the penetration of the visual ray through bodies that break it somewhat, thereby causing refraction (this case we commonly refer to as "penetration of the ray"). The other type involves bodies that prohibit passage, whence there occurs a reflection of the ray from those same bodies prohibiting passage (and we have usually called such bodies "mirrors"). It behooves us, then, to begin by discussing the latter of these two types [of breaking], namely the second type. In the process, we should talk not only about the shapes of reflecting surfaces—i.e., plane, as well as spherical convex and spherical concave[1]—but also about the visual effects created by combining these shapes.

[3] For all cases in which scientific knowledge is sought, certain general principles are necessary, so that postulates that are sure and indubitable in terms either of empirical fact or of logical consistency may be proposed and subsequent demonstrations may be derived from them. We should therefore indicate that three particular principles are needed for the scientific study of mirrors and that, being of the first order of knowledge, they can be understood by themselves.[2] The first of these principles asserts that objects seen in mirrors appear along the extension of the [incident] visual ray that reaches them through reflection, the resulting line-of-sight being determined by the placement of the pupil with respect to the mirror. The second principle asserts that particular spots [on a visible object] seen in mirrors appear on the perpendicular dropped from the visible object to the mirror's surface and passing through it [i.e., the cathetus of reflection]. The third principle, finally, asserts that the disposition of the reflected ray connecting pupil to mirror and mirror to visible object is such that each of the ray's two branches joins at the point of reflection and that both form equal angles with the normal dropped to that point. A line normal to the surface of a sphere is commonly defined as one that forms right angles with all the tangents to the sphere at the common point on its surface [where

---

[1]Here Ptolemy seems to be explicit in limiting the basic reflecting surfaces to three: plane, spherical convex, and spherical concave; see Preface, n. 15 above.

[2]For a general analysis of these principles and their experiential verification, see Lejeune, *Recherches*, pp. 33–41.

the perpendicular is dropped]. Hence, it necessarily follows that all normals to the surface of spheres, when they are extended beyond, pass through the sphere's center.[3]

[4] The implications of the principles just articulated will become evident according to observed facts, as we shall explain. For in all mirrors we find that, if we mark on the surface of any of them the spots where visible objects appear and then cover them, the image of the visible object will no longer appear. But afterward, if we un-cover one spot after another and look toward the areas revealed, the designated spots will appear together with the image of the visible object along the line of sight projected from the vertex of the visual cone.[4] Also, if we stand long, straight objects at right angles to the surfaces of mirrors, and if the distance is moderate, their images will appear [to lie] perfectly in line with those objects [as they] are prop-erly viewed outside the mirror.[5]

[5] From both these instances it follows that a visible object must appear in the mirror at the point where the [extension of the inci-dent] visual ray and the normal dropped from the visible object to the [surface of the] mirror intersect. Moreover, since they intersect, these two lines must lie in the same surface, and this same surface that contains them must be perpendicular to the mirror's surface, because one of the [contained] lines is perpendicular to the mir-ror's surface. Also, the [branch of the whole] visual ray that is re-flected to the visible object will be in the surface just described. Finally, the normal dropped to the point of reflection on the mir-ror's surface must be the common section of all the various sur-faces [of reflection] defined by the reflection of visual rays [at that particular point].

---

[3]The three principles articulated in this passage are easily elucidated by reference to fig-ure 5. Let **XY** represent the surface of a plane mirror or a plane tangent to either a convex or concave mirror. Let **E** represent the eye, **O** a point-object, **I** the image of that point-object, **R** the point of reflection, and **NR** the normal to that point of reflection. **ER** will then be the incident ray and **RO** the reflected ray. Accordingly, the first principle states that the true visual connection between eye and object lies along radial line **ERO**, while the image **I** lies along the rectilinear prolongation of incident ray **ER**. The second principle states that the image **I** lies on the normal dropped to the plane mirror or tangent from generating object **O**. And the third principle states that the angle of incidence **ERN** = the angle of reflection **ORN**. There is implicit in all this an additional principle, one that Ptolemy makes explicit in III, 5 below: namely, that all of the aforementioned lines (**ER, NR, OR,** and **OI**) lie in the same plane (= the surface of reflection), which is itself normal to the plane mirror or tangent.

[4]Lejeune takes this to be an empirical verification of the first principle on the basis of the general directionality of the line-of-sight; see *L'Optique*, p. 89, n. 6, and *Recherches*, pp. 35–36. For a discussion of the possible "Euclidean" and Heronian context for this exercise, see Le-jeune, *Recherches*, pp. 49–50, and Knorr, "Archimedes and the Pseudo-Euclidean *Catoptrics*," pp. 71–74.

[5]Lejeune takes this to be an empirical verification of the second principle; see *L'Optique*, p. 89, n. 7, and *Recherches*, pp. 36–37.

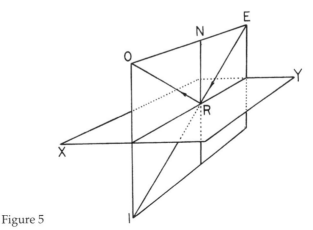

Figure 5

[6] This point is likewise established when the eyes are so situated that each may see the other simultaneously, which obtains when the [axial] visual ray from each falls at the same time on a given point on the mirror. If, however, such is not the case, it follows that neither sees the other, [all of] which is evidence that the [respective] visual rays are reflected equivalently.[6] From these observations it is clear that reflection takes place at equal angles. For the angle will be one and the same according as it measures the incidence of one of the two radial branches upon the mirror or the reflection of the other radial branch from the mirror. If, however, we were to suppose those angles to be unequal on either side, then it is necessary to have the ray from one of the eyes meet the mirror's surface at an angle larger than the one formed by the ray after its reflection from the mirror, while, for the other eye, the situation has to be reversed: i.e., the ray after reflection has to form a larger angle than its incident branch does upon the mirror.

----

[6]As Lejeune explains it in *L'Optique*, p. 90, n. 10, and *Recherches*, pp. 37–41, this empirical verification of the third principle works as follows. Assume that the viewer stands face-on to plane mirror **ACB** (in figure 6), upon which a spot **C** is marked such that the line dropped to it from midpoint **M** between the two eyes E_L and E_R is normal to the mirror. Accordingly, if the viewer focuses both eyes on **C**, he will get a fused image of both eyes along the axial rays, the image I_R of E_R being seen along $E_L CI_R$ and the image I_L of E_L being seen along $E_R CI_L$. Each eye will thus see its counterpart in this fused image, which is located at **F**. Meantime, each eye will see itself along its respective normal, so that the left-hand eye E_L will see its image I_L to the left of **F**, while the right-hand eye E_R sees its image I_R to the right of **F**. When, however, we allow our eyes to resume normal focus on the whole image $I_L FI_R$ in the mirror, then the new focal-point becomes **F**, and the image is seen normally, without the fused image in the center. The reason, of course, is that the respective rays no longer form equal angles at the mirror's surface, as is illustrated by the inequality of angles formed by rays $E_L D$ and $E_R D$ at point **D** on the mirrors.

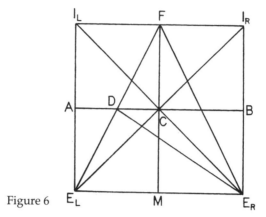

Figure 6

[7] But this point will become clearer and visibly confirmed, and what we have asserted will thus be demonstrated more certainly, by the following experiment:

[8] *[EXPERIMENT III.1]* Let a round, bronze plaque of moderate size, such as the one below [in figure III.1], be set up, and let **A** be its center. Let both faces be planed down as carefully as possible, and let its edges be rounded and polished. Then let a small circle be inscribed at centerpoint **A** on either of its faces, and let it be **BGDE**. On this same face let two diameters, **BD** and **GE**, be inscribed to intersect at right angles; and let each quarter circle be divided into 90 degrees.[7] Finally, let the two points **B** and **D** be taken as centerpoints, and, using **BA** and **DA** as radii, let the two arcs **ZAH** and **TAK** be inscribed.

[9] Now let three thin, small, square, straight sheets of iron be formed. Let one of them remain straight, and let one of its sides be polished so that it appears as a clear mirror. Let the remaining two sheets be curved in such a way that the convex surface of the one and the concave surface of the other [taken together] form a circular section equal to circle **BGDE**, and let the two [respective convex and concave] surfaces of these sheets be polished so that they are made into two mirrors.

[10] Let us cut arcs from each of the [above] two sheets, and let them be represented by **ZAH** and **TAK**. Let line **BA** be drawn in white and **AL** in some other color. Then let a small diopter be mounted upon **AL**, and let the aforementioned plaque be disposed in such a way that [the sight-line of] the diopter passes easily

---

[7] Latin = pars = degree.

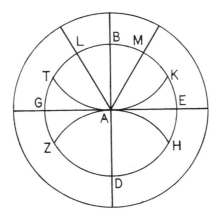

Figure III.1

through point **L** and along line **AL**.[8] Now let the aforementioned plaque be placed with the side upon which the mirrors are [to be mounted] facing up. Of these mirrors, let the plane one lie on **GAE**, the convex one on **ZAH**, and the concave one on **TAK**. Finally, in the middle of the upper edge of each of the mirrors let a pin be attached axially to the mirror so as to keep it in place on point **A**.

[11] Assume, then, that we view with either of our eyes through the diopter, which is placed at point **L** on **AL**, and that we direct our line of sight toward the axial pin of [each of] the mirrors. Accordingly, if we slide a small colored marker [along arc **TBK**] on the plaque's surface until it appears to us to lie on the same line of sight with **A**, then point **L**, point **A**, and the image that is seen in the three mirrors will appear [to lie] upon a single line [of sight].[9] If, therefore, we mark the point at which the colored marker stands on the plaque's surface— i.e., the place from which the marker's image appears in those mirrors (e.g., the place represented by point **M**)—and if we draw out straight line **AM**, we will find that arc **BM** is always equal to arc **BL**. Since that is the case, angle **LAB** will be equal to angle **MAB**, and line **BD** will be normal to all of those mirrors. Line **AL** [thus] delineates the [branch of the] visual ray incident to the mirror's surface from the eye, whereas line **AM** delineates the [branch of the] visual ray reflected from the mirror's surface to the visible object.

---

[8] According to Lejeune, *L'Optique*, p. 93, n. 16, this diopter would probably consist of a narrow vertical sighting-slit mounted on an arm that pivots around **A**.

[9] As Lejeune notes, it would be possible to carry out this experiment using all three mirrors at once, assuming: (1) that the concave mirror were shorter than the plane mirror, which, in turn, would be shorter than the convex one, (2) that the axial pin were common to all three mirrors, and (3) that the small colored marker that is slid along arc **BE** were tall enough to be seen in all three mirrors. Otherwise, of course, the experiment must be carried out for each mirror in turn rather than all at once.

[12] To continue, if we place some moderately long object at **B** and direct our line of sight along **AB**, the entire object will appear along the continuous straight line **AD**,[10] as we established earlier when we claimed that the images of objects erected at right angles upon mirrors are seen along the straight continuation [of the orthogonal] without any deviation, and they seem to move in the same direction as the visible objects.[11]

[13] According to the fundamental principle we accepted in direct vision and upon whose basis we determine the location of a visible object, we proceed likewise to determine the location of an image that is not seen directly.[12] For the composite object is seen as straight according to two components: namely, that which constitutes the image of the visible object and the object itself as it actually appears, and this composite maintains a uniform configuration with respect to each of the two constituents [i.e., image and actual object], just as if all [parts of the image-object combination] were seen as actually facing [the viewer] in the location that it seems to occupy. A straight magnitude [thus] appears straight. But when a true image lies at a moderate distance along a straight line, and when its generating object is stood perpendicularly upon mirrors, it must be the case that, as far as the apparent location is concerned, each component turns out to lie along one [and the same] line. And if the same magnitude appears orthogonal to the mirror's surface while its image lies in a straight line [with it], then that image for its part will appear orthogonal to the same surface, so that there is no discontinuity between it and the composite whole, just as if we set the visible objects up in their actual locations—i.e., those dictated by the relationship within mirrors between visible objects and their images—the relative configuration in each of the two locations being so maintained that [the resultant combination] seems virtually to coincide with the directly viewed object.

[14] Therefore, the principles we have set forth are evidenced by the phenomena we have discussed, and it is not difficult to understand that in these cases reason conforms with sense-experience. It is normal and natural in direct vision for visual radiation to emanate rectilinearly from its origin-point to outlying objects, yet the reflec-

---

[10]The text from this point on to the end of paragraph 13 has been transposed by Lejeune from its original position as paragraph III, 60 in the critical Latin text.

[11]In line with the situation in III, 4 above, the context here seems to dictate that the object in question be laid horizontally along **AB**. For a description and analysis of this notorious experiment, see Lejeune, *Recherches*, pp. 41–46.

[12]Lejeune opines in *L'Optique*, p. 94, n. 22, that this principle was articulated in the now-lost first book and had to do with the rectilinear propagation of visual flux.

tion of the visual flux that occurs from mirrors is not perceived by sight. Therefore, the visual sense must follow its natural and normal inclination by lining the reflected ray up with the initial [incident] ray before reflection and thus judging the resultant [radial line] to be straight, as if nothing happened to it [during reflection] and it kept on rectilinearly. The image of a visible object will thus appear like an object seen without [specular] impediment.[13]

[15] Now the reason that the reflection of the ray is not evident to the visual sense is that it senses nothing whatever without [first making] contact with the objects that generate images of this sort. The impingement of rays upon mirrors occurs in such a way that each ray strikes a particular point on the mirror. However, the place where [any given] ray makes contact [with the mirror] contains nothing of the extent of the sides that form the angle of reflection. By necessity, then, this angle will be imperceptible, since it is not subtended by any ray that strikes the mirror, and so the angle that is formed will be imperceptible. And because only parts of the rays remain on the surface of the mirror itself, then, since its disposition is easily determined, the location marked out by those rays is determined.[14]

[16] Furthermore, because the visual rays [that] emanate from the eye to the mirror [are] normal [to the ocular sphere], the specular image will be conformable with the image derived from direct vision. For one thing, each of their [points] is seen according to rays that are normal to the pupil. Indeed, the rays that pass through the cornea[15] and radiate to the pupil from the origin-point, which lies within the ocular sphere at the centerpoint, all form perpendiculars to the surface of the pupil, which assumes the nature of a convex mirror in terms of its shape and smoothness.[16] This is why [as viewers] we are prone to perceive the images of objects just as the eyes,

---

[13]This is evidently intended as a theoretical justification for the first principle articulated in III, 3 above.

[14]The reasoning in this tortured account seems to be based on the notion that, since each visual ray strikes the mirror at a true point, it fails to sense the incidence and subsequent rebound because it lacks any spatial basis for such a sensation. If, indeed, this is Ptolemy's reasoning, then it seems to be totally incompatible in its basis with his reasoning in II, 50 above, where Ptolemy insists that, were visual radiation to be truly discrete, vision would be impossible, because the rays would terminate at points that, lacking all dimension, would be insensible. Why, then, should the rays take on the character of true discreteness in reflection?

[15]Latin = *aspiciens* = "viewer"; I am following Lejeune, *L'Optique*, p. 96, n. 25, in taking this to mean "cornea" rather than "lens"; this construction is in fact all the more credible in view of the use of *aspiciens* a few lines further on as well as in IV, 4 below, where it can only be taken to mean "cornea."

[16]Here, of course, we are dealing with the cornea; and the likening of it to a convex mirror lends credence to the view that Ptolemy thought of visual imaging in terms of *emphasis*; see n. 35 to II, 23 above.

which face the objects directly, see them. Hence, when the reflection
[of visual radiation] that should have followed a direct, straight
route from visible objects to the pupils is redirected at [the surfaces
of] mirrors, it maintains the original direction from the pupils to
those objects, and that is where the line emanating from the center
of sight coincides with the ray itself that links the pupil and the visible object.[17] The same thing happens in the case of facing images,
when the visual ray falls orthogonally to a mirror's surface and reflects back on itself. For in that case its incidence, from which the object's image derives, will be unique in number and location, and that
location is the one according to which it makes a direct, orthogonal
link between pupil and mirror. Hence, our visual grasp of the object will be twofold and distinct in proportion and power, one stemming from the center of sight to the visible object, the other from the
visible object to the center of sight.[18]

[17] We have already considered that whatever is in line with the
vertex of the visual cone must be determinate and have a definite
order and structure; that is, it must be [spatially referred] to the
origin-point of the visual flux.[19] Likewise, too, the image of visible
objects must be formed according to a unique direction that is determined by the mirror's disposition and that lies along the cathetus
of reflection. Now to any point on a given object there is one and
only one cathetus, whereas any other line, being oblique with respect to this cathetus, is subject to numerous variations.[20]

[18] And since things of this sort [i.e., images and objects] never
exist without the other, objects must lie in the same plane formed by
the reflected ray, because such planes also maintain a normal orientation to the surfaces of mirrors. Within this context, in fact, the
radial path before reflection falls no differently from the path after
reflection.[21] For the arcs of circles constructed upon mirrors around
their centers or axes—arcs, that is, that subtend the angles of incidence and reflection and lie in the same plane with them—have the
same [normal] orientation, along a [given] diameter, to all the
shapes [of reflecting surface] discussed, as long as the plane stands

---

[17]In other words, continuity of radiation is preserved by the continuation of radiation in
the same general direction as well as the connection of the two radial branches at the point of
reflection.
    [18]Ptolemy seems to be implying a perfect symmetry of physical radiation outward from eye
to object and perceptual "radiation" inward, back along the same line, from object to eye. That
symmetry presumably guarantees that the perception will be as geometrically lawful as
physical radiation.
    [19]See II, 26 above.
    [20]The point here is that, given its placement on the cathetus, image-location is unique and
uniquely determined.
    [21]That is, it "falls no differently" insofar as it falls within the very same plane.

normal to the surfaces of the mirrors. That orientation, which comes about variously in those mirrors, is determinate insofar as it lies upon a diameter.[22]

[19] From this, therefore, the relationship of equality between the angles of [incidence and] reflection is easily adduced, as is the fact that it follows a natural course. For projectiles are scarcely obstructed by objects they strike at tangents, whereas they are obstructed to a considerable extent by objects that resist them [directly] along the line of projection. Accordingly, when anything stands directly in the way of these projectiles and stoutly resists them, it interrupts and opposes the line of projection extending to the origin [of motion], just as walls obstruct balls that strike them at right angles. But obstacles [disposed at a tangent] pose no obstruction at all, just as curved bucklers do not obstruct arrows. It should be borne in mind that this explanation extends to all sorts of moving objects, and it should be understood that they all act in such a way.[23]

[20] The action of the visual ray itself must therefore follow this rule, and any of the rays that approaches a mirror and then bounces back from it must maintain the disposition that occurs in the paradigm case—i.e., that the angle formed by the line normal to the mirror at the point of reflection and the line of incidence be equal to the angle formed by the same normal and the line of reflection. If, therefore, the normal and the tangent to which it is normal lie on a plane that cuts the mirror at right angles, it necessarily follows that, within this overall framework, a line that strikes the mirror at some particular inclination and then rebounds from it forms equal angles [with both the normal and the tangent].[24]

[21] It is necessary to stipulate at the outset that the determination of image-locations requires a specific condition: namely, that location be consonant with distance. Now, in the case of direct vision, such locations are often not [properly] apprehended by the visual rays because of their tendency to weaken with distance. This is what usually happens whenever objects that are spread out among themselves and far away from observers appear to be near [to one another] and seem to coalesce because of the decreased capacity of the

---

[22]Clearly, Ptolemy means to say here that, for all spherically convex or concave mirrors, the surface of reflection forms a great circle containing all the diameters within that planar cut.

[23]As Lejeune notes in *L'Optique*, p. 98, n. 32, the likening of optical reflection to physical rebound is found not only in Hero's *Catoptrica*, but also in the Pseudo-Aristotelian *Problemata* XVI, 13.

[24]The Latin text of this entire paragraph is obscure in the extreme; since Lejeune's interpretation in *L'Optique*, p. 99, n. 33, strikes me as plausible, I am following it.

sense for perceiving distant objects clearly. This weakening of visual capacity induces observers to view [distant] objects habitually
as if they were nearby, because in such cases it has greater efficacy
yet hardly avails itself of that capacity in the case of distant objects.
Of course this imperfection in visual capacity is greatly increased in
the event of reflection.[25]

[22] It follows, moreover, that the force of projection in every
moving body weakens when that body does not follow its original
trajectory but goes off in another direction by being deflected before
it makes [full] impact with other bodies.[26] This point becomes evident in the case of visual radiation when mirrors and the objects that
are seen in them are so disposed that some of the visual rays strike
the visible objects without reflection, while some are reflected from
the mirrors, and [the resulting two visual impressions] are made at
the same time.[27] If, under those conditions, the entire image is uniform and near, then the images turn out to be much dimmer and
less distinct than [their generating] objects as seen directly.

[23] Although this weakening ought to be the same for all visible
objects and for all distances at which the image is seen, such [weak-
ening] must somehow [also] depend on the intersection [of the line
of incidence and the cathetus of reflection], since those lines vary in
length. In the case of an image that is common to all types of mirrors this happens on the basis of the previously mentioned weakening that is due to the reflection itself. In the case of an image in a
concave mirror, however, this weakening stems from the quality of
the reflection. And this occurs in two ways at most. According to
one of these, the coherence of the image is greater than it should be,
and the image is bounded and more compacted than it should be,
so that, when an [object] is stood perpendicular to the mirror, not all
of the image is properly aligned with the visible object, because
those points on the image that are farther away from the intersection create a certain deviation in the disposition of the visual radiation.[28] The other way stems from the fact that it sometimes does and
sometimes does not happen that the aforementioned lines—i.e., the
line of incidence and the cathetus of reflection—intersect and that

---

[25]As will be explained in the next paragraph, this is due to the dynamic effect of reflection
upon visual radiation.

[26]Note the recourse to dynamics in the weakening of projection after impact; see II,
19 above.

[27]This obtains, for instance, when the object is stood vertically on the mirror so that object
and image lie in line with one another, as in III, 4 above.

[28]This example is analyzed in IV, 74–76 below. The basic point is that, when stood perpendicularly along the line of sight, a long, thin object creates a sort of horned image on either
side of itself as the composite image goes from virtual to real.

the intersection occurs not in front of the eye but either on the eye itself or behind it.[29] In this situation the image-location will not be clear. In the second case, however, the intersection that should result will be near if there is no visible obstacle to [the formation of] the image itself. Furthermore, if the image-location is indeterminate, the sense is drawn to focus on an intermediate point between the eye and the surface of the mirror that reflects.

[24] For this reason we ought to investigate the ways in which, under these circumstances, we get an image that lies at an appropriate distance and orientation. And we ought to take up the issue of whether the appearance of images in mirrors invariably conforms to the analysis already established [for binocular vision], an analysis that applies as much to reflection as to direct vision. And we ought, first of all, to examine image-location as it relates to the convergence [of visual axes] discussed above, that location being certain only where such a convergence occurs and where it appears. In fact, it often happens that this [sort of] image, while formed according to direct vision, is not seen in the place where the actual body [generating it] proves [to be].

[25] At this point, then, we must consider some issues that we postponed in the second book of this treatise.[30] These issues have to do with the way the apparent locations of visible objects can vary, a point exemplified by the fact that a single object will sometimes appear at two places whereas two objects will sometimes appear in a single place to those viewing them in a given way with both eyes.

[26] *[EXAMPLE III.1]* Let us first discuss the case of convergence in which points **A** and **B** [in figure III.2] represent the vertices of the two visual cones. Let line **AB** be joined, and let it be bisected at point **G**. Then, from that point let **GD** be drawn perpendicular to line **AB**. Let axes **AD** and **BD** converge at point **D**, and let the visible object be placed at **D**. Accordingly, this object will appear single and in its actual location.

[27] So, too, if we draw line **EDZ** through point **D** perpendicular to **GD**, any object placed on this line will appear single and at its true location as long as it is aligned with point **D**.[31]

---

[29]Depending on the relative placement of eye and point-object, then, the image-location can be indeterminate (when the cathetus of reflection and the incident ray do not meet), virtual (when the two intersect behind the mirror's surface), or real (when the two intersect between the mirror and the eye, at the eye, or beyond it). This passage in fact deals only with indeterminate and real images at or beyond the eye; see, e.g., IV, 62–65 below.

[30]See II, 46 above.

[31]This is because the object along line **EDZ** forms a common base for both visual axes; see II, 27 above.

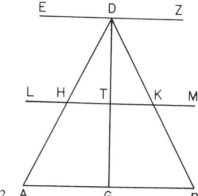

Figure III.2

[28] On the other hand, if line **HTK** is drawn parallel to line **EDZ** while the two axes remain focused on **D**, an object at point **T** will appear at the two locations represented by **H** and **K**. Moreover, two magnitudes placed at those two locations [**H** and **K**] will appear at the three locations represented by points **T, L,** and **M**. At point **T**, of course, both will be seen together at once, as if they constituted a single object. They will be seen separately, however, insofar as **H** will appear at point **L** and **K** at point **M**, and each of the segments **LT** and **TM** will appear equal to **HK**.[32]

[29] But if we focus both axes upon point **T**, we will see point **D** at points **E** and **Z**.[33]

[30] That the number of these images proves to be as we have claimed can be determined experimentally by anyone using the ruler with two pegs standing on it.[34] Whoever wants to ascertain their true locations will do so by placing a finger on the visible object. For the finger actually touches the object when it appears in its true location, whereas, when it does not, the finger does not touch it but withdraws, having encountered nothing there.[35]

[31] *[THEOREM III.1]* We have explained elsewhere the reason for the anomaly that we just discussed, and we have demonstrated it in a variety of ways when we dealt with objects that are seen at

---

[32]The explanation for this set of claims is offered in III, 32 below.

[33]Since the two axial rays **AT** and **BT** converge on it, point **T** will be seen at the center of the visual field. Point **D** will therefore appear to the left of that centerpoint when seen with ray **AD**, which is inclined to the left of axial ray **AT**, whereas it will appear to the right when seen with ray **BD**, which is inclined to the right of axial ray **BT**.

[34]That is, the instrument described in II, 30–33 above.

[35]For a description and analysis of this case, see Lejeune, *Euclide et Ptolémée*, pp. 147–53, or Simon, *Le regard*, pp. 140–142.

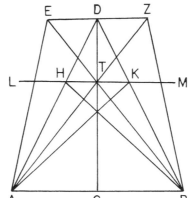

Figure III.3

once with both eyes. Indeed, those objects that are seen with correspondingly arranged rays appear to lie in a single place even if there are actually two of them. However, if they are seen by means of noncorrespondingly arranged rays, such objects, even if they are actually one object, appear to lie in two different locations.

[32] To explain, let us join lines **AE**, **AZ**, **ZB**, **EB**, **TA**, **TB**, **BH**, and **AK** in the preceding diagram [as represented in figure III.3]. Accordingly each of the spots at **E**, **D**, and **Z** will appear [respectively] in one location, since **AD** and **BD** lie on the visual axes themselves, and the rays that fall on **E** and **Z** are correspondingly arranged, since **AE** corresponds to **BE**, while **AZ** corresponds to **BZ**. On the other hand, since **AH** and **BK** lie on the visual axes, **H** and **K** will appear [combined] at single location **T**, but since **BH** and **AK** do not correspond, **H** and **K** will be seen at points **L** and **M**. Finally, because rays **AT** and **BT** do not correspond, point **T** will appear at the two locations **H** and **K**.[36]

[33] Now that we have proposed certain general distinctions concerning the locations at which each of the objects appears, we ought, on that basis, to investigate how visible objects must appear in the places where they really lie. This we will in fact do as soon as we have established another evident principle dealing with the topic under discussion.

[34] *[EXAMPLE III.2]* If line **HTK** is not parallel to line **AB**, as we previously assumed, but if, instead, line **AH** [in figure III.4] is

---

[36]In other words, as **H** shifts rightward to **T** for the eye at **B**, **T** shifts rightward to **K**, while **K** shifts rightward to **M**. Conversely, for the eye at **A**, as **K** shifts leftward to **T**, **T** shifts leftward to **H**, while **H** shifts leftward to **L**. See Lejeune, *Euclide et Ptolémée*, pp. 153–155.

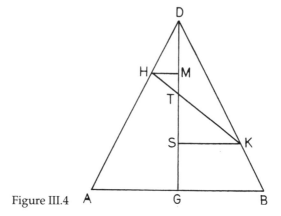

Figure III.4

[assumed to] be longer than line **BK** while the axes remain focused as they were on point **D**, then objects placed at points **K** or **H** will be seen on single line **GD** around point **T**. The two of them will not, however, appear to lie at the same location; rather, **K** will appear to be closer to the viewer than **H**, and the difference in apparent distance between the two will increase as the inclination of **HK** to the [frontal plane **AB** of] the viewer increases. In this case, **H** will be seen at **M** and **K** at **S**, and these two locations depend on where the two perpendiculars [dropped from **H** and **K**] fall on line **GD**.[37]

[35] Since all this is as we have claimed, it follows that nature equalizes the disjunction between the two visual axes and joins them according to the location of the visible object. Therefore, both of them fall upon the object from the Apex,[38] which lies between them and is where the vertices of the visual cones ought to intersect. This *principium* is equidistant from those axes, which have a common sensibility. And this is what keeps the vision tracking in a single direction midway between both sides, because it is impossible for an object facing the visual axes to maintain the same orientation with respect to each of them. In fact, an object directly in front of both eyes is not orthogonal to both axes. How, then, can we determine spatial disposition if each of the axes is inclined to the other, unless, as we said, we do so on the basis of some intermediate reference-line whose distance [from the axes] is proportional

---

[37]See Lejeune, *Euclide et Ptolémée*, pp. 155–156, or Simon, *Le regard*, pp. 142–143.

[38]Latin = *principium*; in this case the *principium* is not the source at the vertex of the visual cone. Rather, it seems to be the source from which the two visual axes are directed toward a common focus. Precisely where this source lies is not clear, but perhaps Ptolemy meant to locate it at the optic chiasma.

[throughout]? Reason dictates that this reference-line be termed the "common axis."[39]

[36] In the case of objects seen in direct vision, we need to provide a general determination of apparent location, and we should note that the displacement of such locations from the actual objects varies with the distance of the common axis from the proper visual axis. And this distance is [measured by] the perpendicular dropped from the visible objects to the common axis.

[37] *[EXAMPLE III.3]* We must still consider whether the determining principle we just adduced actually agrees with observation. Accordingly, let us take the figure [for the first case in this series] as already drawn, so that each of its defining features maintains the same order [as in figure III.5 below]. Let each of the lines **AD** and **BD** represent a visual axis. We will therefore find that whatever lies on line **EDZ** appears where it actually is, while whatever lies on line **HTK** will be encountered at places other than the one it [actually] occupies.

[38] Now it is clear that points **E**, **D**, and **Z** will appear in the places they actually occupy, because the perpendicular dropped from each of [the visual axes] to the common axis **GD** coincides with a single point. Since the displacement of the object's apparent location from its real location toward a given side will depend on the distance between the common axis and the proper axis, the visible objects themselves will be seen at their actual locations on account of the convergence of [all three] axes [at a single point]. It has

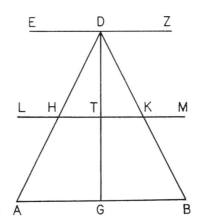

Figure III.5

---

[39]This axis is therefore distinguished from what is termed the "proper axis" (i.e., the axis of the particular visual cone) in the next paragraph.

therefore been demonstrated that each of those objects appears at its true location.

[39] On the other hand, we have explained why **H**, **T**, and **K** do not appear in their true locations: the proper axis of the visual cone emanating from the eye at **A** is **AH**, and its displacement, measured by the perpendicular dropped to common axis **GD**, is **HT**, which is equal to line **TK**. Thus, the spot at which point **T** will be seen will be point **K**. That the same thing also happens if we look from the eye at **B** will be demonstrated in precisely the same way.

[40] Likewise, if two objects are placed at **H** and **K**, and if **LH** and **KM**, respectively, are equal to **HT** and **TK**, then the object placed at **H** will appear to lie at the two points **T** and **L**. On the one hand, it will be seen at point **T** from the eye at **A** along ray **AH**, because its perpendicular distance [from common axis **GD**] is **TH**; on the other hand, it will be seen at point **L** from the eye at **B**, because its distance is **TK**, the perpendicular linking the common axis and the proper axis **KB**, and **TK** is equal to **HL** and lies in a corresponding direction. This is the displacement of the apparent from the true location of **H** as seen from the eye at **B**. By the same token, the apparent location of the object placed at **K** will be at points **T** and **M**.[40]

[41] This analysis corresponds with the general principle we enunciated. So it is clear that what appears in its true location is either what is placed at the point of juncture between the common axis and the proper axis of the visual rays that strike the visible object or what is placed upon the perpendicular dropped to the common axis at the point where it meets the proper axis just mentioned. Objects that are disposed in ways other than those we described are not seen at their true locations, even though they may seem to be.[41]

[42] It is in fact necessary that this [apparent displacement] depend on the distance between the [common and proper] axes when we look at visible objects either with both eyes together or with one or the other [alone]. For, as far as sensible appearances are concerned, when a single object will appear to lie at two locations, if we close either of our eyes, the image that is left after the other has disappeared remains fixed in the place where it appeared before, that is, where it was seen originally [before the eye was closed]. From this it is evident that seeing an object in multiple locations is not re-

---

[40]See Lejeune, *Euclide et Ptolémée*, pp. 156–160, for a description and analysis of this example.

[41]An example can be found in the preceding case, where **H** and **K** appear to form a single fused image that coincides with **T**.

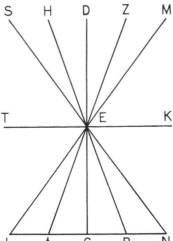

Figure III.6

stricted to vision from two eyes, but applies also to vision from either eye alone.[42]

[43] *[EXPERIMENT III.2]* We can grasp this point even more clearly if we take a board of moderate size, color it black, and mark off on one of its shorter sides a distance **AB** [in figure III, 6] equal to that between the eyes [from pupil to pupil]. Next we drop perpendicular **GD** through midpoint **G** of **AB** and draw out lines **AEZ**, **BEH**, and **TEK**, with **TEK** being parallel to **AB**. Then we paint line **GD** white, **TEK** green, **AEZ** red, and **BEH** yellow. Finally, we fix a tiny marker at **E**, place our two eyes at points **A** and **B**, and focus our sight precisely on point **E**.

[44] Lines **AZ** and **BH** will coincide with the [visual] axes, and line **TEK**, which is green, will appear as a single line because it lies on the point of convergence of the axes. The two lines **AEZ**, which is red, and **BEH**, which is yellow, will appear to us to coalesce along **GD**; yet each will appear separately elsewhere, **AEZ** along **LEM** and **BEH** along **NES**. Meantime, **GED**, which is white, will appear along **AZ** as well as **BH**.

[45] Now if we cover the eye at **B**, none of green line **TEK** will disappear. Of the two white lines, however, the one lying on **AZ** will disappear entirely, and so will the yellow line that is seen along **GD**.

---

[42]That is, when we view the same object alternately with each eye, it appears to shift from right to left or left to right, depending on which eye is closed. Also, as Ptolemy shows in Theorems III.2 and III.3, paragraphs 48–56 below, obliquely disposed objects seen with a single eye are also shifted from their real locations.

Of those lines, meanwhile, at the two extremities, the red line [seen] along **LM** will disappear. The rest of the lines will stay at the locations they occupied before, when both eyes were looking at them. All of this agrees with what we established earlier.[43]

[46] This is why the visual sense, imbued as it is with an amazing propensity to apprehend visible objects accurately, always adjusts its axes until their point of convergence is fixed at the center of the magnitude it apprehends; in that way it assumes a facing orientation, according to which the locations of visible objects are perceived correctly and punctiliously. If, on the other hand, such is not the case but, granting that the object is clear[ly seen], the distance of axial convergence from either side [of the visible magnitude] is unequal, its displacement from the true location will not be very great. When the axes are slanted in any way toward the side, the defining characteristics of objects do not seem clear; instead, they are confused on account of both the sizeable displacement of apparent [from real] locations and the weakness of oblique rays.

[47] Therefore, what happens when we place objects in a so-called facing orientation has been demonstrated. We will discover what happens in the case of an oblique orientation as follows:

[48] *[THEOREM III.2]* Let point **A** [in figure III.7] represent the vertex of either of the visual cones, and let common axis **GB** and visual axis **AB** intersect at point **B**. Then let some line **DEZH** be drawn, and from points **D**, **E**, and **H** let perpendiculars **DTK**, **EL**, and **HMN** be dropped to **BG**. Let **DS** = line **TK**, join **SL**, and extend it until it intersects the prolongation of line **NH** at point **I**.

[49] We say, then, that **DEH** will appear to the eye at point **A** as if it lay along **SLI**. From earlier discussion, it is clear that **E** will appear at point **L**, and **D** will appear at point **S**. Now, however, we will demonstrate that point **H** will be seen at point **I**, a fact that will become apparent if we show that **HI** = **MN**.

[50] Accordingly, since **KL** : **LM** = **TE** : **EN**, and **KL** : **LM** = **KS** : **MI**, and **TE** : **EN** = **TD** : **HN**,[44] then **KS** : **MI** = **TD** : **HN**. By alternation, therefore, **KS** : **TD** = **MI** : **HN**. But **KS** = **TD**, so **MI** = **HN**.

[51] Now let the two perpendiculars **QZR** and **FOP** be dropped to **BG** from points **Z** and **O**. Let us show that point **Z** will be seen at

[43]See Lejeune, *Euclide et Ptolémée*, pp. 160–163, or Simon, *Le regard*, pp. 143–144.

[44]These last two proportionalities follow from the similarity of triangles **KLS** and **LMI** and triangles **TED** and **ENH**.

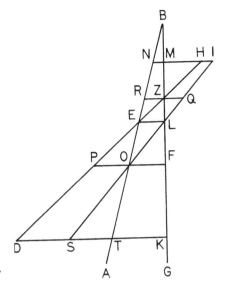

Figure III.7

point **Q** and that point **P** will be seen at point **O**. This is demonstrated if we show not only that lines **QZ** and **ZR** are equal, but also that lines **OF** and **PO** are equal.

[52] Accordingly, since **KL : LZ = KS : QZ**, and **KL : LZ = TE : ER**, and **TE : ER = TD : ZR**,[45] then **KS : QZ = TD : ZR**. By alternation, **KS : TD = QZ : ZR**. But **KS = TD**, so **QZ = ZR**.

[53] Furthermore, since **KS : OF = KL : LF**, and **KL : LF = TE : EO**, and **TE : EO = TD : OP**, then **KS : FO = TD : OP**. By alternation, **KS : TD = OF : OP**. But **KS = TD**, so **OF = OP**.

[54] In addition, because **KT > FO**, while **KT = DS**, then **DS > OP**, and so point **P** will lie closer than point **D** to point **B**, where the axes intersect. Hence, among those objects from which perpendiculars are dropped to the common axis, those whose perpendiculars fall upon it closer to the point of convergence of the two axes appear to lie closer to their true location. Those whose perpendiculars fall farther from the point of convergence of the axes will appear at a single location but farther displaced [from the true location].[46]

[55] From what we have said, it is also evident that the magnitude as we see it at its apparent location has the same shape as the actual object does when seen in its true location, so the difference will be

---

[45]The first proportionality follows from the similarity of triangles **KLS** and **KZQ**, and the third follows from the similarity of triangles **TED** and **ERZ**.

[46]See Lejeune, *Euclide et Ptolémée*, pp. 164–165.

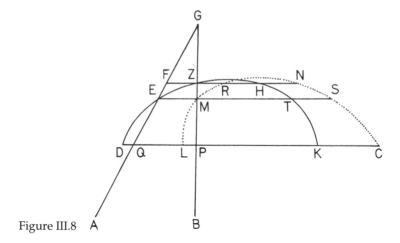

Figure III.8

in location only. What happens in the case of circular objects is as we will now explain.

[56] *[THEOREM III.3]* Let **AG** and **BG** [in figure III.8] represent the [selected proper and common] axes, and let a given circular arc **DEZHTK** be drawn upon them. Then, from points **D**, **Z**, and **E**, let perpendiculars **DP**, **EM**, and **FZ** be dropped to axis **BG**, and let them be extended through and beyond points **H**, **T**, and **K**. Finally, let **DL** and **KC** be marked off equal to **QP**, let **HN** and **ZR** be marked off equal to **FZ**, and let **ST** be marked off equal to **EM**.

[57] The eye at **A**, then, will see **D** at **L**, **K** at **C**, **E** at **M**, **T** at **S**, **Z** at **R**, and **H** at **N**, so it is obvious that the whole of concave line **DEZHTK** will appear to lie along the concave line passing through points **L**, **M**, **R**, **N**, **S**, and **C**. And the same holds for convex lines.[47]

[58] It has therefore been demonstrated that whatever is seen in such a way with both eyes must appear in two places, either separately or conjoined, when we view something face-to-face and focus in the same direction, as long as nothing disturbs the viewing process. For, since its radiation is extremely limited laterally, it is natural for the visual flux to move about the entire visible magnitude and to cause both axes to converge in a very short time with an unrivaled swiftness. And just as with the visual apprehension of distant objects at the very opening of the eye, so with the visual

---

[47]As Lejeune notes in *L'Optique*, p. 115, n. 50, the concave line-image **LMRNSC** cannot possibly be circular—a fact that becomes manifest when one attempts to draw the diagram according to the conditions set forth in the construction. Nevertheless, this point is ignored by Ptolemy and is not reflected in the diagrams supplied by the various manuscripts.

scanning of every visible magnitude by passing the axial pair over various parts of its surface, the operation occurs so quickly that the sensation of the parts occurs simultaneously with that of the whole. But the parts. . . . [48]

[59]. . . . will be concerning distances, insofar as the senses weaken in proportion to the convergence. Generally speaking, in fact, when a visual ray falls upon visible objects in a way other than is inherent to it by nature and custom, it perceives less clearly all the characteristics belonging to them. So too, its perception of the distances it apprehends will be diminished. This seems to be the reason why, among celestial objects that subtend equal visual angles, those that lie near the zenith appear smaller, whereas those that lie near the horizon are seen in another way that accords with custom. Things that are high up seem smaller than usual and are seen with difficulty.[49]

---

[48]As context makes clear, there is a lacuna at this point between II, 58 and II, 59.

[49]This is one of three known Ptolemaic references to the so-called moon illusion: i.e., the perception that celestial objects are larger, sometimes considerably larger, at the horizon than they are at zenith, even though in fact they subtend the same arc at both locations. The other two references can be found in: 1) *Almagest* I, 3, where the illusion is explained in terms of refraction through atmospheric vapors, the resulting image being enlarged in the same way that the image of an object immersed in water is; and 2) the *Planetary Hypotheses*, where the account, which cites perceptual differences arising from differences in distances, is far from clear. For an analysis of these varying interpretations of the illusion, along with an account of Ibn al-Haytham's explanation, see A. I. Sabra, "Psychology versus Mathematics: Ptolemy and Alhazen on the Moon Illusion," in Edward Grant and John Murdoch, eds., *Mathematics and its Applications to Science and Natural Philosophy in the Middle Ages: Essays in Honor of Marshall Clagett* (Cambridge: Cambridge U. Press, 1987), pp. 217–247. Two factors seem to be in play in the account under discussion here. First, there is the perception of distances, which changes as we shift our view upward, so that the apparent distance at zenith will be smaller than that along the horizon (as Sabra points out on p. 226 of the above article, this explanation seems at least basically in accord with that offered by Ptolemy in the *Planetary Hypotheses*). But what is the cause of this difference in apparent distance? Perhaps the clue is to be found in II, 126 above, where Ptolemy makes two interdependent claims: 1) that dimmer objects appear larger than brighter ones of the same size lying at the same distance, and 2) that objects whose color is dimmer appear to be farther away than objects of a brighter hue (the principle underlying atmospheric perspective). Now, in terms of this second claim, it is clear that objects seen toward the horizon appear dimmer than those near the zenith. Thus, by Ptolemy's line of reasoning, they must appear farther away. Add that dimmer objects appear to be larger than brighter ones of the same size, and it follows that bodies such as the moon or stars will appear considerably larger at the horizon than they do at zenith—up to 50 percent larger, in fact; see, e.g., J. O. Robinson, *Psychology of Visual Illusion*, pp. 55–61. What, then, about the second factor cited by Ptolemy: the relative difficulty of seeing elevated things? Interestingly enough, as Robinson points out in the above-mentioned book, the psychologist, Boring, conducted a study some years back that seemed to show that the illusion depends at least in part on the posture of the observer. Thus, when the moon at zenith is viewed from a standing position, with the viewer's neck appropriately craned, the moon appears smaller, whereas when it is viewed from a supine position, with the viewer's neck straight, it does not. More recent study, however, brings Boring's conclusion into doubt . For a discussion of Ptolemy's approach to the moon illusion, see H. E. and G. M. Ross, "Did Ptolemy Understand the Moon Illusion?" *Perception*, 5 (1976): 377–385. For a good general analysis of the moon illusion, see L. Kaufman and I. Rock, "The Moon Illusion," *Scientific American* 207 (1962): 120–130.

[61][50] It has thus been demonstrated how the visual cones attain a perceptual grasp in a completely natural way and how this perceptual grasp leads to a single, primal impression in terms of both sensible effect and location. This happens when we look with both eyes at some spot on an object that is seen with either eye and nothing about the natural operation [of the axes] is altered. Indeed, such a process is invariably followed when both axes, each of which has a sense-capacity peculiar to its eye, are brought to convergence from the Apex toward the visible object by the Governing Faculty. It is this faculty that fixes those axes on one and the same line coinciding with the common axis between the vertices of the visual cones. And it directs the impinging rays according to that line, since it is unique, and the position of the ray is directed upon the visible objects themselves by the power inherent in that faculty. Objects directly intercepted by those axes will clearly seem to occupy a single location, which is [in line with] the common axis, whereas objects that are intercepted by the remaining rays will appear to lie at places whose displacement from the true location depends on the distance of the common axis from the proper axis [of the appropriate visual cone].[51]

[62] As long as we consider visual perception by one eye and a single perceptual impression according to a single origin-point, there is nothing in demonstrations devoted to mirrors that should prevent us from taking into account one of the cases that we have discussed, since it is in the nature of perceptions to arise from a coalescence of the sensation of objects that are seen according to a common apprehension. For things that are seen and perceived by both eyes are seen by means of a single capacity that extends continuously to the origin-point where it is exerted.[52]

[63] So much for what ought to be said about the issues we have raised in the preceding discussion. Let us now discuss images that appear simple according to each of the aforementioned cases, espe-

---

[50]As noted earlier, paragraph III, 60 has been transposed to paragraphs III, 12 and 13 above, where it seems to belong.

[51]According to Lejeune, *Euclide et Ptolémée*, pp. 165–168, the preceding analysis of binocular vision according to horizontal displacement from the common axis is in contradiction with the earlier analysis in book 2, because that latter analysis, focusing as it does on angular displacement, depends on a circular field of view, whereas the former analysis, focusing as it does on horizontal distance, depends on a rectilinear field of view. Cf., however, Pace, *Elementi Aristotelici*, pp. 258–261, and Simon, *Le regard*, pp. 144–146.

[52]Presumably this "single capacity" is exercised at the Apex where the two axes are controlled. The basic point of this passage seems to be to justify Ptolemy's subsequent analysis of reflection in terms of point-objects and a single center of sight, from which a more general analysis of image-formation by extended objects can supposedly be extrapolated; see Lejeune, *L'Optique*, p. 118, n. 57.

cially those that appear in mirrors whose shapes do not affect the visible properties seen in them.

[64] After having spoken first about colors, we offer the basic observation that, in comparison to how the colors themselves appear in direct vision, their images appear somewhat different. The reason for this is the weakening caused by reflection, and we have already pointed out that this is why objects appear dimmer [in mirrors].[53] A change in the colors of such objects [after reflection] also arises from the colors of mirrors. For the visual rays impart something of those colors to the visible objects [seen after reflection]. But this sort of phenomenon should not be attributed to mirrors [alone], because what happens in the case of reflection is common to all bodies that impede visual radiation: the change in color arising from the mirror is common to any and all objects that are seen by unbroken rays, because whatever is seen of colors is always mixed to some extent. So, too, with colors in the air and in transparent media, the flux suffers a sensible coloring. And the same holds for objects that are perceived through transparent bodies.[54]

[65] Therefore, concerning the points that we wish to demonstrate, let us set up geometrical proofs, not indeed according to all of the particular propositions upon which the science of mirrors is ultimately founded—since they would be legion and would require a special treatise—but according to what is possible for us, in order to explain image-formation for everything seen either in direct vision or otherwise.

[66] As far as all the demonstrations that follow are concerned, we ought to bear in mind some points that we will make now so that we need not keep repeating them. First, when we say "eye," we mean the vertex of the visual cone. Second, when we say "a straight line on the surface of mirrors," we mean the common section of the [plane] mirror's surface and the [plane] surface orthogonal to it that contains the broken ray [i.e., the surface of reflection]. Third, when we say "circle inscribed in a spherical mirror," we mean the common section of the mirror's surface with the surface that contains both the broken ray and the center of the sphere [i.e., the surface of reflection]. Fourth, when we say "convex mirror," we mean a spherical mirror whose outward curve faces the eye. Finally, when we say "concave mirror," we mean a spherical mirror whose inwardly curved surface faces the eye.

---

[53]See III, 22 above.
[54]See Ptolemy's discussion of anterior and posterior coloring in II, 108 and 109 above.

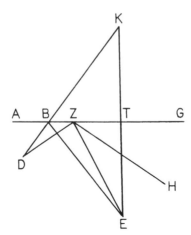

Figure III.9

[67] After having specified these definitions, we should begin by discussing what is observed in plane mirrors for every kind of visible object that is seen in them. In terms of quantity, [we see] the multitude of the objects that we distinguish as well as the sizes and distances of contiguous objects; in terms of quality and the definitive characteristics of things, [we see] differences among motions, the contents of shapes, and spatial locations. It is through these that, in plane mirrors, one eye sees one image of a single object, if it undergoes no alteration, even though it may be quite distant from the mirror.[55]

[68] *[THEOREM III.4]* Let straight line **ABG** [in figure III.9] lie on the surface of a plane mirror, let **D** be the eye, and **E** the visible object. Furthermore, let the visual ray **DBE** emanating from **D** be reflected at equal angles to **E**. We say, therefore, that no other ray emanating from point **D** is reflected at equal angles to **E**.

[69] If such is possible, let the reflecting ray-couple be **DZE**. Then, since angle **ABD** > angle **AZD**, while angle **ZBE** < angle **GZE**, and angle **ABD** = angle **GBE**, it follows that angle **GZE** > angle **AZD**. Therefore, ray **DZ** will not reflect at equal angles along line **ZE**.

[70] From what we have said, then, it is clear that, if we assume that angle **GZH** = angle **AZD**, lines **ZH** and **BE** will not intersect at either of the two points **H** or **E**, nor will they do so in that direction, since angle **GBE** > angle **GZH**.[56]

---

[55]As Lejeune remarks in *L'Optique*, p. 120, n. 59, the distance between object and reflecting surface plays a crucial function in image-location only in concave mirrors.

[56]See Euclid, *Catoptrica*, prop. 4, and Hero, *Catoptrica*, prop. 7.

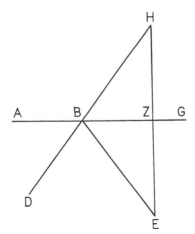

Figure III.10

[71] In like manner it will also be demonstrated that, if we drop perpendicular **ET** to **AG** from visible object **E** and extend lines **ET** and **DB**, they will intersect at point **K**. Accordingly, since angles **DBA** and **EBG** are equal, angle **DBA** will be smaller than a right angle. Therefore, alternate angle **KBZ** will be smaller than a right angle. It therefore follows that, since angle **KTB** is right because it is alternate to angle **GTE**, the sum of angles **TBK** and **BTK** will be less than two right angles. Therefore, lines **ET** and **DB** intersect at point **K**.[57]

[72] Consequently, the image of **E** that the eye at **D** sees lies at point **K**, and so what happens in this mirror is the same as what happens in direct vision, insofar as objects seen by means of a broken ray appear to lie in a single location.

[73] *[THEOREM III.5]* When we look at something in a plane mirror, the distance of the visible object and the distance of its image from the eye are equal.

[74] Let straight line **ABG** [in figure III.10] lie on a plane mirror, let point **D** be the eye and point **E** the visible object, and let ray-couple **DBE** emanating from point **D** reflect at equal angles. Let perpendicular **EZ** be dropped from **E** to **AG**, and let lines **DB** and **EZ** be extended to meet at **H**. We say, then, that line **DH** = **DB** + **BE** and that **EZ** = **ZH**.

[75] Angle **ABD** = angle **ZBE**, and angle **ZBH** = angle **ABD**, since they are alternate angles. But the angles at **Z** are right angles, and line **BZ** is common to the two triangles **BZH** and **BZE**.

---

[57]See Euclid, *Catoptrica*, prop. 15.

Therefore, line **EZ** = **ZH**, and line **BE** = line **BH**. If we take **DB** as common, then the whole of **DBH** will be equal to the sum of lines **DB** and **BE**.[58]

[76] On the basis of what we have said, it will be understood that the image of objects that are farther from the eye lies at a greater [apparent] distance, just as happens in the case of direct vision: the farther removed they are, the farther away they appear to be from the eye according to the amount by which the visual rays are lengthened.

[77] As we will explain, in plane mirrors the sizes of objects seen as they really are in direct vision appear the same as those of their image, and that image appears to occupy the same position. Moreover, unless the particular conditions we will set forth are met, what we see according to direct vision will be nothing like what we see in mirrors, even if they subtend equal angles or one and the same angle, as we have previously shown.[59] Let us therefore agree that the visible objects [assumed in subsequent demonstrations] are always in a facing orientation. For this is the only orientation, even in spherical mirrors, according to which there can be no divergence or error between visible objects and their images on account of the factors we have discussed. Analyzing objects that are not so oriented is difficult, and the results do not jibe with what we have claimed.[60]

[78] Now, when we say "facing orientation," we mean that the visual ray that strikes the midpoint of the straight line joining both ends of the visible object forms two right angles with it; this is a face-to-face position, and it is one of direct opposition [between viewpoint and object]. We say, moreover, that the distances are equal when the rays that emanate from the eye and strike the midpoint of the lines joining the endpoints of the visible magnitudes [i.e., the actual objects and their images] are equal.

[79] *[THEOREM III.6]* Let straight line **ABG** [in figure III.11] lie on a plane mirror, and let **D** be the eye. Now let line **EZ**, connecting both ends of the visible object, be placed in such a way that perpendicular **DB** dropped from **D** to line **AG** bisects it, which defines a facing orientation. Let ray-couples **DH**, **HE** and **DT**, **TZ** emanat-

---

[58]See Euclid, *Catoptrica*, prop. 19, and Lejeune, *Recherches*, pp. 81–82.

[59]Ptolemy is referring here to the analysis in II, 53-62 of size perception according to the relative size of visual angle, obliquity, and radial distance.

[60]As Lejeune points out in *L'Optique*, p. 124, n. 60, by insisting on a facing orientation for the line-objects he will subsequently analyze, Ptolemy is restricting the conditions so that his demonstrations will apply to binocular as well as to monocular vision.

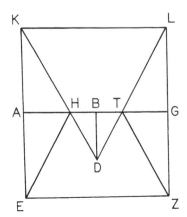

Figure III.11

ing from **D** reflect at equal angles to **E** and **Z**. Then let the two perpendiculars **EA** and **ZG** be dropped to **AG** from points **E** and **Z**, and let these two perpendiculars be extended until they intersect the continuations of lines **DH** and **DT** at points **K** and **L**. Finally, let line **KL** be joined. Accordingly, the image of point **E** will lie at point **K**, and the image of point **Z** will lie at point **L**, and line **KL** will join the two endpoints of the object's image.

[80] From what we have said, it is clear that, since the angles of figure **AGEZ** are right, it follows that **AG** = **EZ**, and **AE** = **GZ**. And we established earlier that **AE** = **AK**, and **GZ** = **GL**. Therefore, the whole of **EAK** = the whole of **ZGL**, **KL** = **EZ**, and the angles at **K** and **L** are right.

[81] The position of the line joining the endpoints of the image of the visible object will thus be as we said. And **EZ** will be congruent with **KL** as far as its appearance to the eye at **D** is concerned. Also, its face-to-face distance and orientation will be the same. And it will subtend the same angle **KDL** as the visible object seemed to subtend according to the way we have set things up.[61] For, when the images of objects subtend precisely the same angles [as the objects themselves], they must always appear equal [to one another].[62]

[82] *[THEOREM III.7]* In plane mirrors, the image of a magnitude appears to have the same shape as that magnitude would if it were to lie where the visual rays would strike unreflected.

[83] Let **ABGD** [in figure III.12] be a right-angled parallelogram, let its sides be bisected at points **E**, **Z**, **H**, and **T**, and let the two lines

---

[61]This, of course, entails shifting the viewpoint **D** to the other side of **AG** an equivalent distance (i.e., = **BD**) when viewing the object **EZ** directly.
[62]See Lejeune, *Recherches*, pp. 87–88.

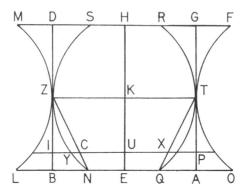

Figure III.12

EKH and ZKT be drawn between them. Let two circular arcs be drawn through point Z such that one of them, LZM, has its convex face toward the mirror and the reflected ray, while the other, NZS, is identical but has its concave face toward them, and let them be positioned in such a way that KZ is perpendicular to them both, as is the case in a face-to-face orientation. Let line EKH lie on the plane mirror, and let the [reflected] visual ray proceed along KZ in the direction of the visible object.

[84] Now let the two lines AE and GH be extended so that lines EO and HF are equal to lines EL and HM [respectively], and let lines EQ and HR be taken equal to lines EN and HS [respectively]. From what we have previously determined, then, it is evident that the image of line BD will appear as a straight line along ATG. Meanwhile, the image of convex arc LZM will appear along FTO, while the image of concave arc SZN will appear along QTR.[63]

[85] Line ATG is straight, because lines EA, KT, and GH are equal, and all the angles are right.

[86] The line passing through points O, T, and F is convex, whereas the line passing through points Q, T, and R is concave. The reason is that the visual ray directed along line KZ to point T is normal to all the images of the lines, whereas the rays that reach the endpoints of the lines and all the remaining spots on them are oblique. With respect to ray [KZT] that reaches T, the rays that reach O and F are proportionately longer than those that reach A

---

[63]Notice that these "appearances" are without reference to a center of sight. Theoretically, of course, Z must represent the center of sight, which will then have to lie on the actual object-surfaces.

and **G**, whereas with respect to ray [**KZT**] that reaches **T**, the rays that reach **Q** and **R** are proportionately shorter than those that reach **A** and **G**. In the case of magnitudes or lines to which visual rays are dropped, when the ratio of oblique to orthogonal rays is greater [than it is for a straight line] from a face-to-face perspective, then those magnitudes will appear convex. When this same ratio is smaller [than it is for a straight line], those magnitudes will appear concave.[64] Line **ATG** will therefore be convex when it is curved to coincide with **OTF**, whereas it will be concave when it is curved to coincide with **QTR**. So line **ATG** [itself] is straight, whereas without exception line **OTF** is convex and line **QTR** is concave.

[87] That each of these types of figure is necessarily seen along a single, continuous line will be obvious from the observed phenomenon itself. It is also naturally evident that the images of continuous magnitudes that are composed of homogeneous parts and that are regularly shaped throughout take the form of a single, simple image when they appear in stationary mirrors under unchanging conditions without being confused. For, given the situation in which objects of this sort have continuous parts, none of the parts is more salient in any respect than the others.

[88] We can, in fact, establish this point mathematically without difficulty on the basis of what we have already described.

[89] To this end, we will draw lines **NZ** and **QT** and will mark point **Y** on concave arc **NZ**. We will then suppose that its image lies on straight line **QT**, and we will draw a line [**XY**] parallel to **AB**. Accordingly, **AQ** : **PX** = **AT** : **TP**, and **BZ** : **IZ** = **BN** : **CI**. In addition, line **AT** = line **BZ**, line **PT** = line **ZI**, and line **AQ** = line **BN**. Thus, line **PX** = line **IC**, and line **PU** = line **IU**. It therefore follows that line **XU** = line **CU**, and line **YU** > line **XU**.

[90] So the image of **Y** will not appear at point **X**. But the distance of **U** from point **X** is the same as that of **U** from point **C**. This latter point will therefore lie above the concave line. Likewise, if we take any line connecting two points on the image, it is impossible for the image of any point on the convex or concave arc to lie on that line.

[91] *[THEOREM III.8]* In plane mirrors, the image lies on the same side as the actual object whose image it is, and if visible objects move in some given direction, their images move in the same direction as far as the eye is concerned.

[92] Let straight line **ABG** [in figure III.13] lie on a plane mirror, and let **D** be the eye and **EZ** the visible object. Let the two

---

[64]See II, 67 above.

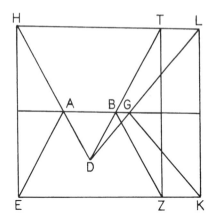

Figure III.13

ray-couples **DAE** and **DBZ** [emanating] from the eye at **D** be reflected at equal angles to points **E** and **Z**. Then let **DA** and **DB** be continued rectilinearly to points **H** and **T** [respectively], where they
intersect the two perpendiculars [**EH** and **ZT**] dropped to **AB** from
points **E** and **Z**. Let line **HT** be joined. Point **E** will therefore be seen
at point **H** and point **Z** at point **T**, and, in fact, those two images appear to lie on the same sides as the two points generating them.

[93] Now, let line **EZ** be continued rectilinearly until **Z** reaches **K**,
and let ray **DGK** [emanating] from point **D** reflect at equal angles
to **K**. Then let line **DG** be extended until it meets point **L** on the perpendicular dropped from point **K** to line **AG**. The image that appeared at **T** has thus been shifted to **L**, and this shift takes the same
direction as that of the visible object itself, i.e., from **Z** toward **K**.

[94] For instance, if we assume that **KZ** and **BG** both lie above
the eye, then image **TL** will appear to lie higher than the eye. And
this image will be seen as the object itself is seen [in direct vision]:
what is higher appears above, and what is lower appears below.
Objects that are seen directly appear in precisely the same way: i.e.,
what is above appears above, and what is below appears below,
for a higher visual ray sees what is above, whereas a lower one
sees what is below.[65]

[95] If, moreover, we assume that both **KZ** and **BG** lie to the right
of the eye, then image **TL** will also lie to the right, and **T** will shift
to the right [when it moves] toward **L**. Nevertheless, the image of **T**
will be judged to lie on the right-hand side [of **TL**], while that of **L**
will be judged to lie on the left-hand side, for when objects are seen
directly, face-on and parallel to our facial plane, their right side does

---

[65]See II, 26 above.

not correspond to [our] right side, nor does their left side correspond to [our] left side; instead, the reverse is true.

[96] This sort of illusion is not, however, due to the mirror but, rather, to perceptual judgment. Accordingly, the object situated at **Z** and seen by means of a right-hand ray appears to lie to our right, and when point **Z** moves, point **T** will appear to move according to the principles we have already articulated. But this is the same as what happens in the case of directly-viewed objects: i.e., what is seen by right-hand rays appears to lie to our right, and what is seen by left-hand rays appears to lie to our left. Yet whatever lies to our right on a facing object is actually situated to the object's left, and the image [**TL**] lies face-on to the plane of the observer. So it necessarily follows that, since what lies toward point **T** is the image of a right-hand object and will appear to our right, the mind judges it to lie to the left, because it is seen by the ray we specified [i.e., the right-hand one] and lies to the left side of an object facing the actual object.[66]

[97] On the basis of what we have shown, it is possible for anyone to understand not only that the images that appear in plane mirrors are no different, in terms of perceptible attributes, from objects that are seen in direct vision, but also that the characteristics of both [kinds of appearance] are similar. Since our discussion to this point leads into the investigation of what happens to images seen in convex and concave mirrors, we should start by saying that in these two types of mirrors several phenomena are different from those that appear in direct vision. As we have already shown, what is seen in plane mirrors appears more or less identical to what is seen directly. Now a straight line happens to be a sort of mean between the other two figures (i.e., convex and concave), and it has a perfectly uniform spatial disposition. The other two figures, though, do not have a perfectly uniform spatial disposition and are opposed not only to one another but also to the straight line. It therefore follows that the straight line should be given precedence over curved[67] lines and arcs, in terms of both nature and dignity. Indeed, among such curved lines, the primary and simple ones, which are perfectly regular, are derived from straight lines.[68]

---

[66]See II, 137–138 above, where Ptolemy discusses the left-right reversal of mirror-images in terms of perceptual illusions. See also Euclid, *Catoptrics*, prop. 19, and Lejeune, *Recherches*, pp. 100–102.

[67]Latin = *curvus*; here we have one of the rare instances in the *Optics* where *curvus* takes the general meaning of "curved" rather than the specific meaning of "convex."

[68]Cf. Proclus' *Commentary* on definition 4 of the *Elements*, in Morrow, trans., p. 87.

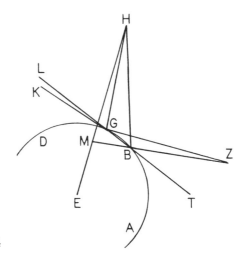

Figure III.14

[98] And just as plane mirrors have a greater capacity than other kinds for preserving the integrity [of images], so also convex mirrors have a greater capacity than concave mirrors for preserving [such] integrity. The reason is that in convex mirrors the reflecting surface invariably lies between the eye and the center of the sphere, whereas the surface of a concave mirror will never, at any time, lie in the position just specified. Also, the divergences between what appears in direct vision and what appears in convex mirrors are less than those between what appears in direct vision and what appears in concave mirrors.[69] That is why we should start with a discussion of convex mirrors, since their study is easier than the study of concave mirrors.

[99] *[THEOREM III.9]* In convex mirrors, [only] one image is ever seen by an observer using one eye, and it appears behind the [surface of the] mirror; moreover, that image will never, at any time, appear to lie on the [surface of] the sphere out of which the mirror is formed.

[100] Accordingly, let **ABGD** [in figure III.14] be the arc of a circle lying on a convex mirror with center **E**. Let **Z** be the eye and **H** the visible object. Let **ZBH** be a ray-couple emanating from **Z** and reflecting at equal angles to **H**. We say, then, that no other ray-couple [emanating from **Z**] reflects at equal angles to that point.

[101] But if such is possible, let the ray-couple be **ZGH**, and draw out line **TBGK**. Then, since angle **ZBT** > angle **ZGT**, and angle

---

[69] In other words, image-distortion is greater and more variable in concave mirrors than in convex ones.

ZGT > [curvilineal] angle ZGA, [curvilineal] angle ZBA will be considerably larger than [curvilineal] angle ZGA. But [curvilineal] angle ZBA = [curvilineal] angle HBD. Thus, [curvilineal] angle HBD > [curvilineal] angle ZGA. Also, [curvilineal] angle HGD > angle HBG. Thus, [curvilineal] angle HGD is considerably larger than [curvilineal] angle HBD. Consequently, ZGH does not reflect at equal angles.[70]

[102] If we posit [curvilineal] angle ZGA = [curvilineal] angle DGL, then it is evident that lines BH and GL will not meet in the direction of H and L. Since angle TGZ > angle KGL, while angle TBZ > angle TGZ, and angle KBH > angle TBZ, then angle KBH > angle KGL. Therefore, lines BH and GL do not meet on the side of points H and L.[71]

[103] Now, let lines HME and ZB be extended to meet at point M. According to the principles we have articulated, then, only at point M will the image of H be seen, and it will always lie between the visible object and E, which is the center of the circle, because the [radial] line connecting E to B [i.e., the normal] bisects angle ZBH. Therefore, when line ZB is extended, it will intersect HE, and the image must lie behind the mirror['s surface], for the mirror['s surface] lies between the eye and the intersection-point of the aforesaid lines.

[104] There is no question that the image of a visible object is formed as we have said and that our reasoning is borne out by actual observation. However, the intersection-point where the image appears does not always lie between the surface of the mirror and the center of the sphere; it is actually possible for the aforesaid two straight lines to intersect on the very surface of the mirror, as well as to intersect between the surface of the mirror and the visible object.

[105] *[THEOREM III.10]* Let ABGD [in figure III.15] be the arc of a circle lying on a convex mirror with center E. Let ZBGH be drawn to cut arc BG on the circle, and let that arc be less than a sixth of the [entire circumference of the] circle. Then let the two lines EBT and

---

[70]Ptolemy's use of curvilineal angles is somewhat idiosyncratic. Normally, the measure of angles with respect to the curved surface would be by means of the tangent to the vertex-point; thus, curvilineal angle ZBA would be virtually equal to the angle formed by ZB and the tangent to point B, the resulting "horn angle" or angle of contingency formed by the tangent and arc BA adding nothing to the measure (see Euclid III, 16).

[71]The point here is to show that, if the reflection takes place according to incident ray ZG rather than incident ray ZB, the reflected ray GL will never reach the visible object H; see Euclid, *Catoptrics*, prop. 4 and Hero, *Catoptrics*, chapter 8.

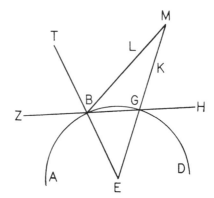

Figure III.15

EGK be drawn, and let **ZB** reflect at equal angles along **ZBL**. We thus say that, when it is extended, **BL** will meet **EGK** in the direction of points **K** and **L**.

[106] Now, since arc **BG** is less than a sixth of the circle, angle **BEG** will be less than two-thirds of a right angle, whereas angle **EBG**, which equals angle **ZBT**, is greater than two-thirds of a right angle. But angle **ZBT** = angle **TBL**, and angle **TBL** > angle **BEG**. Thus, lines **BL** and **EK** intersect on the side of points **K** and **L**. Let them join at point **M**. If, then, we suppose **Z** to be the eye and **M** the visible object, its image will lie at point **G**, which is on the mirror's surface.

[107] *[THEOREM III.11]* Harking back to our previous figure, let line **EK** [in figure III.16] be drawn so that it does not pass through point **G** but, rather, cuts line **HB** at point **K**. Let angle **BEM** be smaller than angle **EBG**, which in turn is greater than two-thirds of a right angle.

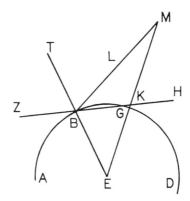

Figure III.16

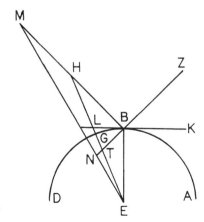

Figure III.17

[108] Accordingly, since angle **EBG** > angle **BEM**, while angle **EBG** = angle **ZBT**, and angle **ZBT** = angle **TBL**, angle **TBL** > angle **BEM**. Hence, when they are extended, lines **BL** and **EK** meet in the direction of **L** and **K**. Let them intersect at point **M**. Therefore, to the eye at **Z** the image of point **M** will appear at point **K**, which lies between the [surface of the] mirror and the visible object.

[109] Moreover, if point **K** could lie between the eye and the [surface of the] mirror, the image would certainly appear in front of the mirror, as is the case in concave mirrors. But, since point **B** invariably blocks point **K**, it follows inevitably that the [surface of the] mirror appears in front of the image, for in that case the eye does not distinguish between the internal and external surface; and so the image must lie behind the surface.[72]

[110] *[THEOREM III.12]* In convex mirrors, the distance of the visible object from the eye, as well as from the mirror's surface, is greater than the distance of the object's image [from the mirror's surface].

[111] Let **ABGD** [in figure III.17] be the arc of a circle lying on a convex mirror, and let the center be at **E**, the eye at **Z**, and the visible object at **H**. Let the visual ray-couple **ZBH** be reflected at equal angles to the visible object, and let the two lines **EB** and **EH** be joined. Then let **BZ** be extended until it intersects **EH** at point **T**. Accordingly, the image of point **H** will lie at **T**. We say, therefore, that lines **BZ** and **BH** [taken together] are longer than line **ZBT**, and that **GH** is longer than **GT**.

---

[72]See Lejeune, *Recherches*, pp. 74–75. Lejeune claims that this somewhat bizarre theorem is included only to provide a parallelism with concave mirrors in terms of the projection of real images; cf., however, Simon, *Le regard*, pp. 159–163.

[112] Let line **KBL** be drawn tangent to the circle at point **B**. Since angle **ABZ** = angle **DBH**, and [horn] angle **ABK** = [horn] angle **DBL**, it follows that angle **KBZ**, which equals angle **LBT**, is equal to angle **LBH**.[73] But angle **LBT** is acute, because angle **EBL** is right, and angle **BLH** > angle **BLT**. Therefore, line **BH** > line **BT**. With line **ZB** taken as common, lines **ZB** and **BH** [taken together] will be longer than line **ZBT**.

[113] Moreover, since **BH** : **BT** = **LH** : **LT**,[74] line **LH** > line **LT**. **GH** therefore surpasses line **GT** by an even greater margin.

[114] On the basis of what we have said, it will also be easily demonstrated that, in the case of objects that withdraw [away from the mirror's surface] or objects that lie farther than others [from that surface], the image lies farther away, and its distance [behind the mirror] appears greater.

[115] Accordingly, let line **BH** be extended to **M**, and let line **EM** be joined. Then let line **ZBT** be extended until it meets line **EM** at point **N**.

[116] If, then, we assume that **H** is distinct from **M** and that **M** is farther away [from the mirror's surface] than **H**, then **N**, which is the image of **M**, will certainly lie farther away from the eye at **Z** than **T**, which is the image of **H**. Moreover, if we assume that **H** coincides with **M** but that its distance increases as it moves from **H** to **M**, then its image will also become more remote, since it has shifted from **T** to **N**.[75]

[117] *[THEOREM III.13]* In convex mirrors, when objects are situated in the way we specified for plane mirrors (where the lines joining the endpoints of the object lie face-on to the mirror), the image appears smaller than the objects themselves [would appear] if they were transposed to where the image lies, with the very same orientation and distance, and were viewed without reflection.

[118] Let **ABG** [in figure III.18] be the arc of a circle on a convex mirror, whose center is **D**, and let **E** be the eye. Let normal **EBD** be dropped from point **E**, let **ZH** be the line joining the endpoints of the visible object, and let the visible object be so disposed that line **EBD** bisects line **ZH** at right angles, as is the case for a facing object. Let the two ray-couples **EAZ** and **EGH** [emanating] from point **E** be reflected at equal angles to **Z** and **H** [respectively]. Then, let lines **DZ** and **DH** be drawn, and let **EG** and **EA** be extended

---

[73]See n. 70 above on Ptolemy's use of curvilineal and horn angles.
[74]This proportionality follows from the fact that angle **HBL** = angle **LBT** within triangle **HBT**; see Euclid, *Elements* VI, 3.
[75]See Lejeune, *Recherches*, pp. 82–84.

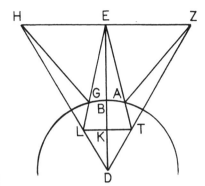

Figure III.18

until they meet lines **HD** and **ZD** at points **L** and **T** [respectively]. Finally, let line **TL** be joined. According to the principles we have established, then, the image of **Z** ought to be seen at **T**, and the image of **H** at **L**. Also, line **TL** will be the line joining the endpoints of the object's image, and its orientation, [being face-to-face,] will be the same as that of **ZH**.

[119] Furthermore, since the distance of point **E**, the eye, from either of the points **Z** or **H** is the same because of their similar disposition, then the angles created by reflection to those points will be equal. Also, the distances of points **T** and **L**, which lie on the image that is seen in the mirror, will be equal with respect to point **E**, so that lines **ET** and **EL** will be equal. Meanwhile, angles **ETK** and **ELK** will be equal, as will angles **TEK** and **KEL**, so triangles **EKT** and **KEL** will be equal and equiangular. But the angles at **K** are right, so line **TL** is parallel to line **ZH**. Consequently, **DH** : **DL** = **ZH** : **TL**. But line **DH** > line **DL**, so line **ZH** > line **TL**.[76]

[120] Now, if line **ZH** were transposed to the location of line **TL**, and if the eye at **E** were to view it directly, it would be seen under a larger visual angle than **TEL**, which [image **LT**] subtends. And this accords with what we proposed.[77]

[121] *[THEOREM III.14]* In convex mirrors, facing straight lines appear convex. In the case of circular arcs, however, those whose convex curvature faces the mirror and the reflected ray appear convex, whereas those whose concave curvature faces the mirror sometimes appear convex, sometimes straight, and sometimes concave.[78]

---

[76]As Ptolemy demonstrates in the next theorem, line **TL** does not define the actual image but, rather, its rectilinear cross-section, which indicates only its relative size (see II, 47 above); the image itself will be convex.
[77]See Euclid, *Catoptrics*, prop. 21.
[78]This latter claim is in fact treated not in this theorem but in the one following.

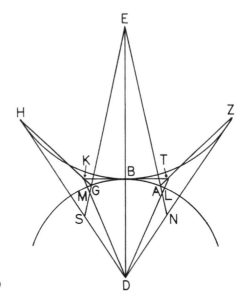

Figure III.19

[122] Let **ABG** [in figure III.19] be the arc of a circle lying on a convex mirror, whose center is **D**, and let **E** be the eye. Let normal **EBD** be dropped from the eye to the mirror, and let two equal arcs **BA** and **BG** be marked off on either side of **B**. Let the two ray-couples **EAZ** and **EGH** [emanating] from the eye at **E** be reflected at equal angles from points **A** and **G** [respectively]. Let two lines be drawn tangent to the circle at point **B**, and let one of them, **TBK**, be straight, while the other, **ZBH**, is a circular arc with its convex curvature facing the mirror and the reflected ray. Then, let lines **TD**, **KD**, **ZD**, and **HD** be drawn to intersect [the extensions of] lines **EA** and **EG** at points **L**, **M**, **N**, and **S** [respectively]. Accordingly, the image of point **T**, which is one endpoint of the straight line, will lie at **L**, and the image of point **K** [the other endpoint] will lie at **M**. Meantime, the image of **Z**, which is one endpoint of the convex line, will lie at **N**, whereas the image of **H** [the other endpoint], will lie at **S**.

[123] It is therefore evident that for both [line-segments] common point **B** will appear at **B**, since this is where the mirror's surface and the visual ray intersect. However, the convexity of arc **ABG** lies toward the eye at **E**, and the images of simple lines, which comprise a single line, themselves comprise a single line of a unique type.[79] Hence, the eye will see the two lines passing through points **L**, **B**,

---

[79]In other words, we can assume that the entire line-image, from **M**, through **B**, to **L** will be completely continuous and regular in its curvature, as indeed will the line-image from **S**, through **B**, to **N**; see III, 87 above.

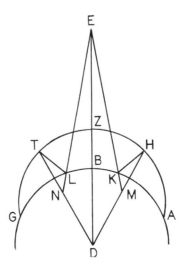

Figure III.20

and **M** and points **S**, **B**, and **N** as convex, since they are more sharply curved than arc **ABG**. Still, arc **SBN** will be more sharply curved than **LBM**. Since the objects face the eye, and since the oblique lines produced from [the endpoints of] the diameter are proportionately longer with respect to the normal [in the case of the image] than the corresponding oblique rays when the object is seen directly, the line passing through points **L**, **B**, and **M**, which defines the image of straight line **TBK**, will be convex. By the same token, the line passing through points **N**, **B**, and **S**, which defines the image of convex line **ZBH**, will also be convex.[80]

[124] *[THEOREM III.15]* If the line that is seen is concave, it can be demonstrated as follows how it is possible, using the distance along the normal between the two arcs [of the concave object-segment and the convex mirror], that the image might sometimes appear convex, sometimes concave, and sometimes straight:

[125] Let **ABG** [in figure III.20] be the arc of a circle lying on a concave mirror whose center is **D**, let the eye be at **E**, and let the normal be **EBD**. Let circular arc **AZG** be drawn from points **A** and **G** so that it cuts perpendicular **BE** [at **Z**]. Let two equal arcs **ZH** and **ZT** be marked off on each side of **Z**, and let the two ray-couples **EKH** and **ELT** [emanating] from point **E** be reflected at equal angles to **H** and **T**. Let **DT** and **DH** be drawn, and let them intersect the extensions of lines **EK** and **EL** at points **M** and **N**.

---

[80]See II, 67 above; see Lejeune, *Recherches*, pp. 92–93.

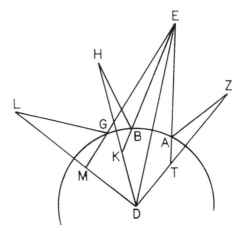

Figure III.21

Thus, **H** will be seen at **M**, and **T** at **N**, while points **A** and **G** will be seen where they actually lie.

[126] Depending on how sharply curved concave arc **AZG** is, it is possible for points **M** and **N** sometimes to lie between arc **ABG** (which is an arc on the mirror) and the straight line joining points **A** and **G**, sometimes to lie on that straight line itself, and sometimes to lie between that straight line and **D** (which is the sphere's center). Now it is clear that, when images of this kind lie on one line, and points **M** and **N** lie between the mirror and the straight line joining **A** and **G**, then the eye at **E** will see their convexity directed toward it. But if points **M** and **N** lie on that same straight line [joining **A** and **G**], then the image will appear straight, whereas if the image is farther removed from **E** [than that straight line], it will appear concave. This will be obvious through a construction of the image on the basis of the ratios between oblique and normal rays.[81]

[127] *[THEOREM III.16]* In convex mirrors, the image of objects is seen on the same side as the actual object, and, if the visible objects move in a given direction, their image moves in the same direction.

[128] Let **ABG** [in figure III.21] be the arc of a circle lying on a convex mirror whose center is **D**, let **E** be the eye, and let the visible object occupy points **H** and **Z** on both sides of the eye. Let the two ray-couples **EAZ** and **EBH** be reflected at equal angles to those points, let lines **DZ** and **DH** be drawn, and let **EA** and **EB** be extended to meet them at points **T** and **K** [respectively]. Accordingly,

---

[81]See n. 79 above.

Z will be seen at **T**, and **H** at **K**, and each of those images [will lie] on the same side as the actual object of which it is an image.

[129] Let **H** be moved toward **L**, and let ray-couple **EGL** be reflected to it at equal angles. Then let line **DL** be drawn so that line **EG**, when extended, intersects it at **M**. The image that lies at point **K** is thus transposed to **M**, so it is transposed to the same side as the object itself; and this follows necessarily from the reasons that we gave in our discussion of the image seen in plane mirrors.[82]

[130] If we place points **H** and **L** above the eye, then points **K** and **M**, which are the images of **H** and **L**, ought to appear above the eye, and the magnitude that lies between them will appear above the eye. And if we place **H** and **L** to the right of the eye, their images **K** and **M** will lie to our right, and so [the magnitude that lies between them] will be judged to lie to the left. Still, this image will not be seen in the same way as an object that faces us directly,[83] but[what lies to the right] will be judged to lie to the left, and what lies to the left will be judged to be right according to what is usual, as we have already said, in the case of facing objects.[84]

[131] From what we have established, it follows necessarily that, in convex mirrors, variations in regard to number, location, and movement are much like those that occur in the case of direct vision. For it has been demonstrated that, in convex mirrors, the image of a single object appears single, that the images of visible objects are equivalently oriented with the objects themselves, and that, when those objects are transposed in a given direction, their images are transposed along with them in the same direction.

[132] On the other hand, in regard to variations in sizes, distances, and shapes, the greater part of those involving images in convex mirrors do not correspond to those involving objects viewed directly—and this for reasons that hark back to general principles already articulated. For one thing, we find that the images of magnitudes situated at an equal distance and orientation appear smaller than the actual objects [generating them], because the angles those images subtend are smaller. For another, we find that images do not always maintain the same internal organization as the actual objects [that generate them]. Indeed, the ratio of oblique rays to the normal does not necessarily correspond to that which obtains in the case of objects that are viewed directly. Also, while some

---

[82]See Theorem III.8, paragraphs 91–96 above.
[83]Because this phrase contradicts the claim that follows, Lejeune was led to speculate in *L'Optique*, p. 145, n. 92, that it may represent a later interpolation.
[84]See Euclid, *Catoptrics*, prop. 20.

images of objects may have the same overall shape as the visible object, they are still not alike in terms of intensity [of curvature], as is the case with the images of convex objects which always appear different from the objects themselves; and the same holds for concave objects. Moreover, distances appear smaller [than they really are], for the radial line that reaches the image is shorter than the one reaching the actual object. And, as one has been used to judging sizes, shapes, and distances in direct vision, so also, all other things being equal, one will judge them [in reflection].

BOOK 4
REFLECTION FROM CONCAVE MIRRORS

Topical Resume

Introductory Section [1]: [1] summary of book 3 and statement of goal for book 4

Finding the Points of Reflection in Concave Mirrors [2–107]
[2] depending on relative location of center of sight and point-object, reflection can occur from all, or from several, or from one, or from no points on a concave mirror, [3–5] **Theorem IV.1**, demonstrating that if the center of sight and the object-point coincide, reflection will occur from every point on the mirror, [6–25] **Theorems IV.2–IV.9**, analyzing conditions under which reflection can occur either from a great circle on the mirror's surface or from one of the smaller circles parallel to that great circle, [26–38] **Theorems IV.10–IV.14**, analyzing conditions under which reflection can occur from three points on the mirror, [39] **Theorem IV.15**, analyzing conditions under which reflection can occur from two points on the mirror, [40–61] **Theorems IV.16–IV.21**, analyzing conditions under which reflection can occur from only one point on the mirror, [62–65] **Theorem IV.22**, showing that the cathetus of reflection and the incident visual ray can meet behind the mirror, between the mirror and the eye, at the eye, or beyond the eye, [66–68] **Theorem IV.23**, analyzing image-location according to place of object-point on the mirror's radius, [69–70] how sight adjusts for indeterminate images, [71–73] **Experiment IV.1**, confirming various appearances depending on location of eye and object with respect to mirror's surface, [74–80] **Experiment IV.2**, showing empirically how an image of a straight rod normal to the mirror's surface is formed, [81–96] **Theorems IV.24 and IV.25**, analyzing conditions under which only one reflection will occur, [97] **Theorem IV.26**, demonstrating that, under the conditions just analyzed, the image appears between the eye and the mirror's surface, [98–107] **Theorems IV.27–IV.29**, analyzing conditions under which no reflection can occur

173

Image-Formation and Image-Distortion [108–155]

Image-Distance [108–119]: [108] transitional statement, [109–113] **Theorem IV.30**, showing that, when the image lies behind the mirror, it will appear more distant than the object, [114–119] **Theorem IV.31**, demonstrating that, when the image lies between the mirror and the eye, its distance can be equal to, less than, or greater than the object's distance

Image-Size [120–129]: [120–122] **Theorem IV.32**, showing that, when the image lies behind the mirror, it appears larger than the object, [123–124] capsule summary, [125–129] **Theorem IV.33**, demonstrating that, when the image lies between the mirror and the eye, it can appear larger than, equal to, or smaller than the object

Image-Shape [130–141]: [130–141] **Theorem IV.34**, analyzing shape-distortion depending upon whether image lies behind or in front of the mirror

Image-Orientation [142–155]: [142–146] **Theorem IV.35**, showing that, when the image lies behind the mirror, it is inverted right-to-left, [147–151] **Theorem IV.36**, showing that, when the image lies in front of the mirror, it is inverted top-to-bottom, [152–155] general conclusions about image-formation in concave mirrors

Composite Mirrors [156–182]

[156–157] general observations on image-formation in such mirrors, [158–160] convex and concave cylindrical mirrors, [161–162] mirrors in the shape of a torus, [163] general statement about distortion in composite mirrors, [164–170] **Theorems IV.37** and **IV.38**, analyzing images in conical mirrors, [171–173] **Theorem IV.39**, analyzing images in polyhedral mirrors, [174–177] **Theorem IV.40**, analyzing image-formation in multiple plane mirrors set at angles to each other, [178–182] **Theorem IV.41**, analyzing image-reversal in reflection from multiple even-numbered and odd-numbered mirrors

The Fourth Book of Ptolemy's *Optics*

[1] In the preceding book we sufficiently discussed the principles and the determinate conditions that are necessary to a scientific account of mirrors, and we took into account the properties of images that appear in plane and convex mirrors. We also discussed what must happen in those [images] as far as all the visible properties are

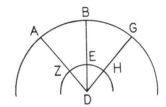

Figure IV.1

concerned. In the present book, then, we must pursue the remaining points that we promised to explain about mirrors. So let us speak of those phenomena that occur and appear in the last kind of simple mirrors that are not composite: i.e., mirrors for which the surface facing the eye is concave; and let us proceed in this analysis as we have proceeded in the others.[1]

[2] Let us start by demonstrating that it is possible for one and the same visual ray to be reflected from a concave mirror [in several ways]: sometimes from all points on the mirror, sometimes from all points on one of its great circles, sometimes from three points, sometimes from two only, or sometimes from none. It is therefore necessary for us to distinguish [among these cases] and to explain the circumstances under which each of these cases occurs and where the image must be formed for each type of reflection according to whatever kind of situation we have proposed.

[3] *[THEOREM IV.1]* Accordingly, we say that reflection from the entire surface of the mirror occurs as follows:

[4] Let **ABG** [in figure IV.1] represent the arc of a circle lying on a concave mirror whose center is **D**. Let **BD** be drawn normal to the mirror. Let **D** be the center of sight and **E** the midpoint [where the axis of the visual cone intersects the surface] of the cornea.[2] In this plane, on centerpoint **D**, let arc **ZEH** be drawn through the corneal surface. Then, let the two lines **AZD** and **GHD** be drawn. They will therefore be perpendicular to the mirror. Hence, all rays emanating from point **D** to arc **ABG** are reflected back on themselves to point

---

[1]It is clear that, like his analysis of convex mirrors, Ptolemy's analysis of concave mirrors was ultimately based on empirical evidence, albeit sometimes misconstrued. If Experiments IV.1 and IV.2, paragraphs 71–80 below, are trustworthy guides, that evidence derived from observations made using the basic apparatus described in Experiment III.1, paragraphs 8–10 above: i.e., the circular bronze plaque graduated into degrees upon which the variously-shaped mirrors can be stood upright. It is probable, then, that the concave mirror Ptolemy used for his observations comprised a semicylindrical section rather than a spherical section. Thus, his conclusions, which are applied to spherical concave mirrors, were doubtless extrapolated from the more limited case of semicylindrical concave mirrors.

[2]Latin = *aspiciens*; see III, 16, n. 15 above.

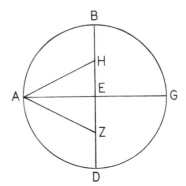

Figure IV.2

D. Moreover, they are apprehended and sensed according to the spots at which rays like **DA**, **DB**, and **DG** strike [as they are] projected from the eye, and they are actually apprehended and sensed according to points **E**, **Z**, and **H**.[3]

[5] The same happens for all parts of the mirror that are sensed by the base of the visual cone. For if we assume that surface **ADG** forms a cone when it is rotated about axis **BD**, then what will appear in the whole surface of the mirror will be the image of the cornea, and it will appear continuous upon the entire surface. For the rays that are projected from point **D** to the mirror's surface and the normals that are reflected back upon them from the mirror meet at a point common to them all as well as to the reflecting surface, and this surface is the one that defines the locations of the image.[4]

[6] *[THEOREM IV.2]* Reflection from [a single] circle on the circumference of such a mirror generally takes place according to the following example:

[7] Let circle **ABGD** [in figure IV.2] be drawn with diameter **BED**. Let two points **Z** and **H** be marked upon **BED** on each side of centerpoint **E**, and let **EZ** = **EH**. Then, through point **E**, let diameter **AEG** pass orthogonally to diameter **BED**. Finally, let lines **ZA** and **AH** be drawn.

[8] Thus, since **ZE** = **EH**, while line **AE** is common, and since the angles at **E** are right, angle **ZAE** = angle **EAH**. Hence, lines **AH** and **AZ** will be reflected at equal angles. Likewise, too, a reflection at

---

[3]Evidently, then, **ZEH** represents the pupil (projected onto the corneal surface) through which the visual cone **DAG** radiates from **D** to the reflecting surface. Once reflected, the rays within that cone are projected back to **ZEH** where, by *emphasis*, they are visually sensed. As a consequence, the image that is perceived is of the pupil rather than of vertex/center-of-sight **D**.

[4]See Euclid, *Catoptrics*, props. 5 and 24, and Hero, *Catoptrics*, chapter 9.

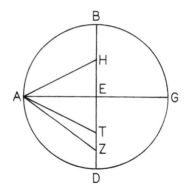

Figure IV.3

equal angles occurs at point **G**. Now if we assume **ABG** to be the arc of a circle lying on a concave mirror, and if we suppose that surface **AZH** revolves about diameter **BD**, then points **A** and **G** certainly describe a circle on the mirror, and all the rays projected to that circle are reflected in the same way as they are along **ZAH**.[5]

[9] *[THEOREM IV.3]* If, now, we suppose that **ZE** is longer than **EH**, we assert that it is impossible for a ray to be reflected from point **A** to points **Z** and **H** at equal angles. For if we cut line **EZ** [in figure IV.3] at point **T** [to form segment **ET**] equal to line **EH**, and if we join **TA**, then angle **EAH** = angle **EAT**, according to what we have just demonstrated. However, angle **EAZ** > angle **EAH**. Therefore, it is impossible for a ray to be reflected [at equal angles to **Z** from **H**] via point **A**.

[10] *[THEOREM IV.4]* We furthermore assert that it is impossible [under the same conditions] for such reflection to occur from some point, such as **K**, between **A** and **D** [in figure IV.4]. For if we join lines **KH**, **KE**, and **KZ**, then, since angles **EKH** and **EKZ** are supposed to be equal, **ZE** : **EH** = **KZ** : **KH**.[6] But **ZE** > **EH**. Therefore, **ZK** > **KH**, which is [to posit a shorter line] longer than a longer one, and that is false. Therefore, no ray between **Z** and **H** is reflected at equal angles from point **K**.[7]

[11] *[THEOREM IV.5]* We wish to demonstrate that on the arc between points **A** and **B** there is a point from which reflection takes place at equal angles. This demonstration is as follows:

---

[5]Lejeune, *L'Optique*, p. 150, n. 6, cites *Catoptrics*, prop. 27 as the appropriate Euclidean counterpart to this theory. In fact, the two theorems bear only the most superficial of resemblances.
    [6]Euclid, *Elements* VI, 3.
    [7]Although this theorem seems related to Euclid, *Catoptrics*, prop. 27, Euclid's analysis is predicated on the equality of **EH** and **EZ**.

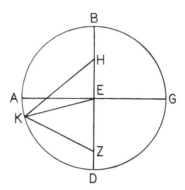

Figure IV.4

[12] Suppose [in figure IV.5] that the amount by which **ZE** exceeds **EH** (i.e., **ZT**) : **EH** = ZH : **KH**, **KH** being longer than **HB**. Let line **KL** be drawn tangent to circle **ABG** at point **L**, and let lines **LZ** and **LH** be connected. We therefore assert that **HLZ** reflects at equal angles.

[13] Let line **MEN** pass through point **E** parallel to line **KL**, and let it meet the extension of line **LH** at **N**. Then let [normal] **EL** be joined, and cut off **ET** equal to **EH**. Therefore, since **ZH** : **KH** = [**ZT** : **EH** =] **ZT** : **ET**, then, by composition,[8] **KZ** [i.e., **KH** + **ZH**] : **KH** = **ZE** [i.e., **ZT** + **ET**] : **ET**, while **ET** = **EH**. By alternation,[9] **KH** : **EH** = **KZ** : **ZE**. But **KH** : **EH** = **KL** : **EN**,[10] and **KZ** : **EZ** = **KL** : **EM**. Thus, the ratio of lines **EM** and **EN** to one another is one of

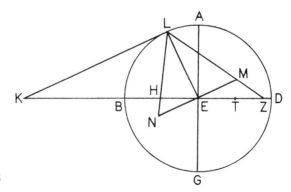

Figure IV.5

8Latin = *coniunctim*.
9Latin = *permutim*.
10These are proportional sides of similar triangles **KLH** and **ENH**; see Euclid, *Elements* VI, 3. The same applies to the next proportionality.

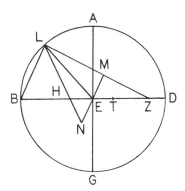

Figure IV.6

identity; so they are equal.[11] Moreover, line **LE** is common, and both angles of the two triangles [**ELN** and **ELM**] at **E** are right, because angle **KLE** is right. Hence, angles **ELN** and **ELM** will be equal.

[14] *[THEOREM IV.6]* On the other hand, if the proportion is not as we have stated, but if instead we suppose that **ZT** : **TE** = **ZH** : **BH**,[12] we assert that no ray reflects from **H** to **Z** at equal angles from any point between **A** and **B**.

[15] If such is possible, though, then let the reflection occur from point **L** [in figure IV.6]. Thus, according to the previously-drawn example, if we draw line **LB** and then draw line **MEN** parallel to it, it clearly follows from what we have demonstrated that **ME** = **NE**. Therefore, since angle **MLH** is bisected by line **LE**, **ML** = **LN**. But line **LE** is common, and angles that are subtended by equal sides are equal. Hence, angles **MEL** and **LEN** will be right, and so will angle **BLE**, which is impossible.[13]

[16] Such a conclusion will be even more absurd if the endpoint of the line upon which the proportion is based lies between points **B** and **H**.[14]

[17] *[THEOREM IV.7]* In general, then, it is clear that, in the case of all concave mirrors centered on **E** with a radius greater than **EH** but less than **EK**, if **EZ** > **EH**, and **KZ** : **ZE** = **KH** : **EH**, then a ray projected between **Z** and **H** must be reflected at equal angles at the

---

[11]Since **KH** : **EH** = **KZ** : **ZE** and **KH** : **EH** = **KL** : **EN**, then **KZ** : **ZE** = **KL** : **EN**. But **KZ** : **EZ** = **KL** : **EM**. Thus, **KL** : **EN** = **KL** : **EM**, whence it follows that **EM** = **EN**.

[12]In this case, then, point **K** in the previous theorem has been moved from outside the circular section to its circumference so that it coincides with point **B**.

[13]**BLE** can only be a right angle if **BL** is tangent to the circle at point **L**; see Euclid, *Elements* III, 16.

[14]Here the point is fully generalized to the case in which point **K** in theorem VI, 5 is moved to within the circular section.

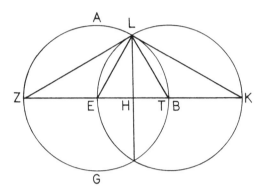

Figure IV.7

point of intersection between the mirror and a circle drawn upon diameter **KE**.

[18] Let some point **B** [in figure IV.7] be taken on line **KH** [and let circle **KLE**, with radius **BK**, be circumscribed about it]. Then let circle **ABG**, which lies on the concave mirror, be drawn upon centerpoint **E** with radius **EB** so that it will cut circle **KLE** at point **L**. We therefore assert that there will be a reflection at equal angles between points **Z** and **H** from point **L**.

[19] For if we join the lines **KL** and [normal] **EL**, angle **KLE** will be right, because it is inscribed in a semicircle, and its vertex touches the circumference of circle **ABG**.[15] In fact, line **KL** is tangent to circle **ABG**, because it is orthogonal to line **EL**, which emanates from the centerpoint **E**. It is therefore obvious that either of the two rays projected between points **Z** and **H** from the point of tangency [**L**] reflects one upon the other at equal angles as [is the case along] **ZLH**.[16]

[20] *[THEOREM IV.8]* Moreover, from what we have said, it necessarily follows that, if **LH** is perpendicular to **EK**, then line **ZL** is tangent to circle **KLE** at point **L**, and angle **ZLH** is the largest of all the angles at which reflection takes place.

[21] Let point **T** [in figure IV.7] be the center of circle **KLE**, and let line **LT** be joined. Thus, since **ET** = **TL**, angle **TLE** = angle **TEL**. But angle **TEL** = angle **LZE** + angle **ELZ**.[17] Thus, angle **TLE** = angle **EZL** + angle **ELZ**. Also, angle **ELZ** = angle **ELH**, so it follows that angle **TLH** = angle **LZH**. If angle **HLZ** is taken as common, then the whole of angle **TLZ** = angle **EZL** + angle **HLZ**. But these two angles together form a right angle, since the angle at **H** is right.

---

[15]Euclid, *Elements* III, 31.
[16]This follows from theorem IV.5 above.
[17]Euclid, *Elements* I, 32.

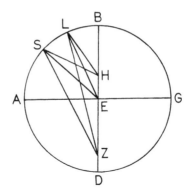

Figure IV.8

Thus, angle **TLZ** is right, and line **LZ** is tangent to circle **KLE**. Of all the angles at which reflection takes place, then, angle **ZLH** is the largest.[18]

[22] *[THEOREM IV.9]* Accordingly, it will be shown as follows that, if reflection takes place at equal angles from some given point on a semicircle, it is impossible for it to take place from any other point on that semicircle:

[23] If it is possible, let the reflecting ray-couples be **ZL, LH** and **ZS, SH** [in figure IV, 8], and let normals **EL** and **ES** be joined. Therefore, since both angles [**ZLH** and **ZSH**] are bisected, **ZL : LH = ZE : EH** and **ZS : SH = ZE : EH**, so that **ZL : LH = ZS : SH**. By alternation, **ZS : ZL = SH : LH**, but line **ZS** < line **ZL**.[19] Therefore, line **SH** < **LH**, which is false.

[24] What we have said will also be demonstrated in like manner for any other semicircle, because, if the ray is reflected at equal angles from, say, point **L**, and if we assume that point **B** lies on a circular arc within the concave mirror and that surface **HZL** is revolved about diameter **DB** [as axis], then from every point on the circle formed by the circumscription of point **L**, the ray-couple connecting points **Z** and **H** will reflect at equal angles.

[25] Moreover, in such a situation, the images of visible objects will lie on the mirror's surface, for the line passing from the visible object through the center of sight to the mirror is normal to that mirror. As a result, the intersection of lines that defines the image always lies at the center of sight itself, no matter whether we place

---

[18]This follows because, were angle **ZLH** to increase beyond this point, **ZL** would no longer be tangent to circle **LEK**, so the original conditions for reflection at equal angles would no longer obtain.

[19]Euclid, *Elements* III, 7.

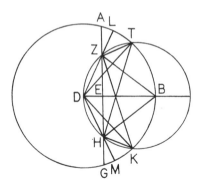

Figure IV.9

that viewpoint at **Z** or **H**.[20] Because it is difficult to see at this point, the visual sensitivity inclines toward the surface from which the reflection occurs and, touching the mirror in that way, makes a mutual interchange, taking the image from the place where it was [actually formed] to the place where it appears [to lie]. Consequently, the colors of such images, when they lie at a great distance, are either indistinguishable or barely distinguishable from the colors of the mirrors. For, since those images are not perceived at the proper place, and since the visual sense is barely capable of determining their place, it inclines to a place nearer the mirror where the image seems [to lie].[21]

[26] *[THEOREM IV.10]* After this, however, we ought to investigate the cases in which the appearances are created by three reflections from one arc while the eye lies at predetermined points.

[27] Let **ABG** [in figure IV.9] be a circle with center **D**, and let **AEG** be drawn. Then let perpendicular **DEB** be drawn, and let arc **ABG** be marked off on a concave mirror. Let there be two points **Z** and **H** on line **AEG** such that circle **TDK**, which passes through those two points as well as **D**, cuts both arcs **AB** and **BG**. Finally, let lines **DZL** and **DHM** be drawn, and let **ZE** first be supposed equal

---

[20]In other words, since the image-location is determined by the intersection of the cathetus of reflection (**BD** in this case) with the incident ray, that location will perforce lie at the center of sight, whether it be at **Z** or **H**.

[21]Ptolemy's point here is that, since the eye has undue difficulty in perceiving such an image that coincides with the center of sight, it is compelled to transpose that image to a more convenient and "natural" location, that location being the actual surface of reflection. Thus transposed, the image blends with the reflecting surface, taking on its color. As a result, the image is either "indistinguishable or barely distinguishable"; see II, 23 above. In fact, no image whatever is visible in such a case, but, as Simon remarks in *Le regard*, pp. 163–164, Ptolemy is constrained by his theoretical principles to assume that, whenever a visual ray reaches a visual object—no matter the circumstances under which it does so—some sort of visual perception must ensue. Accordingly, when no such perception exists, it must somehow be invented.

to **EH**. We say, then, that reflection at equal angles [between **Z** and **H**] takes place at three points: i.e., at point **B**, point **T**, and point **K**.

[28] Let rays **ZB** and **BH**, **ZT** and **TH**, and **ZK** and **KH** be the reflecting couples, then, and let normals **TD** and **KD** be drawn. Therefore, since **ZE** = **EH**, and line **EB** is common, while the angles at **E** are right, angle **EBZ** = angle **EBH**. So too, angle **ZTD** = angle **DTH**, and angle **DKZ** = angle **DKH**, because arc **DZ** = arc **DH**, and points **T** and **K** lie on the circumference of the circle.[22]

[29] *[THEOREM IV.11]* We say, however, that no other ray is reflected [at equal angles] from any point lying between points **L** and **M** on arc **LM**. For, since the angles at points **T** and **K** on arc **KDT** are subtended by [equal] arcs **ZD** and **DH**, there is a reflection [at equal angles] from these two points. Moreover, the ray to **B** is reflected [at equal angles], because line **BD** bisects line **AG** [and thus line **ZH**, which subtends angle **ZBH**]. On the other hand, no rays are reflected from any points other than these three on the arc between **L** and **M**, because they lie neither on arc **DTK** nor on its center.

[30] It is evident, moreover, that those rays that are [supposedly] reflected between [arcs] **AL** or **GM** do not do so at equal angles, because they do not encompass centerpoint **D**.[23]

[31] *[THEOREM IV.12]* If such [reflection at equal angles] is possible between **M** and **L**, though, then let ray-couple **ZNH** in the previous figure [redrawn as figure IV.10] be reflected [from somewhere] between points **T** and **B**, and let ray-couple **ZSH** be reflected [from somewhere] between points **K** and **M**. Let normals **DONF** and **DQCS** be drawn, and let lines **ZS** and **ZN**, lines **NH** and **SH** [and lines **ZC** and **CH**] be joined.

[32] Accordingly, since arc **DZ** = arc **DH**, then angle **ZFD** = angle **DFH**, and angle **ZCD** = angle **DCH**.[24] But angle **ZND** was [supposed to be] equal to angle **DNH**, and angle **ZSD** was [supposed to be] equal to angle **DSH**, so the angles of triangle **ZNF** are equal [respectively] to the angles of triangle **HFN**, and the angles of triangle **CSZ** are equal [respectively] to the angles of triangle **SCH**. Meantime, sides **FN** and **SC** are common, so line **ZN** = line **HN**, while line **SZ** = line **SH**. But lines **NOD** and **SQD** bisect the angles

---

[22]Euclid, *Elements* III, 21.

[23]The logic here is that, for all of the possible reflecting ray-couples from **ZLH** to **ZAH** or from **HMZ** to **HGZ**, the normal from centerpoint **D** will invariably fall outside the angles formed by those ray-couples. There will thus be no way for the angles of incidence and reflection to be equal relative to that normal. Ptolemy applies this principle of disproof throughout the rest of this book.

[24]Euclid, *Elements* III, 21.

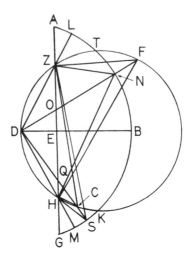

Figure IV.10

at **N** and **S**. Therefore, line **ZO** = line **OH**, and line **ZQ** = line **QH**. Each of these [equalities], however, is impossible, since line **ZE** = line **EH**.

[33] In the same way it will also be demonstrated that reflection [at equal angles] cannot occur between **B** and **K** or between **T** and **L**.

[34] *[THEOREM IV.13]* Reflection [at equal angles] occurs at three places, even if line **EZ** is longer than line **EH**.

[35] Let the circle passing through points **D**, **Z**, and **H** [in figure IV.11] cut each of the arcs **LB** and **BM** [such that **EZ** > **EH**]. Ac-

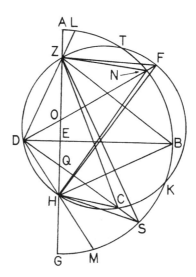

Figure IV.11

cordingly, as we have just shown, no ray will reflect from points **T** and **K**, because arc **DZ** > arc **DH**. Nor will it reflect from point **B**, since **ZE** > **EH**, and the two angles at **E** are right. However, in lieu of the two reflections that took place from points **B** and **K**, two rays will reflect from the arc that lies between points **B** and **K**. Moreover, in lieu of the reflection from point **T**, a ray will reflect between points **L** and **T**.

[36] *[THEOREM IV.14]* No ray reflects from the arc that lies between **T** and **B**, nor from the arc between **K** and **M**, for if a reflection from any of these arcs were supposed, the result would be impossible.

[Suppose that such a reflection does occur from points **N** and **S** on arc **ABG**. According to this supposition, then:]

[37] Since angle **ZSD** = angle **DSH**, and angle **ZND** = angle **DNH**, then angle **ZNF** = angle **FNH**. On the other hand, angle **ZFN** > angle **NFH**, while angle **ZCD** > angle **HCD**, because arc **ZD** > arc **DH**. From this it follows that angle **SCH** > angle **SCZ**, whence **ZN** > **NH**, and **SH** > **ZS**.[25]Thus, angle **ZHN** > angle **NZH**, and angle **SZH** > angle **SHZ**. It follows, then, that angle **NOH** < angle **NOZ**, and angle **ZQS** < angle **SQH**; this means that obtuse angles must be smaller than acute angles, which is impossible.

[38] If, however, we suppose a reflection [at equal angles] from the arc that lies between **T** and **L** or the one between **B** and **K**, then the result is not impossible. This we would demonstrate in the same way using a similar figure, for the obtuse angles should be larger if we assume the figure to conform to this condition.[26]

[39] *[THEOREM IV.15]* In order not to prolong this discussion, we ought to stipulate that, if circle **ZDH** does not cut some arc other

---

[25]Since angle **ZNF** = angle **FNH**, while angle **ZFN** > angle **NFH** (because arc **ZD** > arc **DH**), then angle **FHN** > angle **FZN**, and angle **NFH** > angle **ZFN**. Thus, side **HN** of triangle **FHN** subtends a smaller angle (**NFH**) than does size **ZN**, which subtends angle **ZFN** of triangle **ZNF**. From this it follows that **ZN** > **HN**. The same line of reasoning applied to triangles **ZSC** and **HSC** leads to the conclusion that **SH** > **ZS**.

[26]Let figure 7 represent the situation supposed, so that reflection at equal angles between points **Z** and **H** does occur from 1) point **F** on arc **TL** and 2) from point **S** on arc **BK**. CASE 1: angle **ZFD** = angle **HFD** (by supposition); angle **ZND** > angle **HND** (because arc **ZD** > arc **HD**); thus angle **ZNF** < angle **HNF**; and so it follows that **ZF** < **HF** (because it subtends a smaller angle). But **ZF** : **HF** = **ZO** : **HO** (by Euclid, *Elements* VI, 3), so it follows that normal **DOF** cuts **ZH** such that oblique angle **FOH** > acute angle **FOZ**. CASE 2: angle **ZSD** = angle **HSD** (by supposition); angle **ZCD** > angle **HCD** (because arc **ZD** > arc **HD**); thus angle **ZSC** < angle **HSC**; and so it follows that **SZ** > **SH** (because it subtends a larger angle). But **SZ** : **SH** = **ZQ** : **QH** (by Euclid, *Elements* VI, 3), so it follows that normal **DQC** cuts **ZH** such that oblique angle **CQZ** > acute angle **CQH**.

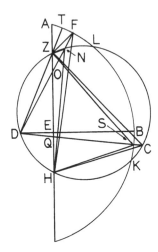

Figure 7

than **BM**, and if point **Z** passes through point **A**, as in the present figure [IV.12], or beyond it [along line **EA**], then the reflections that occurred at equal angles from points **B** and **K** [in theorem IV.10 above] will occur only from two places between **B** and **K**—as long as we exclude arc **TL**—because the whole of arc **AB** will lie within circle **DZH**.[27] In the same way it will be demonstrated that none of the other rays reflects at equal angles.

[40] *[THEOREM IV.16]* If, however, we assume that circle **ZDH** cuts arc **LB** only, and if we suppose that its other intersection-point is **B**, as represented in the present figure [IV.13], or somewhere between points **A** and **B**, then the reflection takes place between points

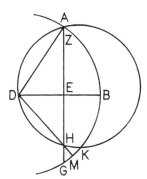

Figure IV.12

---

[27]By the previous theorem, no reflection can occur from point **Z** to point **H** from anywhere within arc **AB** nor from anywhere outside arc **BK**. Hence, by default, the reflection must occur from somewhere within arc **BK**.

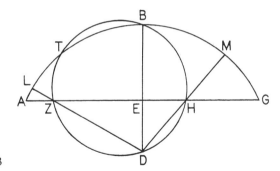

Figure IV.13

T and **L**, because arc **BK** is out of the account, and therefore reflection does not occur at equal angles from it, for arc **BG** will lie outside the circle. Also, according to what we have shown, it will be demonstrated that no reflection occurs at equal angles from any point other than the one we have specified.[28]

[41] *[THEOREM IV.17]* Nearly the same thing that we have shown also happens when circle **ZHD** does not cut circle **ABG**, and when point **E** lies between points **Z** and **H**. For in that case only one reflection takes place at equal angles.

[42] Let circle **DZH** [in figure IV.14] not cut circle **ABG**, and, for a start, let **ZE = EH**.

[43] It is obvious from what we have shown that there will be a reflection at equal angles from point **B**. Furthermore, it will be evident that there is no reflection from arcs **AL** and **GM**. Now, it will be demonstrated as follows that it is impossible for a reflection [at equal angles] to occur from the arc that lies between **L** and **B** or the one that lies between **B** and **M**.

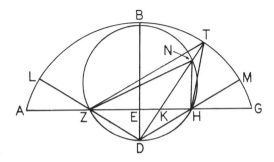

Figure IV.14

---

[28]See theorem IV.13 above.

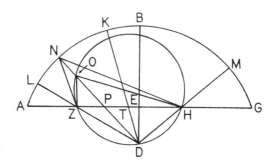

Figure IV.15

[44] If such a reflection is possible, then let it take place according to **ZT** and **TH**. After [normal-]line **DKNT** is drawn, lines **ZN** and **NH** will be joined. Since angle **ZND** = angle **DNH**, insofar as arc **DZ** = arc **DH**, it follows that angle **ZNT** = angle **HNT**. But angle **ZTN** was [supposed to be] equal to angle **DTH**. Therefore, triangles **ZNT** and **THN** are equiangular, and line **TN** is common. Consequently, **ZT** = **TH**, so it follows that **ZK** = **KH**. This entails a longer [line] being equal to a shorter [line], which is impossible.

[45] The same thing will also be demonstrated if we assume the reflection to occur between points **L** and **B**.

[46] *[THEOREM IV.18]* Let **ZE** [in figure IV.15] be longer than **EH**, let **ZH** be bisected at point **T**, and let line **DT** be drawn to **K**. We say, then, that reflection [along the ray-couple joining] **Z** and **H** can only take place [at equal angles] between points **K** and **L**.

[47] Now it has been demonstrated several times that [such reflection] does not occur from arc **AL** or arc **GM**. Furthermore, it will be demonstrated from the present figure that it is impossible for reflection to occur between points **B** and **M**. For, in such cases, line **EH**, which is shorter, must be longer than line **EZ**, which is longer than it.[29] Nor is it possible for reflection to occur at equal angles from arc **KB**, because, if [such] a reflection is supposed at point **K**, then angle **ZTK** = angle **HTK**: an acute equal to an obtuse angle.[30] And if [the reflection] is assumed to occur at point **B**, then angle **ZEB** > angle **HEB**, although both are right.[31] Finally, if the reflection is assumed to occur between **K** and **B**, an acute angle be-

---

[29]This follows because, under the conditions specified, with arc **ZD** > arc **DH**, the normal from **D** that bisects the angle formed by the ray-couple reflecting from arc **BM** will cut line **ZH** to the right of **E** but with the angle on the side of **E** acute. Therefore, the normal itself must cut **ZH** in such a way that the distance between it and **H** is greater than half **ZH**, and since **E** lies to its left, then **EH** must exceed even that. But **EH** is shorter than half **ZH** by construction.

[30]Since normal **DTK** bisects **ZH**, and since it is supposed that angle **ZKT** = angle **HKT**, it follows that the angles at **T** will be right angles and, thus, equal to one another.

[31]The presumed inequality of angles **ZEB** and **HEB** follows because arc **ZD** > arc **DH**.

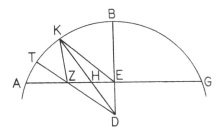

Figure IV.16

comes greater than an obtuse angle. But all these conclusions are impossible.

[48] But if we assume that the reflection at equal angles takes place between **L** and **K**, then what ought to occur actually does. Accordingly, let the reflection occur along **ZN** and **NH**. Let [normal] **DPON** be drawn, and let **OZ** and **OH** be joined.

[49] Angle **ZND** was [posited] equal to angle **DNH**, while angle **ZOD** > angle **DOH**, and arc **ZD** > arc **DH**. Meantime, supplementary angle **NOH** > supplementary angle **NOZ**, so line **NH** > **ZN**. Therefore, line **PH** > line **PZ**, which is in fact the case. So too, angle **NZH** > angle **NHP**. It follows, then, that obtuse angle **NPH** > acute angle **NPZ**, which is as it should be.

[50] *[THEOREM IV.19]* Now, let circle **ABG** [in figure IV.16] be erected on centerpoint **D**, and let line **BD** be drawn so that it intersects line **AEG** at right angles. Let arc **ABG** of the circle lie within a concave mirror, and let one of the [end]points [of the reflected ray-couple] lie at **Z** and the other at **E**. Accordingly, we say that, from arc **ABG**, only one reflection at equal angles can occur between points **E** and **Z**.

[51] Let **ZE** be bisected at point **H**, then, and let lines **DHK** and **DZT** be drawn. Thus, the rays that [might be supposed to] reflect from arcs **AT** and **BG** do not encompass centerpoint **D**.[32] Moreover, if we suppose the reflection to occur from point **K**, then the angles will be unequal.

[52] But, assuming that this is not the case, let the reflection occur along **ZK**, **KE**. Then, since **ZH** = **EH**, line **EK** = **KZ**, for the ratio of the one to the other is as that of the corresponding two lines **ZH** and **EH**, since angle **ZKE** is [supposed to be] bisected by line **KD**.[33] Moreover, since line **HK** is common, then angle **KHZ** = angle **KHE**; so an acute angle is equal to an obtuse one, which is impossible.

---

[32]See n. 23 above.
[33]This follows from the necessary equality of the two triangles **ZHK** and **HKE**; see Euclid, *Elements* I, 38.

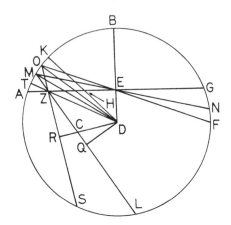

Figure IV.17

[53] This conclusion will be all the more absurd if we suppose the reflection to occur between points **B** and **K**, for in that case an acute angle will be greater than an obtuse one.[34] But if we suppose the reflection to occur between points **T** and **K**, the converse holds: i.e., the obtuse angle will be larger than the acute one. Nor can the desired result occur in any other way.

[54] *[THEOREM IV.20]* So too it will be demonstrated that it is impossible for two reflections at equal angles to occur simultaneously between points **E** and **Z** from two points between points **K** and **T**. The demonstration is as follows:

[55] If such reflection is possible, then, using a figure like the previous one [recast in figure IV.17], let the two reflecting ray-couples be **EMZ** and **EOZ**. Let the rays be extended to points **S**, **L**, **F**, and **N**, and let [normal] lines **DO** and **DM** be drawn. Then, let the reflection [be assumed to] occur at equal angles.

[56] Therefore, since **DZ** > **ED**, the line that bisects angle **EDZ** will cut **EZ** between **E** and **H**.[35] On this basis, then, angle **EDO** will be much larger than angle **ZDO**. Hence, because angle **EOD** is posited equal to angle **ZOD** on account of reflection, the sum of the two angles **EDO** and **EOD**, which equals angle **DEF**, is greater than the sum of the two angles **ZDO** and **ZOD**, which equals angle **DZS**. But angle **DEF** is acute, so angle **DZS** in turn will be acute. If, how-

---

[34]In the previous case, with **ZE** bisected, the angles at **H** would necessarily be right because of their supposed equality. Now, if the point of reflection is moved from **K** toward **B**, point **H**, where the normal and **ZE** intersect, will also move toward **E**. Hence, the new angle **ZHK** will be greater than its previous counterpart, but it will still remain smaller than angle **ZEB**, which is a right angle.

[35]This follows from the equality of **ZH** and **HE** and the fact that **DZ** > **ED**; see Euclid, *Elements* VI, 3.

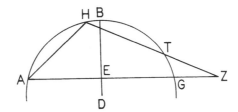

Figure IV.18

ever, from centerpoint **D** we draw **DQ** perpendicular to **MZL** and **DCR** perpendicular to **OZS** and intersecting **ML** at **C**, then line **DC** will be longer than **DQ**, since the angle at **Q** is right. From this it follows that **DR** will be much longer than **DQ**, and line **ML** will be longer than **OS**.[36] Hence, angle **ZMA**, which is [posited] equal to angle **BME**, will be larger than angle **ZOA**, which is posited equal to angle **EOB**. It therefore follows that **MEN** must be nearer centerpoint **D** than **OEF**,[37] which is false, since angle **DEN** is acute. Thus, ray-couples **ZOE** and **ZME** do not reflect at equal angles.

[57] *[THEOREM IV.21]* Again, let **ABG** [in figure IV.18] represent a circle with centerpoint **D**, and draw [normal] line **BD**. Then draw a line cutting it at right angles, and let that line be **AEG**. Suppose that arc **ABG** lies within a concave mirror. Let each of the two [end]points [of the reflected ray-couple] be located on straight line **AG**, but let one of them not lie within the circle. We say, then, that between those two points only one ray is reflected at equal angles from arc **ABG**.

[58] Of course, if the two [end]points are equidistant from point **E**, then the reflection occurs from the single point **B**. This is obvious, because the two lines that are drawn to **B** from these [end]points subtend equal arcs.

[59] Yet if one of the [end]points lies at point **A** and the other at point **Z**, both of which lie on the extension of line **EG**, then no ray reflects at equal angles from arc **BG**. For, given that arc **AB** = arc **BG**, it follows that the arc subtended by the line joining point **A** and any of the points lying on arc **BG** is greater than the arc subtended by the line joining that [given point on **BG**] and point **Z**.

[60] On the other hand, from the arc lying between **A** and **B**, a ray, such as **AH** and **HZ**, can be reflected at equal angles, and arcs **AH** and **HT** can then be equal. But it is impossible for another ray to be so reflected from this arc. For, of the two lines connecting the

---

[36]Euclid, *Elements* III, 15.
[37]According to the posited conditions of angular equality, **MEN** = **ML**, while **OEF** = **OS**, but **ML** > **OS**, so **MEN** > **OEF** (in fact, **MEN** < **OEF**).

aforesaid [end]points and some point other than **H**, one cuts off an arc larger than either of the two previous arcs [**AH** and **HT**] while the other cuts off an arc smaller than those two.

[61] The same thing would also happen if point **A** lay outside the circle and line **EZ** were longer than line **EA**.

[62] What we have discussed concerning the position [of the radial endpoints] involves cases in which the line joining the viewpoint and the visible object lies between the mirror and the center of the sphere. It is now time for us to discuss reflections occurring at equal angles and to show that, in certain such cases, there is no possible intersection of the lines at which we said the image of the visible object is formed. In certain cases, however, that intersection occurs behind the mirror, whereas in others that intersection occurs behind the eye.

[63] *[THEOREM IV.22]* In the case where the line-segment reflected from the mirror to the visible object is equal to the line connecting the visible object and the center of the sphere, it necessarily follows that the [incident] ray-segment emanating from the eye to the mirror and the normal dropped from the visible object to the mirror are parallel. But in the case where the two aforesaid lines are unequal, if the line connecting the center of the sphere and the visible object is longer than the reflected line-segment, then the intersection-point we have cited will lie behind the mirror. If, however, the reflected line-segment is longer than the line joining the center-point [of the sphere] and the visible object, then the intersection-point will lie behind the eye.

[64] Let **ABG** [in figure IV.19] represent the arc of a circle lying on a concave mirror with centerpoint **D**. With the eye placed at point **E**, let the reflection be assumed to occur at equal angles along **EB** and **BZ**. Then, from point **D**, let line **DH** be drawn equal to **HB**. Likewise, from point **D**, let line **DT** be drawn longer than **TB**, let line **DZ** be drawn shorter than **ZB**, and let **H**, **Z**, and **T** represent the visible objects. Finally, let each of the lines drawn from **E** to the visible objects fall, like **EZ**, between the center of the sphere and the mir-

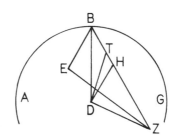

Figure IV.19

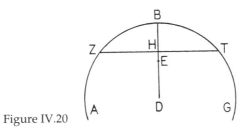

Figure IV.20

ror. We say, then, that **DH** is parallel to **EB**, that **DT** intersects it behind the mirror, and that **DZ** intersects it behind the viewer.

[65] Now, since **DH** = **BH**, angle **DBH**, which is equal to angle
DBE, will be equal to angle **BDH**. Thus, angle **BDH** = angle **DBE**,
and so line **EB** will be parallel to line **DH**. Again, since line **BT** <
DT, angle **BDT** < angle **DBT**, which is equal to angle **DBE**. Thus,
lines **EB** and **DT** intersect on the side of **B** and **T**, which lie beyond
the mirror, since **BZ** is always reflected from the other side of line
BD. By the same token, if angle **BDZ** > angle **DBZ**, which is equal
to angle **DBE**, lines **BE** and **DZ** must meet on the side of points **E**
and **D**, and it is evident that the intersection occurs behind the eye,
because **ZE** lies between **B** and **D**.[38]

[66] *[THEOREM IV.23]* Along with what we have said, it will become evident that, generally speaking, if the visible object lies upon
the normal dropped to the mirror, and if the section [of the normal]
that is cut off on the side of the center is not longer than half the
radius, then the reflected ray will be longer than the line connecting
point **D** and the visible object.

[67] Let **ABG** [in figure IV.20] represent the circumference of a circle with centerpoint **D**, draw normal **BD**, and let it be bisected at **E**.
Let the visible object lie at **E**. Thus, all of the lines drawn from **E** to
the circumference will be longer than **EB**, which is equal to **ED**.
Moreover, those lines will be even longer in comparison to **ED** if
they are drawn from some other point lying between **E** and **D**.

[68] But if the visible object is placed [at a point] between **E** and
B, the line drawn from that point to the circle can sometimes be
equal to the line between that point and **D**, sometimes longer, and
sometimes shorter. Let line **ZHT** be dropped at right angles to **DB**,

---

[38]See Lejeune, *Recherches*, pp. 75–78, for an analysis of this theorem. Knorr, "Archimedes,"
pp. 55–58, makes a strong case for connecting this theorem to Euclid's *Catoptrics*, particularly
to prop. 28, where each of the instances cited by Ptolemy is dealt with: case 1 = image-location behind the eye; case 2 = indeterminate image-location; case 3 = image-location behind
mirror; case 4 = image-location in front of mirror between reflecting surface and eye; cf.,
however, Lejeune, *Recherches*, pp. 131–133.

and let **ZT** be the side of an [inscribed] square, so that **ZH = DH** and so that arc **ZBT**, which is equal to a quarter of the circle['s circumference], subtends the whole of the mirror's width. Then let lines be drawn to the arc lying between **Z** and **T** from some given point on line **HB**. These lines will then be shorter than the lines connecting point **D** and the points selected [on arc **ZT**]. . . . [39]

[69]. . . . But if the aforesaid two straight lines are parallel, and the image has no determinate location where it may be seen, then the sight fixes on a location common to both the image and the mirror. The same thing happens when the two straight lines just mentioned intersect in such a way that the image appears at the center of sight.[40] But when the point of intersection lies behind the mirror, the image-locations preserve their proper disposition, for they appear behind the mirror. When, however, the distances [of such intersection-points] become inordinate, they are always [perceptually] shortened.[41]

[70] In addition, when the point of intersection lies behind the eye, the visible object is not seen behind the viewpoint, for such a perception is impossible. Rather, it appears in front of the mirror, even though the impression arising in the sense does not render the image in its proper place but transfers it to a location that is not its own, just as we have already pointed out. For the location, which is not properly assigned, renders the visible object on the very surface of the mirror.[42] But when the location is properly determined, the image appears where it ought to according to the eye. And this is how images are formed in a mirror: namely, that objects whose images are formed beyond the eye and the mirror appear behind the mirror, while those whose images are formed behind the eye appear in front of the mirror.

[71] *[EXPERIMENT IV.1]* But we can examine the cases of image-location that we have discussed, each in turn, using the bronze template we described earlier.[43] As before, in figure IV.19 [redrawn as IV.21], we draw lines **EB** and **ZB** [reflected from **ABG**]

---

[39]As Lejeune observes in *L'Optique*, p. 172, n. 23, there is an obvious lacuna at this point. According to his analysis of the theorem in *Recherches*, pp. 78–80, the point of the incomplete theorem is clear enough: to show that, when the object lies at point **E** or on **ED**, the image will lie in front of the mirror (i.e., it will be real), whereas, when it lies between **E** and **B**, it may be real, virtual, or indeterminate, depending on the counterplacement of the eye.

[40]This point echoes the one made in IV, 25 above about the melding of image with mirror-surface.

[41]See IV, 72 and IV, 78 below.

[42]See n. 40 above.

[43]See III, 8–10 above.

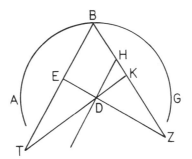

Figure IV.21

at equal angles and **DH** parallel to **EB**. We then draw **EDZ**, position the eye at **E**, and place the concave mirror on arc **ABG** while setting ourselves to observe the images of small colored pegs placed upon line **BZ**.

[72] The image of the peg placed at **Z** or **H** will seem to coalesce [with the surface] where point **B** lies on the mirror, and it will take on both its position and color. For there is no proper, determinate location for the image of anything located at **Z** or **H**.[44] In fact, nothing lying at point **H** has a proper intersection-point, whereas what lies at **Z** shares a common location with the eye, since the intersection lies at **E**.

[73] The images of objects lying between these two points [**Z** and **H**] do not appear on the mirror itself, because those images have determinate locations. But whatever lies between **B** and **H** appears to lie beyond the mirror at the point of intersection, whereas whatever lies between **H** and **Z** appears to lie in front of the mirror for the reason appropriate to it. For if we draw line **BET**, and lines **DK** and **DZ** as well, and if their intersections lie at points **E** and **T**, the image of point **Z** will lie at point **B**. Accordingly, the image of the intersection at point **T** will appear toward **K**, which lies in front of **B** with respect to **D**, because the location of **T** with respect to **E** lies in that direction.[45]

[74] *[EXPERIMENT IV.2]* On the basis of the mirror as laid out, it is possible for us to gain a general understanding of these points concerning the intersection and displacement that happen to image-locations, which we have said incline toward a position nearer [the reflecting surface]. Assume that we draw line **BDE** [in figure IV.23 below] and place a long, thin rod, such as **BZ**, along **BD**.

---

[44]As Lejeune remarks in *L'Optique*, p. 173, n. 36, this perceptual shortening of inordinate distances plays a part in the observational analysis outlined in Experiment IV.1, paragraphs 71–73 below (esp. 73) and Experiment IV.2, paragraphs 74–80 below (esp. 78).

[45]See Lejeune, *Recherches*, pp. 104–108, for an evaluation of these experimental results; cf., however, Simon, *Le regard*, pp. 160–165.

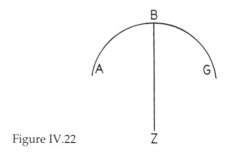

Figure IV.22

[75] If we place either eye at point **E** and look at **BZ**, we will see its image coincide with point **B** on the mirror, and it will lie between two equal segments on the mirror's surface. Moreover, what the viewer will see as the composite of the rod and its image looks like what is represented in this figure [IV.22], which is composed of line **BZ** and arc **ABG**. The reason is that, when the visual ray is reflected to magnitude **BZ**, the intersection that determines image-location has no proper, determinate position but will instead lie at point **E** where the eye is also situated. And the images that are in the mirror will appear continuous.[46]

[76] If the midpoint between the eyebrows is at **E** and the eyes are placed at **H** and **T** [in figure IV.23], the image of **BZ** curves in the direction of the mirror. And when it lies at a reasonable distance, it will appear in front of the eyes. Subsequently, it tends to curve gradually toward each of the eyes in the form of the Greek letter alpha, which looks like Υ.[47] And it does not stop curving until it reaches the visual rays that are parallel to line **EB**. After that, it tends to curve back toward the mirror.

[77] Among the rays that emanate from **H** and are reflected to **BZ**, only one (e.g., line **KH**) is parallel to the aforesaid line [**EB**]. The rest of the rays lying between points **B** and **K** intersect line **DB** beyond the mirror so that, when all of them are extended, they intersect it as lines **MH** and **HN** do. But those lines lying between **A** and **K** intersect it behind the eye, as do lines **SH** and **OH**.

[78] This is what happens in such a transposition of distances. From the intersection formed at point **N** the image is transferred toward **F**, whereas, in the case of the remaining intersection-points,

---

[46]Since the object itself lies along the cathetus of reflection, all of the intersections between the incident rays and that cathetus will lie at **E**, which, according to Ptolemy, will then transpose the image to the actual point of reflection on the mirror's surface. As a result, the entire image of **BE** will appear to coalesce with arc **ABG** on that surface.

[47]See Lejeune, *L'Optique*, p. 16*, as well as p. 176, n. 24, for a discussion of this orthographic peculiarity.

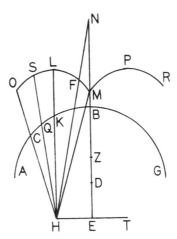

Figure IV.23

much the same thing happens (as should be the case according to the reasoned account) until we arrive at the point [where the line of sight] is parallel to **EM**. The [resulting] image will be formed along **MFL**. From then on, the image will turn back on account of the weakness of the visual power, which judges parallels to extend to infinity and assumes that the distance of any intersection in the case of the subsequent, nearly parallel rays extends unimaginably far, as if it reached to the Pleiades or whatever other distance one might conceive. Because of this, then, the visual power reverts to moderate limits. And, because the particular intersection occurs behind the eye, whereas the visible object forms a continuous whole, the images of [such] objects must be similar to them in kind.[48]

[79] As far as any intersection formed beyond **L** is concerned, the [image created by] reflection will always be near the mirror, e.g., along line **LSO**, because the point on the [visible] magnitude grasped by [radial] line **KH**, which is parallel to line **MBZ**, would appear at point **K**, if it were isolated. But what is seen by rays **SH** and **OH** will appear on the mirror—e.g., at **Q** and **C**—because the intersection of those rays with line **BE** occurs behind the eye. On the other hand, the part of the magnitude that appears along line **KCQ** is not disposed as it actually appears, because it is neither in a determinate place nor disjoined from image **MFL** but appears continuous with it, even though it lies in front of the mirror—the point being to insure that a continuous object not be perceived as discontinuous. Therefore, since the intersection formed on the side of the

---

[48]In other words, whatever specular discontinuities there may be in the images, they will be perceptually adjusted by the eye so that the resulting image appears to be as continuous as its generating object.

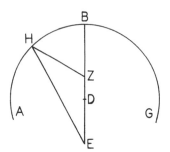

Figure IV.24

mirror is specified by point **L**, the place of **K** is transposed potentially to point **L**, so that its image is joined to the remainder and
made continuous with it. In other words, line **LSO**, which is transposed to the same side and will have its image correspond to **KQC**,
reverts to what is nearer, so that the image, lacking its proper location, is transposed to an appropriate place for something with a determinate location.

[80] In order not to repeat ourselves, we say that on the other side
the same thing happens in the case of the [radial] lines emanating
from point **T** to arc **BG**, the [base of the] visual cone here being represented by line **MPR**, which is similar to line **MLO**.[49]

[81] *[THEOREM IV.24]* Having, therefore, elucidated these
points, let us now demonstrate that at point **E** only one reflection
can be made at equal angles when the position [of the object] is as
has been specified.

[82] Let **ABG** [in figure IV.24] be the arc of a circle lying on a concave mirror with center **D**, and let line **BDE** be drawn. We say, then,
that if the eye lies on **EB**, but not at point **D**, the ray directed at point
B reflects back onto itself, a fact that is evident, since the angles at **B**
are equal. However, those rays that are reflected to a point between
B and **D** or between **D** and **E** from any point lying between **A** and
B or between **B** and **G** (e.g., **EH** and **HZ**) do not fall at equal angles
from any side of centerpoint **D**. For they form unequal angles on
each side, e.g., angle **BHE**, which is greater than angle **AHE**.[50]

[83] Again, since the case is as previously explained, it is clear
that, when the ray that is reflected to point **Z** passes through points
B and **E**, the image of the eye appears on the actual surface of the
mirror toward the [front side of] the mirror. For, as we have said,

---

[49]See Lejeune, *Recherches*, pp. 109–111.
[50]In short, **EZ** is not normal to the mirror. This theorem is peculiar in that the angles of incidence and reflection are measured in reference not to the normal **DH** but rather to the curvi-
linear surface of reflection itself.

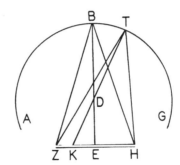

Figure IV.25

the image tends toward one of the locations where the intersection that is common to it and the mirror lies, and that is at point **B**.[51]

[84] *[THEOREM IV.25]* We must also demonstrate that, on the whole, when the center of the sphere lies between the mirror itself and the line joining the eye and the visible object, reflection at equal angles takes place from only one point.

[CASE 1]

[85] Let **ABG** [in figure IV.25] be the arc of a circle lying on a con-cave mirror with center **D**, and let normal **BDE** be drawn with line **HEZ** cutting it at right angles. Let one of the [end]points [of the re-flecting ray-couple] be **Z** and the other **H**. First, let **EZ** = **EH**. We say, then, that between points **Z** and **H** there is no reflection at equal angles except from point **B**.

[86] Let **BZ** and **BH** be joined. Accordingly, since lines **EB** and **EZ** are equal to lines **BE** and **EH**, respectively, and since the angles at **E** are right, the two triangles will be equiangular. Thus, angle **BZE** = angle **BHE** and angle **ZBE** = angle **HBE**.

If it is possible, let another ray-couple, such as **ZTH**, reflect at equal angles, and let normal **TDK** be drawn. Therefore, since line **KT** bisects angle **ZTH**, then **KH** : **KZ** = **TH** : **TZ**.[52] But **KH** > **KZ**, so **TH** > **TZ**. Thus, angle **TZH** > angle **THZ**, and so angle **BZH** will be even greater than angle **BHZ**, which cannot possibly be the case, since we have just shown them to be equal. Consequently, **TZ** and **TH** do not reflect at equal angles.

[87] What we have established will also be evident if we assume some other line.

---

[51]Given its idiosyncrasies, Lejeune raises the possibility in *L'Optique*, p. 180, n. 38, that this theorem represents a later interpolation into the text.

[52]Euclid, *Elements* VI, 3.

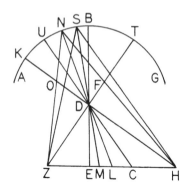

Figure IV.26

[CASE 2]

[88] Let a diagram [figure IV.26] be constructed similar to the previous one, but let **EH** > **EZ**. Then let lines **ZDT** and **HDK** be drawn, let line **ZH** be bisected at point **L**, and let angle **ZDH** be bisected by line **DM**. Let **EZ** = **EC**, and let lines **LDN**, **MDS**, and **CDU** be drawn. We say, then, that the reflection at equal angles of the ray-couple joining **Z** to **H** will occur between points **N** and **S**.

[89] But from arc **KA** and arc **GT** no ray-couple will reflect at equal angles, for the ray-couple reflected from either of them does not encompass centerpoint **D**.[53]

[90] So let us demonstrate first that it does not reflect from point **S**. If it could, let it reflect along **ZSH**. Therefore, since angle **ZDM** = angle **HDM**, angle **SDZ** = angle **SDH**. However, angle **ZSD** was [supposed] equal to angle **DSH**. Therefore, triangles **ZDS** and **HDS** will be equiangular, with line **DS** common. Accordingly, line **ZD** = line **DH**, so angle **DZH** = angle **ZHD**. But angle **DZH** = angle **DCZ**, because sides **CE** and **EZ** are equal, as are the angles at **E** in triangles **CED** and **EZD**, and line **ED** is common. Consequently, in triangle **CDH** interior angle **ZHD** = exterior angle **DCZ**, which is false.

[91] If, moreover, we assume a reflection [at equal angles] from some point between **S** and **T**, the result becomes even more absurd. For in that case **ZD** ought to be longer than **DH**. Also, line **MH**, which is the longer, will be shorter than **ZM**, which is the shorter, and interior angle **DHZ** > angle **DZH**, which is equal to exterior angle **DCZ**, all of which is impossible.

[92] Again, assuming that it is possible, let the reflection occur along **ZNH**. Thus, since line **ZL** = line **LH**, then **ZN** = **NH**, and an-

---

[53]See n. 23 above.

gle **NZH** = angle **NHZ**. It therefore follows that angle **NLZ** = supplementary angle **NLH**, which is impossible.

[93] This will be even more absurd, however, if we assume the reflection [at equal angles] to occur between **K** and **N**. Such an assumption leads to the conclusion that a line shorter than **ZN** is longer than a line longer than **NH**, while angle **NLZ**, which is more acute than angle **NLH**, is larger than an angle that is more obtuse than angle **NLH**.[54]

[CASE 3]

[94] But if there is a reflection at equal angles, it will certainly fall between points **S** and **N**, in which case nothing results that is contrary to what we have proposed. That only one ray-couple reflects at equal angles from this arc will be obvious, as we shall now explain. But in order not to confuse the figure, let us assume that the two points **S** and **N** lie on the arc from which we have claimed that it is possible for reflection [at equal angles] to occur. Moreover, assuming that it is possible, let the two ray-couples [**ZSH** and **ZNH**] be reflected at equal angles from those two points.

[95] Accordingly, **ZS** : **SH** = **ZM** : **MH**, and **ZN** : **NH** = **ZL** : **LH**. But **ZL** : **LH** > **ZM** : **MH**, so **ZN** : **NH** > **ZS** : **SH**. By alternation, **ZN** : **ZS** > **NH** : **SH**, but **NH** > **SH**, because it is nearer to centerpoint **D**. Therefore, **ZN** will be much longer than **ZS**, which is impossible, since **ZS** > **ZN**, insofar as it is nearer the centerpoint.

[96] But if we transpose **N** to point **U** and **S** to a position between points **S** and **B**, the arc from which the reflection occurs is defined by the line joining the centerpoint and the midpoint of the line joining the two points [**Z** and **H**] along with the line that bisects the angle subtended by this line [**ZH**]. And there is only one ray that reflects at equal angles from these points.

[97] *[THEOREM IV.26]* To continue, let us now demonstrate that whenever the disposition is as just discussed [i.e., when the centerpoint lies between the mirror and the eye-object line], the image is always formed between the eye and the mirror [and appears] in its proper place. For the normals dropped from points **Z** and **H** [in figure IV, 26 above] and passing through centerpoint **D** always

---

[54]If, for instance, we choose **U** as the point of reflection, then, given that angle **ZUH** will be bisected by normal **UDC**, it follows that **ZU** : **UH** = **ZC** : **CH**. Thus, **ZU**, which is shorter than **ZN**, will be longer than **UH**, which is longer than **NH**. By the same token, angle **UCZ**, which is subtended by the supposedly longer side **UZ**, should be greater than angle **NLZ**, which is subtended by the supposedly shorter side **NZ**.

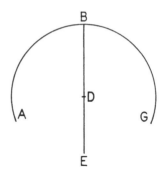

Figure IV.27

intersect the incident ray (which is the one by means of which the image is perceived) in front of the mirror. Moreover, the normal invariably passes through centerpoint **D** in the triangle formed by **ZH** and the reflected ray-couple (e.g., lines **ZN** and **NH**), for in such a reflection, when the eye is situated at point **Z**, point **H** appears as if it lay at point **O**, and if we place the eye at point **H**, we will see point **Z** at point **F**. In addition, the image will lie in front of the mirror between the eye and the mirror itself according to previously established principles. It also happens that certain distances shorten and appear as if they were nearer the mirror.

[98] *[THEOREM IV.27]* In cases of the sort we are going to discuss, there will be no reflection at equal angles.

[99] Let **ABG** [in figure IV.27] be the arc of a circle lying on a concave mirror with centerpoint **D**, and let line **BDE** be drawn. We say, then, that, if the eye and the visible object are distinct and lie on straight line **BE**, and if point **D** does not lie between them, there is no reflection at equal angles between them, whether one of them lies at point **D** or not.[55]

[100] This point is easily grasped, because no ray whatever is reflected at equal angles from the arc lying between **A** and **B** or between **B** and **G**, for in that case the rays do not encompass centerpoint **D**.[56] From point **B**, moreover, there is only a potential, not an actual, reflection, for one of the posited entities [i.e., eye or object] blocks the reflection. If, in fact, we assume the eye to be closer to the mirror, it itself is what blocks the visual ray reflected at point **B** from reaching the object, insofar as it lies in the way of the ray and blocks it. And if we assume that the visible object is closer to the mirror, then it blocks the visual ray from reaching point **B**, insofar

[55]See Euclid, *Catoptrics*, prop. 26.
[56]See n. 23 above.

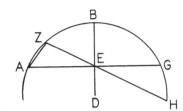

Figure IV.28

as the visible object is located between the eye and **B**, and blocks this latter point.

[101] *[THEOREM IV.28]* Now, let **ABG** [in figure IV.28] be a circle with centerpoint **D**, let line **AG** be drawn, and let **DEB** be dropped perpendicular to it. Let arc **ABG** be the segment of a circle inscribed in a concave mirror. Then, of the two points [eye and object] under discussion in the preceding figures, let one be placed at point **A** and the other at point **E**. We say, then, that there is no reflection at equal angles between them from arc **ABG**.

[102] It is clear that no such reflection occurs from arc **BG**, since it is impossible for the rays reflected between the aforesaid [two points] to encompass point **D**. From arc **AB**, as well, it is obvious that no reflection at equal angles can occur. For if such a reflection could occur, for instance, along **AZ** and **ZE**, then arc **HGZ** would be equal to arc **AZ**, which is false, because arc **HGZ** is much larger than arc **ZA**, given that arcs **AB** and **BG** are equal.

[103] *[THEOREM IV.29]* Likewise, too, it will be evident that a reflection [at equal angles] cannot occur if one of the points lies within the circle on line **AG** while the other lies at **E**, nor [can it occur] if one of the points lies at **A** while the other lies between **A** and **E** or if one of them lies outside the circle while the other lies between **A** and **E**.

[104] Again, let circle **ABG** [in figure IV.29] with center **D** be constructed, let line **AG** be drawn, and let perpendicular **DEB** be dropped to it. Let arc **ABG** be the arc of a circle inscribed in a

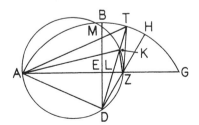

Figure IV.29

concave mirror. Let one of the points—e.g., **A**—lie between **A** and **E**, and the other—e.g., **Z**—between **E** and **G**. Then let circle **ADZ** be drawn to pass through points **A** and **Z**, but let it not cut arc **BG**. Finally, let lines **AD** and **DZH** be drawn. We say, then, that between **A** and **Z** no ray reflects from arc **ABH** at equal angles.

[105] As was shown in the preceding figure, it is clear that no such reflection takes place from arc **AB**, because of what results from the [size-]difference in the arcs. Moreover, that [such] a reflection does not take place from arc **GH** is obvious, because the ray-couple reflected at this very location does not encompass the centerpoint **D**. But let us now show that no reflection takes place at equal angles from the arc between **B** and **H**.

[106] Assuming that such [a reflection] is possible, let **ATZ** be the reflecting ray-couple, and let lines **TKLD**, **KA**, and **KZ** be joined. Then, since arc **AD** > arc **DZ**, angle **AKD** > angle **DKZ**. If follows, therefore, that angle **ZKT** > angle **AKT**. Now, if we suppose that angle **KTZ** = angle **KTA**, then, given that the angles at **T** are [assumed] equal, while side **KT** is common to the triangles **KTA** and **KTZ** [it follows that **TZ** = **TA**. But] **TZ** will also be longer than **TA**.[57] Thus, angle **TAZ** > angle **TZA**. For this reason, and because the angles at **T** are [assumed] equal, if follows that obtuse angle **TLA** < acute angle **TLZ**, which is false. Therefore, **ATZ** does not reflect at equal angles. In the same way it will also be demonstrated that, if point **A** were to lie outside the circle, no reflection [at equal angles] would take place.

[107] From this it will be easily understood that, because no reflection at equal angles can possibly take place between points that are situated in such a way [with respect to the mirror's centerpoint], the eye does not apprehend the posited object, and, since it is not apprehended by the sense, the object will have no image.

[108] As far as what happens in particular cases of reflection and as far as the apparent locations of images are concerned, let the distinctions we have drawn suffice. We must subsequently examine all possible variations among visible objects, each in turn (not only of those whose image lies behind the mirror, but also of those whose image lies between the mirror and the eye), not according to particular aspects, but according to the essential and true nature of the phenomenon.[58]

---

[57]This follows from the fact that it subtends a larger angle.

[58]As Lejeune remarks in *L'Optique*, p. 189, n. 49, Ptolemy is acknowledging here that there can be a distinct variance between what is predicted by pure geometry (the "essential and true") and what appears to be the case by observation ("particular aspects"). The prime ex-

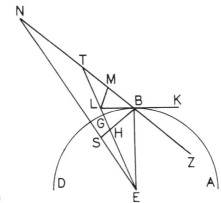

Figure IV.30

[109] *[THEOREM IV.30]* Accordingly, in concave mirrors, when the image lies behind the mirror, the distance of the visible object [along the reflected ray-couple] will be smaller than that of its image [along the incident ray] if the visual radiation were to continue behind the mirror.

[110] Let **ABG** [in figure IV.30] be the arc of a circle inscribed in a concave mirror with centerpoint **E**. Let point **Z** be the eye, and point **H** the visible object to which ray-couple **ZBH** is reflected at equal angles. Let [cathetus] **EHGT** be drawn, and let **BZ** and **HE** be extended to intersect at point **T** beyond the mirror. Therefore, the image of **H** will lie at point **T**. We say, then, that **BZ** + **BH** < **ZT**, whereas **GH** < **GT**.

[111] Let [normal] **EB** be drawn, and let line **KBL** be drawn tangent to the circle at point **B**. Therefore, [curvilinear] angle **ABZ** = [curvilinear] angle **GBH**, and [horn] angle **ABK** = [horn] angle **GBL**, so that the whole angle **KBZ**, which is equal to angle **TBL**, is equal to angle **LBH**. Now angle **BLH** is acute, since angle **LBE** is right. Therefore, angle **BLH** < angle **BLT**. And if we posit angle **BLM** = angle **BLH**, then, since angles **MBL** and **LBH** in triangles **MBL** and **LBH** are equal, while side **BL** is common, it follows that **MB** = **BH**. Consequently, **BH** < **BT**. And if we take **BZ** as common, then **ZB** + **BH** < **ZBT**.

[112] Moreover, since **LH** : **LT** = **BH** : **BT**, while **BH** < **BT**, then **LH** < **LT**, and **GH** will be much smaller than **GT**.

[113] From what we have established, it is evident that, if the distance between the objects and the same viewpoint increases, or if

ample, of course, can be found in the perceptual adjustments made to images in concave mirrors when the locations are indeterminate or inconvenient (i.e., at or behind the eye); see Experiments IV.1 and IV.2, paragraphs 71–81 above.

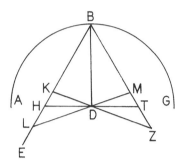

Figure IV.31

that distance is greater, then the distance between the images and the eye increases, or that distance will be greater. For if we extend **BH** to **S** and continue [cathetus] **ES** until it meets the prolongation of **ZBT** at point **N**, then the image of **S** will lie at point **N**, and so **BS + BN > BH + BT**.[59]

[114] *[THEOREM IV.31]* If the image that appears in concave mirrors lies between the eye and the mirror, the distance of the visible object from the eye [along the reflected ray-couple] will be greater than the distance of its image [along the incident ray]; nevertheless, the image's distance from the mirror will sometimes be smaller, sometimes greater, and sometimes equal.

[115] Let **ABG** [in figure IV.31] be the arc of a circle inscribed in a concave mirror with centerpoint **D**, and let normal **DB** be drawn. Let point **E** be the eye, and let ray-couple **EBZ** emanating from point **E** be reflected at equal angles. Then, let [cathetus] **TDH** be drawn through point **D** perpendicular to **BD**, and let the two catheti **KDZ** and **LDM** intersect one another at corresponding angles.

[116] If, therefore, we place [magnitude] **ZTM** at the location that is observed, then, according to principles established earlier, the image of **Z** will lie at point **K**, the image of **T** at **H**, and the image of **M** at point **L**. Indeed, they fall between **E** and **B**, and their distance from **E** will be less than **EB**. Moreover, that distance will be much less than the [eye's] distance from the object-points reached by reflection, e.g., [along] **EB** and **BM**.

[117] It is also evident that, if angle **EBZ** is bisected by line **BD**, then **BT = BH**, and **DT = DH**. But line **BZ > BK**, while **ZD > DK**. Also, line **BM < line **BL**, and **MD < LD**. Hence, the image of **T** will lie at point **H**, and the distance of both [points] from point **B** will be equal. But the distance[from point **B**] of **Z**'s image, which appears at **K**, will be less [than that of **Z** itself], and the distance

[59]See Euclid, *Catoptrics*, prop. 28, part 3.

[from **B**] of **M**'s image, which appears at **L**, will be greater [than that of **M** itself].

[118] So it is clear that, when the distance of objects from the same eye increases, or when it is greater, the distance of their images from the eye increases or will be greater. For, since the distance of point **Z** from the eye is greater than that of the remaining points, its image appears farther away, and since the point-object at **M** lies nearer the eye, its image, which is at **L**, is formed nearer [the eye] than the remaining images.

[119] On this basis, then, it is evident that, when the distance of the visible object is no greater than the distance of the line [**HDT**] passing through the center of the sphere, but when, instead, the object itself and the mirror lie on the same side of this line, the actual object will always lie a shorter distance from the mirror than its image. For, when the situation is as we have just stated, the image of the visible objects will lie farther from the sphere. But if the visible object lies behind the line passing through the center of the sphere, the result will not be as claimed.[60]

[120] *[THEOREM IV.32]* In concave mirrors, if the image of a visible object appears behind the mirror and the location of the generating object is what we have called "facing," then, according to what we have already said in regard to other mirrors, the lines joining the endpoints of the images of visible magnitudes appear longer than the lines joining the endpoints of the visible magnitudes themselves, assuming that those magnitudes were observed directly from the [same] fixed viewpoint at the same distance as that of the image and with the same orientation.

[121] Let **ABG** [in figure IV.32] be the arc of a circle inscribed in a concave mirror with centerpoint **D**. Let **E** be the eye, and let normal

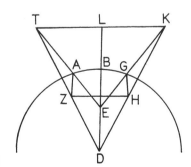

Figure IV.32

[60]See Lejeune, *Recherches*, pp. 84–86.

DEBL be drawn from **E**. Let **ZH** be the line joining the endpoints of the visible object, which is oriented so that line **BD** bisects it at right angles, for it must be placed in a directly facing position. Then, let the two rays **EA** and **EG** be reflected from **E** to **Z** and **H**. Let those two lines be extended to intersect the continuations of catheti **ZD** and **DH** at points **T** and **K**, which lie behind the mirror. Let line **KT** be joined, and let line **DB** be extended to cut it at point **L**. Therefore, the image of **Z** will be seen at **T**, and the image of **H** at **K**. But line **TK** joins the endpoints of the object's image, and its orientation is the same as that of **ZH**.

[122] Therefore, since points **Z** and **H** lie the same distance from the eye at **E**, and since the angles of reflection are equal, image-points **K** and **T** will lie the same distance from point **E**. Also, since both lines **EL** and **ET** are equal to lines **EL** and **KE**, respectively, and since angle **TEL** = angle **KEL**, it necessarily follows that the [respective] angles of both triangles are equal. The angles at **L** will be right, and **KT** will be parallel to line **ZH**. Moreover, **KT** : **ZH** = **TD** : **ZD**. But **TD** > **ZD**. Therefore, **TK** > **ZH**. *And if ZH is transposed to location KT and maintains the same orientation, visual angle KET will be larger than the visual angle subtended by ZH, assuming that it is looked at directly and that its distance and orientation are the same as that of TK. Thus, to the eye at E, KT will appear longer than KH.*[61]

[123] and [124] See note 64 below for the text of these paragraphs.

[125] *[THEOREM IV.33]* In concave mirrors, when the image of a visible object appears between the mirrors and the eye, and when it is oriented [in a facing position] as we specified, the straight lines joining the endpoints of the visible magnitudes sometimes appear to be equal to [those of] their images, sometimes longer, and sometimes shorter, when those visible magnitudes are observed directly and maintain the same disposition and distance as their images, assuming that the eye remains fixed.

[126] Let **ABG** [in figure IV.33] be the arc of a circle inscribed in a concave mirror with centerpoint **D**, let line **BD** be drawn, and let **E** be the eye. Let line **ZH**, which joins the endpoints of the visible object, be positioned in such a way as to be bisected by line **DB** at right angles, according to what must obtain in the case of things that have

---

[61]The text in italics represents an interpolation by Lejeune to complete the theorem. That the text has become somewhat mangled at this point is evidenced by paragraphs 123 and 124 in the Latin text, which simply recapitulate conclusions drawn subsequently in IV, 127–129.

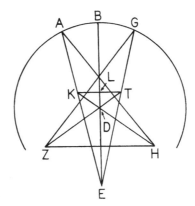

Figure IV.33

a facing orientation. Let the two ray-couples **EAH** and **EGZ** emanating from point **E** be reflected at equal angles to points **Z** and **H**.
Let [catheti] **Z[D]T** and **K[D]H** be drawn, and let them cut lines **EA**
and **EG** in points **T** and **K**. Finally, let line **KLT** be joined. The image of **Z** will therefore appear at point **T**, and the image of **H** at point
K. **KT** will be the line joining the endpoints of the image, and its orientation will be the same as that of **ZH**.

[127] Therefore, since the distances of the eye at **E** from points **Z**
and **H** must be equal, as we have previously said, and since the angles of reflection are equal, the distances of **K** and **T** from point **E**
are equal. Moreover, the two lines **KE** and **EL** are equal to lines **ET**
and **EL**, and angle **KEL** = angle **TEL**. Therefore, triangles **ETL** and
EKL will be equiangular, and the angles at **L** will be right, while line
TK will be parallel to line **ZH**. Also, **ZH** : **KT** = **DZ** : **TD**. But we
have already shown in the previous discussion that it is possible for
DZ sometimes to be equal to **DT**, sometimes shorter, and sometimes longer.[62] Thus, it is possible for **ZH** sometimes to be equal to
KT, sometimes longer, and sometimes shorter.

[128] However, if **ZH** is transposed to the position of **KT**, while
it maintains the same orientation, angle **KET** will sometimes be
equal to the angle subtended by **ZH**, if its disposition and distance
are the same as that of **KT** and if it is observed directly, and it will
appear equal to the eye at **E**. But sometimes it will be greater and
will appear greater, whereas sometimes it will be smaller and will
appear smaller.[63]

[129] Moreover, all objects whose distances from point **E** are less
than half the radius of the sphere that forms the mirror maintain a

---

[62]See theorem IV.31, paragraphs 114–119 above.
[63]See Lejeune, *Recherches*, pp. 89–90, for an analysis. See Knorr, "Archimedes," pp. 58–60,
for the relationship between theorems 32 and 33 and proposition 28 of Euclid's *Catoptrics*.

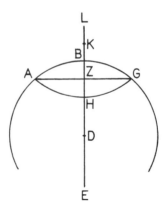

Figure IV.34

distance between the endpoints of the magnitude that is greater than the distance between the endpoints of the image.[64]

[130] *[THEOREM IV.34]* When the image of an object that is seen in concave mirrors lies behind them, or between them and the eye, straight objects that face them directly appear concave. Furthermore, circular objects whose concave side faces the mirror and the reflected ray appear concave, but those whose convex side faces the mirror sometimes appear concave, sometimes straight, and sometimes convex.

[CASE 1]

[131] First, then, let the image of the visible object lie behind the mirror. Let **ABG** [in figure IV.34] be the arc of a circle inscribed in a concave mirror with centerpoint **D**, and let line **BDE** be drawn. Let line **ABG** be bisected at point **B**. Then, through points **A** and **G**, let straight line **AZG** be drawn along with circular arc **AHG**, which has its concave side facing the mirror and the reflected ray. Let **DH** > **HB**,[65] and let the eye be at point **E**.

---

[64]The following text, transposed from its original position above in the critical Latin text, provides a sort of capsule summary of the conclusions drawn to this point:

*[123] Now, as we have demonstrated in preceding theorems, **ZD** can sometimes be equal to **ZT**, sometimes longer, and sometimes shorter, so **ZH** can sometimes be equal to **KT**, sometimes longer, and sometimes shorter. But if **ZH** is transposed to location **KT** and maintains the same orientation, then angle **KET** will sometimes equal the angle subtended by **KH**, when its disposition and distance are the same as that of **TK**, and when it is seen directly; [thus] to the eye at **E** the image will appear the same size as the object. Sometimes, however, the angle subtended by **ZH** will be greater, and the object will appear larger, whereas sometimes the aforesaid angle becomes smaller, and the object will appear smaller.*

*[124] Furthermore, in the case of identical objects lying less than half the radius of the sphere away from point **D**, the length [of the line] joining both endpoints of the magnitude remains greater than the length [of the line joining both endpoints] of the image.*

[65]Under these conditions, then, the image will always lie behind the mirror.

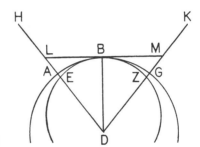

Figure IV.35

[132] The image of points **A** and **G** on these two lines will thus appear at points **A** and **G** themselves. Meantime, the image of points **Z** and **H** will appear behind the mirror, **Z** at point **K** and **H** at point **L**, for it has been demonstrated that the farther objects lie from the same viewpoint, the farther their images will lie from that viewpoint. But the distance of **H** from the eye at **E**, as judged along the ray-couple [**EB, BH**] reflected from point **B**, is greater than the distance of **Z** from the eye at **E** [along ray-couple **EB, BZ**]. Point **L** is also farther away than point **K**, and both of those points lie on the image.

[133] So the image of line **AZG** will appear on the line that passes through points **A**, **K**, and **G**, whereas the image of arc **AHG** will appear on the line passing through points **A**, **L**, and **G**. Moreover, the concavity of arc **ABG** will appear to be turned toward center of sight **E**, and the lines passing through points **AKG** and **ALG** will be more pronounced in their concavity than **ABG**, since the rays falling perpendicular to those arcs are proportionately longer than the ones falling more obliquely. Generally speaking, though, the images of straight and concave objects will appear concave.[66]

[134] To continue, let a convex arc **EBZ** [in figure IV.35] be drawn through point **B**, tangent to the circle and [thus] to arc **ABG**. Let [normal] lines **DEH** and **DZK** be drawn, and let **DE** and **DZ** be longer than **EA** and **ZG**. Let tangent **LBM** be drawn through point **B** of the circle.

[135] Accordingly, the image of **EZ** can sometimes be seen between **L** and **A** and between **M** and **G**, sometimes on **L** and **M** themselves, and sometimes beyond points **L** and **M** (e.g., at points **H** and **K**), all depending on the distance of **E** from **A** and of **Z** from point **G**. But the image of **B** will lie at point **B** itself.[67]

---

[66]See Lejeune, *Recherches*, pp. 94–95.

[67]Although Ptolemy does not specify the fact, it is clear that the center of sight must in this case coincide with the center of curvature **D**. This, of course, poses an extreme limitation on the resulting conclusion.

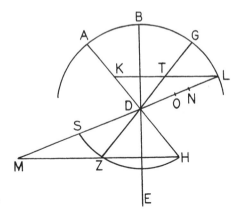

Figure IV.36

[136] If the images of **E** and **Z** lie at points **L** and **M**, then the image of convex [arc] **EBZ** will appear along straight line **LBM**. But when the images of **E** and **Z** lie between **A** and **L** and between **G** and **M**, then the image will appear concave, for when oblique rays falling on objects are proportionately smaller than perpendicular ones, those objects appear concave. Finally, if the image[s] of **E** [and] **Z** lie beyond points **L** and **M**, then the image will appear along a convex line, for in this case the opposite to what we just specified happens insofar as the oblique rays are proportionately longer than the perpendicular ones. Thus, when something is concave in comparison to straight line **LBM**, it will be absolutely concave, and when it is convex in comparison to that [same straight line], it will be absolutely convex.[68]

[CASE 2]

[137] Now, let the image lie between the eye and the mirror. Let **ABG** [in figure IV.36] be the arc of a circle inscribed in a concave mirror with centerpoint **D**, let line **EDB** be drawn normal to the mirror, and let the eye be at point **E**. Let **HZ** be the line joining the endpoints of the visible object, and let line **ED** bisect it at right angles.

[138] Accordingly, when catheti **ZDG** and **KDH** are extended, the image of **Z** will appear at point **T**, and the image of **H** at point **K**. Let cathetus **TK** be drawn, and let it be extended to point **L** on the arc. Let line **LDM** be drawn, and let it intersect the extension of line **ZH** at point **M**. Thus, the image of point **M** will appear between points **D** and **L**, since none of the rays emanating from the

---

[68]See Lejeune, *Recherches*, pp. 95–97.

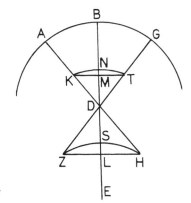

Figure IV.37

eye is reflected at equal angles from point **L** to point **M**.[69] But if we suppose that the image of point **M** lies at point **N**, then the image of straight line **MH** will be seen along the [curved] line passing through points **K**, **T**, and **N**, and the concave side of this image will face the eye.

[139] The same thing holds if we assume that the line passing through points **Z** and **H** is circular, with its concave side facing the mirror. For, if we construct the arc to which the visual ray is reflected and have its concave side facing the mirror, as is represented by arc **HZS**, then the image of point **S** will lie between points **D** and **N** (e.g., at point **O**). And the line passing through points **K**, **T**, and **O** will be even more concave [than the previous one through **K**, **T**, and **N**]. But this very line [**KTO**] lies on the visible object's image.[70]

[140] Let us now suppose that arc **ZSH** [in figure IV.37] has its convex side facing the mirror, let us draw straight line **ZH** joining the endpoints of the visible object, and let **KNT** represent its concave image. When the rays are shorter, the image of the object seen by them lies closer [to the eye]. Therefore, since the distance of **S** from the eye at **E** is less than that of **L** [from **E**], as measured by the ray-couple[s **EB**, **BS** and **EB**, **BL**] reflected from point **B**, and since the distance [**EM**] of its image, which lies at **M**, is less than that of **L** [along **EN**], then point **S** will be seen between **D** and **N**.

[141] So the image of **S** can sometimes appear at **M**, and the image of **ZSH** will lie on straight line **TMK**. But sometimes [it can appear] between **M** and **N**, and the line will appear concave. Finally, the image can sometimes appear between points **D** and **M**, and the line will appear convex, depending on how pronounced the

[69]This follows because the resulting ray-couple **MLE** would not encompass centerpoint **D**; see n. 23 above.

[70]See Lejeune, *Recherches*, pp. 97–98.

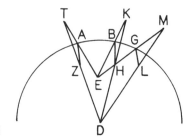

Figure IV.38

concavity [of the visible object] is. Still, the image of [end]points **Z** and **H** will remain fixed at points **T** and **K**.[71]

[142] *[THEOREM IV.35]* In concave mirrors, when the image of a visible object lies behind the mirror, it is perceived to be on the same side as the actual object; and when visible objects are moved in any direction, their images appear to move in the same direction.

[143] Accordingly, let **ABG** [in figure IV.38] be the arc of a circle inscribed in a concave mirror with centerpoint **D**. Let the eye be at **E**, and let **Z** and **H** represent two visible objects flanking it. Let catheti **DZT** and **DHK** be drawn, and let the two ray-couples **EAZ** and **EBH** be reflected at equal angles from **E** to **Z** and from **E** to **H**. Then, let lines **EA** and **EB** be extended so as to intersect lines **DT** and **DK** at points **T** and **K**. Therefore, the image of **Z** will lie at point **T**, and the image of **H** at point **K**. Moreover, those images will be perceived to lie on the same side as the actual objects.

[144] Let **H** be moved to **L**, and let [ray-couple] **EGL** be reflected to it at equal angles. Then, let [cathetus] **DLM** be joined, and let it intersect the extension of **EG** at point **M**. Therefore, the image at **K** will be moved to point **M**, which lies in the same direction as that toward which the actual object has been moved.

[145] Again, if **H** and **L**, which represent the visible objects, are above eye-level, then their images **K** and **M** will be above eye-level and will appear above [the original location]. For the upper parts of objects directly facing the viewer are apprehended by higher rays.

[146] If, however, we suppose **H** and **L** to lie to the right of the eye, then their images, which are at **K** and **M**, happen to lie to our right. But those images are not judged to lie to the right, because the part of what we see facing us that is touched by right-hand rays and that appears to the right is actually on the left-hand side of what we see.

[71]See Lejeune, *Recherches*, pp. 98–100.

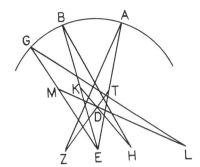

Figure IV.39

And [so] the left-hand side of a facing visible object is apprehended by the right-hand rays in direct vision. But in the case of reflection, it is the right-hand side that is apprehended, and the facing sides of the images are directly opposite us, so the images that lie to the right are judged to lie to the left according to the normal arrangement of the visual rays in regard to position.[72]

[147] *[THEOREM IV.36]* In concave mirrors, when the image of a visible object lies between the mirror and the viewer, it is perceived as the reverse of the actual object, and when the actual object is moved in a given direction, its image appears to move in the opposite direction.

[148] Let **ABG** [in figure IV.39] be the arc of a circle inscribed in a concave mirror with centerpoint **D**. Let the eye be at **E**, and let **Z** and **H** represent two visible objects on either side of the eye. From point **E** let the two ray-couples **EAZ** and **EBH** be reflected at equal angles to those [two point-objects]. Let catheti **ZDT** and **HDK** be drawn, and let lines **EA** and **EB** intersect them at points **T** and **K**. Thus, the image of point **Z** will lie at point **T**, and the image of point **H** at **K**. Consequently, the images will appear on the opposite sides of their actual [generating] objects.

[149] By the same token, if point **H** is moved to point **L**, if ray-couple **EGL** is reflected to it, and if cathetus **LDM** is extended to meet line **EG** at point **M**, then the image of **H**, which is at **K**, will move in the opposite direction from that in which the actual object has been moved.

[150] Furthermore, if **H** and **L** lie above eye-level, their images, which are represented by **K** and **M**, will lie below us, and the upper parts will appear lower. For things that are apprehended by lower

---

[72]See II, 137–138 and III, 96 and 130 above. For the Euclidean counterparts to this theorem, see *Catoptrics*, props. 11, 12, and 28, part 3.

visual rays appear at the lower side of the visible object, but what is apprehended in such reflections by lower rays appears at the upper side.

[151] And if we suppose that **H** and **L** lie to the right of the eye, then their images (i.e., **K** and **M**) will appear to our left. And so right-hand objects will appear to the right, for the parts of facing objects that are apprehended by right-hand rays appear to the left, while those that are apprehended by left-hand rays appear to the right. Among objects seen by means of such reflections, moreover, those that move toward the right appear to move toward the left, because [in this case] our hand does not seem to move as a mirror-image, but in the opposite direction, as if it were a right hand [rather than a left]. Indeed, facing objects that are seen in direct vision are disposed so that their right sides lie to our left.[73]

[152] From what we have explained, then, it necessarily follows that, in concave mirrors, images appear in a different way from the actual objects that are seen in them; sometimes the result is the same as that in direct vision, but sometimes not.

[153] As in the case of objects that are seen directly, variation [in the image] depends on a face-to-face position and on a back-to-front displacement only, just as we have shown is the case with convex and plane mirrors. For the images of objects that lie farther from a given viewpoint always lie farther from that same viewpoint. Indeed, reflection has no effect on the displacement of rays in a back-to-front direction.[74]

[154] But side-to-side displacement does cause a difference, because the visual rays are changed in their fundamental nature. Thus, for example, we have shown that a high position will sometimes appear high in concave mirrors—just as we observe when the object is so placed that its image appears behind the mirror, in which case it will be as if the object were seen directly. Sometimes, however, what lies toward the top will be seen toward the bottom, so that [the image] is inverted in comparison to what is seen directly. Moreover, right-hand things sometimes appear to the left, as is the case [in direct vision] with objects that face us. Sometimes, though, the opposite happens: i.e., [the right-hand side appears to lie] to the right.

[155] Also, in this sort of mirror, the eye sometimes sees a single image and sometimes several images of the same object. And when

---

[73]See Lejeune, *Recherches*, pp. 102–103.
[74]That is, reflection causes no reversal in this direction.

the eye remains fixed and unmoved, some images will lie farther [from the eye] than the actual object [that generates them], whereas others will lie closer [to the eye than their generating object]. Furthermore, the images of magnitudes will at times appear to be the same size as the magnitudes themselves, at times larger, and at times smaller. The shapes of some objects (e.g., concave ones) appear similar to those of their images, whereas the shapes of other objects (e.g., straight ones) appear dissimilar, and the shapes of yet other objects (e.g., convex ones) sometimes appear similar and sometimes dissimilar. But the way in which these variations are perceived is governed by the set of principles we have established.

[156] What we have said to this point provides almost everything that is necessary for an analysis of images appearing in simple, non-composite mirrors. However, anyone who wants to understand not only what happens when the three primary shapes—i.e., plane, convex, and concave—are juxtaposed but also what results from such a composition of shapes can do so using the reasoning that has already been set forth.

[157] If one considers what has been said about plane, convex, and concave mirrors, and if one considers the diagrams pertaining to each type according to what we have shown, and if one applies to these composite mirrors specific properties of their [constituent] cross-sections and the cross-sections of the objects that appear facing them, then no further discussion will be necessary for such an analysis, unless one wishes to repeat the same points. For, on the basis of earlier analyses, it will be easy for the student to understand the specific types of image-variation that are created in each kind of composition. Still, it does no harm to talk briefly about these variations.[75]

[158] Take, for instance, mirrors composed of a straight and a convex section, such as those that are cylindrical. When the eye faces the convex surface of such a mirror, and when the visible object stands upright in front of the mirror, which also stands upright, then the object will appear undistorted[76] along the vertical [i.e., along its own length], its upper edge being above and its lower edge below. It will, however, appear compressed along the horizontal [i.e., along its own width], and its right-hand side will appear to the left, while its left-hand side will appear to the right. On the other hand, when the object faces the [upright] mirror sideways, it will

---

[75]As Lejeune points out in L'Optique, p. 208, n. 71, by including a discussion of composite mirrors, Ptolemy is following a tradition established by Euclid's and, more to the point, Hero's Catoptrics.

[76]Latin = moderata = "equal"; see Lejeune, L'Optique, p. 208, n. 74.

appear undistorted along the vertical [i.e., along its own width], although, again, its right-hand side will appear to the left and its left-
hand side to the right. It will, however, appear compressed along
the horizontal [i.e., along its own length], and the upper edge will
appear above and the lower edge below.

[159] We turn now to mirrors composed of a concave and a
straight section, which take the form of a cylinder with its concave
surface facing the viewer. When the images formed in such a concave surface are behind the mirror, and when the mirror and the
object facing it are both upright, then the object will appear undistorted along the vertical [i.e., along its own length], and the upper
edge will appear above while the lower edge will appear below.
It will, however, appear distended along the horizontal [i.e., along
its own width], and the right-hand side will appear to the left,
while the left-hand side will appear to the right. Yet when the
object faces the mirror sideways, it will appear undistorted along
the vertical [i.e., along its own width], and, again, the right-hand
side will appear to the left and the left-hand side to the right. But
it will appear distended along the horizontal [i.e., along its own
length], and the upper edge will appear above, while the lower edge
will appear below.

[160] If, however, the image appears [to lie] between the concave
surface of the mirror and the eye, then, when the object and the mirror face one another upright, the object will appear undistorted
along the vertical [i.e., along its own length], and the upper edge
will appear above, while the lower edge will appear below. Along
the horizontal [i.e., along its own length], however, the object will
sometimes appear undistorted, sometimes distended, and sometimes compressed, and the right-hand side will appear to the right
and the left-hand side to the left. Yet, when the object faces the mirror sideways, it will appear undistorted along the vertical [i.e.,
along its own width], and the right-hand side will appear to the left
and the left-hand side to the right. Along the horizontal [i.e., along
its own length], though, it will sometimes appear undistorted,
sometimes distended, and sometimes compressed, and the upper
edge will appear below and the lower edge above.

[161] Finally, in mirrors composed of a concave and a convex
section,[77] if the image of the visible object ought to be seen behind

---

[77]As Lejeune remarks in *L'Optique*, p. 210, n. 76, this composite mirror is in the form of a
torus—that is, it is shaped like a doughnut with the center of concavity at the center of the
hole. Thus, the observations of image-distortion will depend on whether the line of concavity is vertical to the vertically-standing viewer (i.e., the surface of reflection coincides with the

the mirror, on account of the concavity,[78] and if the object stands upright, facing the [horizontally positioned] convex section, the object will appear compressed along the vertical [i.e., along its own length], and the upper edge will appear above while the lower one appears below. It will appear distended along the horizontal [i.e., along its own width], though, and the right-hand side will appear to the left while the left-hand side will appear to the right. Yet, if the object stands upright toward the [vertically positioned] convex section, it will appear compressed along the horizontal [i.e., along its own width], while, again, the right-hand side will appear to the left, and the left-hand side will appear to the right. Along the vertical [i.e., along its own length], however, it will appear distended, and the upper edge will appear above, while the lower edge will appear below.

[162] But if the image ought, on account of the concavity, to be seen between the mirror and the eye, then, if the visible object stands upright in front of the [horizontally positioned] convex section of the mirror, it will appear compressed along the vertical [i.e., along its own length], and the upper edge will appear above, while the lower edge will appear below. Along the horizontal [i.e., along its own width], however, it will sometimes appear distended, sometimes compressed, and sometimes undistorted, and the right-hand side will appear to the right while the left-hand side will appear to the left. But if the upright object faces the [vertically positioned] convex section, it will appear compressed along the horizontal [i.e., along its own width], and the right-hand side will appear to the left, while the left-hand side will appear to the right. Along the vertical [i.e., along its own length], though, it will sometimes appear undistorted, sometimes distended, and sometimes compressed, and the upper edge will appear below and the lower edge above.[79]

[163] Generally speaking, in the case of all images formed in composite mirrors, forward and backward motions are reversed, insofar as certain parts are nearer and certain farther [from the mirror's surface]. Also, in this case, the apparent motions are not the same as in direct vision, nor are they the same as in the case of objects whose images appear in uniform spatial order.[80]

---

plane of the "hole") or whether it is horizontal to that same vertically-standing viewer (i.e., the surface of reflection is orthogonal to the plane of the "hole").

[78]In other words, the viewer must stand between the concave surface and the center of curvature in order to insure that his image is projected behind the reflecting surface.

[79]In this case, then, the viewer must stand beyond the center of curvature, the amount and type of distortion depending on how far beyond.

[80]Latin = *nec sicut fit in rebus quarum omnes partes apparent secundum maius et minus*; I take this somewhat cryptic statement to mean that, whereas images in regular mirrors (i.e., plane

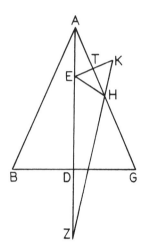

Figure IV.40

[164] A mirror can also be conical, and when the visual flux strikes the interior surface of such a mirror, the resulting image will have an acute angle, for it is composed of straight and concave lines. However, the outer surface of the cone is composed of straight and convex lines, as we have explained.

[165] *[THEOREM IV.37]* Let a conical mirror be constructed, let a section of it be represented by triangle **ABG** [in figure IV.40], and let the visual ray strike its interior surface. Let its axis be **AD**, and let point **E** lie on **AD**. Let the eye lie on the extension of axis **AD** at point **Z**, and let ray-couple **ZHE** reflect from **AG** at equal angles between **Z** and **E**. Then let [cathetus] **ET** be drawn normal to **AG**, and let it intersect the extension of **ZH** at **K**. The two therefore intersect behind the mirror. Since angle **KTA** is right, angle **ZHG**, which equals angle **THK**, is less than a right angle.[81]

[166] An image of **E** happens to be formed along the entire circle that **K** describes when plane **EZK** is revolved about axis **AZ**. When this reflection reaches **AD** and [the incident ray **ZH**] is extended, then, because of the mirror's concavity, the image will appear along the entire circle passing through point **H**, assuming that axis **AD** remains fixed and the plane revolves about it. For all the planes passing through the axis are normal to the surface of the cone, just as the planes passing through the centers of spheres are normal to their surfaces.

_____

and spherical) are distorted in a fundamentally regular way throughout, they are distorted irregularly in composite mirrors—hence, the irregular variations in apparent motions because of the irregularity in relative spatial dispositions along the vertical and horizontal.

[81]In short, **AG** is treated as a section of a plane mirror, and the image-location is determined accordingly.

[167] In the same vein, if the mirror is shaped like a pyramid, with its base not circular but formed of straight lines, and if the observer's eye is directed toward its interior as represented in the preceding figure, then the resulting reflection does not take place along a circle, and the image will not appear continuously around [the interior surface]. Instead, the reflection takes place from one point on each of the mirror's sides. Moreover, that point will lie on the normal dropped from the pyramid's vertex between the [flanking] edges [of the given side] to the [edge at the] base [of that side].[82] Only those planes that pass through these sections in the direction of each reflecting lateral surface and that contain the reflecting ray-couple can pass through the axis as well as the center of sight and the visible object. The number of such planes will be equal to the number of the [pyramid's] sides.

[168] *[THEOREM IV.38]* In cases of this sort [i.e., of concave conical mirrors], a single image is formed, as long as the object that generates the image is not taken at the outset to be large. But when the sight falls on the [outside] surface of a conical mirror, what happens is as follows. We start by describing a circle around the mirror's base, we color that circle brightly, making it somewhat larger than the base circle of the cone, and we place its center upon point **D** with great care. Then, if we direct our line of sight toward the cone's vertex along its axis, we see the circle described about the base [within the mirror].

[169] Accordingly, let the mirror itself be [represented by section] **ABG** [in figure IV.41], with **AD** being its axis. Let **DAE**, **BGZ**, and **AGH** be drawn, and let **E** be the eye and **Z** the visible object.[83] Then, among the rays emanating from **E**, let one (e.g., [ray-couple] **ETZ**) reflect at equal angles to **Z**. Let [cathetus] **ZH** be drawn perpendicular to **AH**, and let it intersect the extension of line **ET** at **K**.[84] Point **K** will therefore lie behind the mirror, for angle **KHT** is right, while angle **HTK** is acute.

[170] Hence, the image of point **Z** will lie at **K**. And when this plane [i.e., the plane of the diagram] is revolved about axis **AD** so that a cone is thereby formed [from section **ABG**], point **Z**, in the process of revolving, describes a circle on the cone's base-plane

---

[82]In other words, the point of reflection for each triangular side of the pyramid will lie on the line bisecting the angle at the side's vertex, that line being normal to the side's lower edge.

[83]**Z** thus represents the point of intersection between the circle drawn about the cone's base and the cutting plane of section **ABG**.

[84]Again, **AG** is being treated as a section of a plane mirror, so the image-location at **K** is determined accordingly.

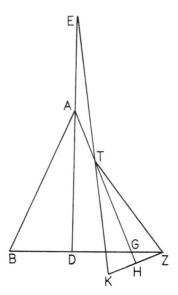

Figure IV.41

according to the [placement of the] visible object. Point **K**, for its part, will lie on its image, and point **T** will designate the spot on the mirror through which the image is seen, just as we discover from observations as well as from previously-determined principles.

[171] *[THEOREM IV.39]* In the case of mirrors that are formed in such a way as to appear spherical to the viewer, if they are actually composed of [separate] mirrors all having the same shape, multiple images of a single object are formed. This happens because each [constituent] mirror is capable of providing only one, single point [of incidence/reflection] for the visual flux.

[172] Let **ABGD** [in figure IV.42] be the arc of a circle with centerpoint **E**, and let lines **AB**, **BG**, and **GD** of equal length be inscribed. In this particular case, let the eye be at point **E**. Let plane **ABGD** be continuous, and let it pass through the sphere as well as through the plane of reflection. Let **EZ**, **EH**, **ET**, **EK**, **EL**, and **EM** represent all the possible rays for each mirror. Accordingly, everything lying between **H** and **T** as well as between **L** and **K**, and thus everything that the ray strikes at an angle, falls outside the figure of the mirror when it is reflected. On the other hand, what lies between **Z** and **H**, **T** and **K**, and **L** and **M** falls inside the figure of the mirror when it is reflected. The number of images will be equal to the number of mirrors.[85]

---

[85]As Lejeune observes in *L'Optique*, p. 215, n. 84, this "demonstration" is vague in the extreme, in great part because of the lack of specification for the representative rays. For in-

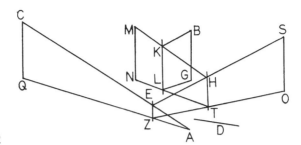

Figure IV.42

[173] In fact, each of the rays reflected from the mirror contains the whole.[86] Still, not all the intersections that are thus made will be continuous; some will be discontinuous at small intervals.[87] And reflection will occur from objects that lie outside the figure of the mirror. And, just as in the case of objects seen in direct vision, so too here, things that are separated appear more clearly, and a single object subtending a discrete angle and visual ray-bundle does not appear continuous in several locations whose number is equal to the number of angles.[88] When there is absolutely no difference in the number of images, when the shape of the basic component mirror is plane, when the [composite] mirror itself takes on a concave or convex shape, and when the angles at the junctures [of the component mirrors] necessarily create discontinuities, still, variations occur rather in its essence and in the estimation of distances and in the judgment of the location of each of the images.[89]

---

stance, it is unclear whether **EZ** and **EM** are perpendicular as the figure suggests. Moreover, it is unclear what Ptolemy means by asserting that all the reflections that occur between **H** and **T**, as well as **L** and **K**, fall outside the mirror. Again, the figure suggests that any ray emanating from **E** to a point on segment **HB** will reflect to mirror **BG** at such an angle that it will then reflect from **BG** along a line that will take it past point **D** through the opening at arc **AD**. Meantime, any ray emanating from **E** to a point on segment **BT** will reflect to mirror **AB** at such an angle as to cause the second reflection to take it past point **A** through the opening at arc **AD**. Presumably, then, the same reasoning will apply to the area between **K** and **L**. On the other hand, reflections between **Z** and **H**, as well as between **L** and **M**, will cause at least three internal reflections. Whatever the case, however, only the reflections along the normals will create images; hence, the number of images will equal the number of reflecting surfaces. See Euclid, *Catoptrics*, prop. 14 and Hero, *Catoptrics*, props. 14 and 17 for possible sources for this theorem.

[86]Lejeune takes this to mean that all the rays, taken as a whole, are actually continuous, an interpretation that is perhaps borne out by the very next sentence. Another possible meaning here is that each of the rays represents an entire set of rays that share the same fundamental reflecting characteristics.

[87]Perhaps these "small intervals" refer to the points of juncture **B** and **G** of the reflecting surfaces.

[88]Perhaps the point here is to contrast convex or concave mirrors composed of a definite number of discrete plane mirrors to convex or concave mirrors composed of an indefinite number of such mirrors. In the latter, the images formed are continuous.

[89]The meaning of this last sentence escapes me entirely. Indeed, the entire theorem is so cryptic and vague as to suggest either that it was interpolated into the original by a badly muddled editor or that it suffered considerably in translation.

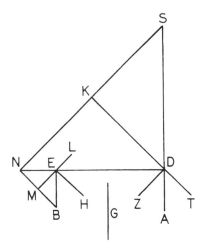

Figure IV.43

[174] Or else this does not happen, and the contrary occurs, so that the image of one object is formed in several mirrors according to the number of [reflecting] surfaces, but chiefly among plane mirrors in which images are formed. This is what happens when the eye and the visible object are on separate sides and an object is interposed to block them from one another. In such a case, in fact, several images are formed according to the continued reflection of the ray from one or more plane mirrors to the visible object, and the images within these mirrors are formed along the line of incidence. Accordingly, let the following serve as an example to clarify our point:

[175] *[THEOREM IV.40]* Let the eye be placed at point **A** [in figure IV.43] and the visible object at **B**. Let some object **G** be placed between them to screen one from the other. Let **AD** and **BE** be drawn from **A** and **B** to opposite points, and let line **DE** be drawn. Let angle **ADE** be bisected by line **DZ** and angle **BED** by line **EH**. Let the two perpendiculars **KDT** and **LEM** be dropped to **DZ** and **EH** [respectively], and let them each represent a straight line on a plane mirror, according to the principles we have established.[90] Accordingly, **AD** will reflect at equal angles along **DE**, and **DE** will reflect [at equal angles] along **EB**, for these particular lines form equal angles with the normals dropped to the mirrors. Indeed, the visual ray emanating from **A** is reflected along **ADEB**.

[176] Also, in conformance with the fundamental principles [of reflection], the visible object at **B** is seen at that point in the mirror

---

[90]In other words, **LM** and **KT** represent the common sections of the two mirrors with cutting-plane **ASNB**.

where the cathetus dropped from the visible object to the mirror and the incident ray intersect. But the incident ray, represented by **AD**, is unique, whereas the catheti dropped from the visible object are multiple, their number being equal to the number of mirrors. It is impossible for all of them to converge on the initial line of sight.[91] On the contrary, in the case of the first mirror, only the cathetus falling to **S**, being the last of all the catheti, meets at a common intersection with the original line of sight on the mirror, according to the principle we have established for visual rays.[92] Also, when cathetus **BM** is dropped from **B** to **LE** until it intersects the extension of **DE** at point **N**, then point **N** will be the first endpoint of the cathetus that is the last in the succession of reflections from point **E**. Furthermore, if a cathetus, such as **KN**, is dropped from that same point [**N**] to **TK** (this is possible because angle **EDK** is acute) so that it intersects the extension of [incident] visual ray **AD** at **S**, then point **S** will be the endpoint of the last cathetus that is the first in the succession of the reflections at **D**.

[177] In addition, the image of **B** that is seen by the eye **A** will be at point **S**, according to the principles governing plane mirrors; these principles are perfectly evident here, since **ADS** = **AD** + [**ED** +] **EB**. Thus, since **BE** = **EN**, and **ND** = **DS**, then, with line **AD** forming a common segment, the whole of **AS** = **AD** + **ED** + **EB**.[93]

[178] *[THEOREM IV.41]* Much the same thing as we established [in the case of single reflections] happens in the case of multiple reflections: i.e., the sizes of the images and of the actual objects are equal, which is invariably the case with plane mirrors, but the orientation is not the same. In fact, the orientation appears to be the same only when the number of mirrors is uneven. If they are even in number, then right-hand objects will appear to the left and left-hand objects to the right.

[179] Accordingly, let the eye be at **A** [in figure IV.44], let **BG** be the visible object, and let **D** be some body screening [eye and object from one another]. Let there be a reflection from three mirrors: namely, **EZ**, **HT**, and **KL**. Let a visual ray be reflected toward **B** at equal angles from every mirror, and let it form [successive

---

[91]Thus, no normal to mirror **MEL** will intersect the extension of incident ray **AD** behind mirror **KDT**.

[92]In this case, **N**, which is where the image of object-point **B** lies in mirror **ML**, is itself taken as an object-point for mirror **KT**, its image in that mirror being located along cathetus of reflection **NS**.

[93]See Euclid, *Catoptrics*, props. 13 and 14, esp. the latter. Chapter 17 of Hero's *Catoptrics*, which Lejeune cites in *L'Optique*, p. 219, n. 90, is inapposite, since it has nothing to do with multiple reflections among mirrors.

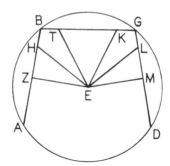

Figure IV.44

segments] **AE**, **EH**, **HK**, and **KB**. Toward **G**, meantime, let a visual ray [be reflected along segments] **AZ**, **ZT**, **TL**, and **LG**. Let [all] the radial segments be extended to intersect the normals. According to the first distance [from mirror **HT** to mirror **KL**, the image of] magnitude **BG** will appear at **MN**. According to the second distance [from mirror **EZ** to mirror **HT**, the image of **MN** will appear] at **SO**. And according to the third distance [from the eye to mirror **EZ**, the image of **SO** will appear] at **CQ**.

[180] From the foregoing analysis, then, it is obvious that the [apparent] distance of the image is equal to the sum of the lengths of the radial segments [along one reflected ray-set]. For, since the juxtaposed angles [i.e., of incidence and reflection] are equal and the extensions of the reflected rays are always straight, it necessarily follows that **KB** and **GL** are congruent with **KM** and **LN**,[94] while **HM** and **TN** [are congruent with] **HS** and **TO**, [as are] **ES** and **ZO** with the whole of **EC** and **ZQ**. By the same reasoning, the eye at **A** perceives **BG** in the mirrors as if it were seen in direct vision at the same location with the same orientation—i.e., at **CQ**, where in this case the objects are perceived according to [visual] angle **EAZ**, and they appear equal.

[181] It is also evident that, when the visible object faces the visual ray that reaches it by way of the three mirrors, point **B** of magnitude **BG** will appear to the right insofar as **M** and **S** are [projected] to **C**, which will [in turn] appear to lie toward the right-hand side of the eye at **A**. Point **G**, meanwhile, insofar as [it is projected via] **N** and **O**, will appear at **Q**, which lies to the left-hand side of the eye at **A**. But if we posit two [instead of three] mirrors—i.e., **EZ** and **HT**— then point **M**, which is on the left-hand side of magnitude **MN**, will appear through mirror **TH** at **C** and will appear to lie to the right from the perspective of **A**, while **N**, which is actually on the right-

---

[94]That is, figure **KLMN** is perfectly congruent with **KBGL**.

hand side, will appear at **Q**, which lies to the left [from the perspective of **A**].

[182] Thus, as we said, the same happens whenever the mirrors are even in number, because at each, single reflection the visual ray switches direction, just as is the case in things that are seen directly, insofar as their right-hand portions, which face us, are perceived by left-hand rays, and their left-hand portions by right-hand rays. When the mirror is single, and the reflection is too, then, on the basis of the first reflection, right-hand objects are seen by right-hand rays, in contradistinction to the previous case and in contradistinction to the case of an object seen directly. The same also holds in the case of several uneven-numbered mirrors. Indeed, in the case of even-numbered mirrors, reflections occur according to successive switches in direction, just as in the case of things that are seen directly, and they cause a reversal in the true position [of the visible object]. On the other hand, when the number of mirrors is odd, the opposite happens, according to the way that is appropriate for the first reflection.[95]

---

[95]See Knorr, "Archimedes," pp. 61–64, for a discussion of Theorems IV.40 and IV.41 and their possible relationship to both Euclid's and Hero's treatments of the same topics.

BOOK 5
REFRACTION

Introductory Section [1–4]: [1] summary of book 4 and statement of goal for book 4, [2] review of claims previously made about refraction and its physical cause, [3–4] principles common to both reflection and refraction

Experiments on Refraction [5–22]
[5] simple experiment to show how refraction affects where the object is seen, [6] **Example V.1**, illustrating this experiment, [7–12] **Experiment V.1**, determining the angle of refraction from air to water, [13–18] **Experiment V.2**, determining the angle of refraction from air to glass, [19–21] **Experiment V.3**, determining the angle of refraction from water to glass, [22] concluding statement

Atmospheric Refraction [23–30]
[23–24] description of effect of atmospheric refraction on astronomical observation, [25–26] **Theorem V.1**, analyzing that effect in two dimensions, [27–29] **Theorem V.2**, analyzing the effect in three dimensions, [30] possible practical ramifications

General Considerations on Refraction [31–45]
[31] principle of reciprocity, [32–34] **Experiment V.4** to determine a general law of angular relations in refraction, [35–37] theoretical considerations on the inequality of angles in refraction, [38–45] **Theorem V.3**, intended to demonstrate the necessity of such angular inequality

Image-Location [46–63]
[46] introductory paragraph, [47–54] **Theorems V.4–V.7**, analyzing conditions under which the cathetus of incidence and the refracted ray do or do not meet when the surface of refraction is plane or spherical, [55] summary of results, [56–61] **Theorems V.8–V.10**, analyzing conditions under which the cathetus of refraction and the refracted ray do or do not meet when the surface of refraction is plane or spherical, [62–63] summary of results

Image-Distortion [64–87]
[64–66] preliminary conditions for the analysis, [67–68] **Experiment V.5**, basic description of three glass vessels to be used; observation

228

using cubical vessel for refraction through a plane interface, [69–74] **Theorems V.11** and **V.12**, analyzing distance-distortion in refraction through a plane mirror, [75–78] **Theorems V.13** and **V.14**, analyzing size-distortion in refraction through a plane mirror, [79–82] **Theorem V.15**, showing that no shape-distortion occurs in refraction through a plane interface, [83] preliminary observations on refraction through a convex interface using the cylindrical glass vessel, [84–87] **Theorems V.16** and **V.17**, analyzing distance-distortion in refraction through a convex interface

The Fifth Book of Ptolemy's *Optics*

[1] There are two ways in which the visual ray is broken. One involves rebound and is caused by reflection from bodies that block the [visual ray's] passage and that are included under the heading of "mirrors." The other way, however, involves penetration and is caused by a deflection in media that do not [completely] block the [visual ray's] passage, and those media are included under the single heading "transparent." In the preceding books we have discussed mirrors; and, insofar as it is possible for it to be demonstrated, we have explained not only variations in the images of visible objects according to the principles laid out for the science of optics, but also what happens with each of the visible properties [in reflection]. It thus remains for us at this point to analyse what sorts of variations occur in such objects when we look at them through transparent media.

[2] It has been claimed earlier that this sort of bending of the visual ray does not occur [the same way] in all liquids and rare media; what happens, instead, is that in each one of these [media] the amount of deflection is determined solely by the way in which the medium allows penetration. It has also been claimed that the visual ray radiates rectilinearly, and such rays break only because of an impedance posed by the surfaces separating media of different consistency. It has also been claimed that refraction occurs not only in the passage from rarer and more tenuous to denser media—as happens in the case of reflections—but also in the passage from a denser to a rarer medium. And it has been claimed that this breaking does not take place at equal angles; however, the angles [of incidence and refraction] do bear a certain consistent quantitative relation to one another with respect to the normals.[1]

---

[1]Latin = *sed habent similitudinem quandam et quantitatem que sequitur habitudinem perpendicularium*. This paragraph is rife with implications. First, it is evident that Ptolemy sees a

[3] At this point we ought to investigate the quantitative relationship between the angles [of incidence and refraction] according to specific intervals. But we should start by discussing the phenomena that such refractions have in common with reflections. First, in either case, whatever is seen appears along the continuation of the incident ray—i.e., along the continuation of the ray that emanates from the eye to the surface at which it is broken—[second, the object appears] on the straight line dropped perpendicularly from the visible object to the surface where the breaking occurs. It therefore follows that, just as was the case for mirrors, so in this case, the plane containing the broken ray-couple must be perpendicular to the surface where the breaking occurs.[2]

[4] We have already shown in the place where we laid out the principles governing mirrors that the above points are in the nature of observable phenomena and that what happens [in the breaking of rays] is quantifiable.[3]

[5] That this is clear and indubitable we can understand on its own terms by means of a coin that is placed in a vessel called a *baptistir*.[4] For, if the eye remains fixed so that the visual ray passing over the lip of the vessel passes above the coin, and if water is then poured slowly into the vessel until the ray that passes over the edge of the vessel is refracted toward the interior to fall on the coin, then objects that were invisible before are seen along a straight line extended from the eye to a point higher than the true point [at which the coin lies].[5] And it will be supposed not that the

---

fundamental relationship (direct proportionality?) between the density of a given transparent medium and the amount of impedance it poses to the passage of visual rays. Second, Ptolemy seems to believe that the amount of deflection or refraction of such rays is fundamentally related (in direct proportionality?) to the amount of impedance posed by the given medium. Thus, we are led to conclude a fundamental relationship (direct proportionality?) between density and amount of refraction. The problem, of course, is how to gauge "density" in this account. Should we somehow take it as physical, so that it measures the relative amount of matter contained in a given volume? If so, then what is the relative density, for instance, of glass to water? Worse, why are so many opaque objects less dense than transparent ones? In view of these kinds of issues, we should be chary of assuming that Ptolemy has a true model for refraction in mind. Perhaps all he means to convey is a vague sort of analogy. For previous references to density as an optical factor, see paragraphs II, 19 and 25 above; see also V, 33 below.

[2]See III, 5 above.

[3]See III, 14–17 above; suffice it to say, the implication is that there exists a quantitative relationship between the angles of incidence and refraction that is analogous to that between the angles of incidence and reflection. It will not, of course, be one of equality, but it will presumably be simple and straightforward.

[4]See n. 17 to the Preface. From later discussion in book V, we know that the *baptistir* to which Ptolemy refers consists of a hollow semicylinder whose ends are closed off so that it can hold water.

[5]As Lejeune remarks in *L'Optique*, p. 225, n. 9, this "experiment" harks back to at least the third century B.C.; see, e.g., Euclid, *Catoptrica*, definition 6.

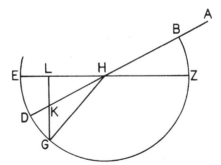

Figure V.1

ray is refracted toward those lower objects but, rather, that the objects themselves are floating and are raised up to [meet] the ray. For this reason, such objects will be seen along the continuation of the [incident] visual ray, as well as along the normal dropped [from the visible object] to the water's surface—all according to the principles we have previously established.[6]

[6] *[EXAMPLE V.1]* Now, let us suppose that point **A** [in figure V.1] is the eye, **ZHE** the common section of the plane containing the refracted ray-couple and the surface [of the water] in the vessel, and **ABD** the ray passing over the vessel's lip at **B**. Let us also suppose that there is a coin at **G**, which lies toward the bottom of the vessel. Then, as long as the vessel remains empty, the coin will not be seen, because the body of the apparatus at **B** blocks the visual ray that could proceed directly to the coin. Yet, when just enough water is poured into the vessel so that its surface reaches line **ZHE**, ray **ABH** is deflected along line **GH**, compared to which **AH**[**D**] is higher. In that case, then, the coin will appear to be located along the cathetus dropped from point **G** to **EH**—i.e., cathetus **LKG**, which intersects line **AHD** at point **K**. Moreover, its image-location will lie on the radial line passing from the eye and continuing rectilinearly to point **K**, that radial line being higher than the actual ray [**HG**] and nearer the water's surface; so the image will appear at point **K**.

[7] The amount that the ray is refracted in water below the [original line of] sight is determined according to the following experiment,

―――――――――

[6]Thus, image-location in refraction is determined in precisely the same way as it is determined in reflection; see III, 5 above. In fact, Ptolemy's determination of image-location in refraction is incorrect, and it ultimately leads him into false conclusions about image-distortion later on in the book; see Lejeune, *Recherches*, pp. 167–169 for an analysis of Ptolemy's "laws" of refraction.

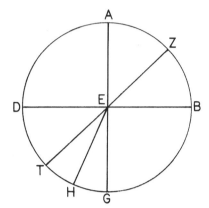

Figure V.2

which is conducted by means of the bronze plaque that we constructed for analysing the phenomena of mirrors.[7]

[8] *[EXPERIMENT V.1]* Let circle **ABGD** [in figure V.2] be described on that plaque about centerpoint **E**, and let the two diameters **AEG** and **BED** intersect one another at right angles. Let each of
the [resulting] quadrants be divided into 90 equal increments. At
the centerpoint let a small marker of some color or other be attached, and let the plaque be stood upright in the small vessel [dis-
cussed in the previous experiment]. Then let a suitable amount of
water that is clear enough to be seen through be poured into that
vessel, and let the graduated plaque be placed erect at right angles
to the surface of the water. Let all of semicircle **BGD** of the plaque,
but nothing beyond that, lie under water, so that diameter **AEG** is
normal to the water's surface. From point **A**, let a given arc **AZ** be
marked off on either of the two quadrants that lie above the water.
Furthermore, let a small, colored marker be placed at **Z**.

[9] Now, if we line up both markers at **Z** and **E** along a line of sight
from either eye so that they appear to coincide, and if we then move
a small, thin peg along the opposite arc **GD** under water until the
end of the peg, which lies upon that opposite arc, appears to lie directly in line with the two previous markers, and if we mark off the
portion of the arc **GH** that lies between **G** and the point at which the
object would appear unrefracted, the resulting arc will always turn
out to be smaller than **AZ**. Moreover, if we join lines **ZE** and **EH**, angle **AEZ** > angle **GEH**, which cannot be the case unless there is refraction—that is, unless ray **ZE** is refracted toward **H** according to
the excess of one of the opposite angles over the other.

---

[7]See III, 8 above.

[10] Furthermore, if we place our line of sight along normal **AE**, we will find the image directly opposite along its rectilinear continuation, which will extend to **G**; and this [radial line] undergoes no refraction.

[11] In the case of all the remaining positions, when arc **AZ** is increased, arc **GH** in turn will be increased, and the refraction will be greater. When arc **AZ** is 10 degrees out of the 90 into which quadrant [**AB**] is divided, then arc **GH** will be around 8 degrees.[8] When **AZ** is 20 degrees, then **GH** will be 15.5. When **AZ** is 30 [degrees], then **GH** will be 22.5. When **AZ** is 40 [degrees], then **GH** will be 29. When **AZ** is 50 [degrees], then **GH** will be 35. When **AZ** is 60 [degrees], then **GH** will be 40.5. When **AZ** is 70 [degrees], then **GH** will be 45.5. And when **AZ** is 80 [degrees], then **GH** will be 50 [see Table V.1 below for a synopsis].[9]

TABLE V.1

AIR TO WATER

incidence	refraction
10	8.0
20	15.5
30	22.5
40	28.0
50	35.0
60	40.5
70	45.5
80	50.0

---

[8]As Lejeune notes in *L'Optique*, p. 229, n. 14, the qualifier "around" (*ad prope*) in this context suggests that Ptolemy was self-consciously rounding his results off—to the nearest half degree as it turns out; see also *Recherches*, pp. 159–163. Quite clearly, the bronze plaque used in this experiment is modeled after equivalent sighting-instruments that Ptolemy applied to astronomy (e.g., the graduated disk for the meridianal armillary described in *Almagest*, I, 12). If, as Lejeune claims, Ptolemy was able to calibrate such astronomical disks to one-sixth of a degree, then the plaque he used in his refraction-experiments must have been considerably smaller, no doubt to accommodate the limitations of the *baptistir*; see *L'Optique*, p. 227, n. 12, and *Recherches*, pp. 156–157. For a discussion of the later Greek attitude toward observational accuracy, see G. E. R. Lloyd, "Observational error in later Greek science," in Jonathan Barnes, et al., eds., *Science and Speculation: Studies in Hellenistic theory and practice* (Cambridge: Cambridge U. Press, 1983), pp. 128–164.

[9]The half-degree intervals that I have given decimally are expressed verbally in the Latin, so that my rendering of "15.5 degrees" translates "quindecim partium et dimidie," and so forth. Ptolemy, however, would have tabulated the results in an entirely different way, according to the sexagesimal format used by him in the *Almagest*. On that basis, the same "15.5" tabulation would have taken the form "15 30" (or "15;30" with modern notation). Given the evident limits of observational accuracy, to within around half a degree, we can draw some general conclusions about Ptolemy's procedure in drawing up this and the subsequent two tables of refraction. First, in comparison to the proper values for the angles of refraction (r) as computed on the basis of the sine law, Ptolemy's values are relatively accurate for all values of i except for the last value—i.e., $i = 80$. Second, the structure of the tables indicates that, along with recognizing the inexactness of his brute observations, Ptolemy had a specific algorithm in mind: namely, that the progression of values for r depends on constantly decreasing increments, the rate of decrease being constant at half a degree. Thus, if r varied constantly with i, then the progression of values would be as follows: when $i = 10$, $r = 8$; when $i = 20$, $r = 16$; when $i = 30$, $r = 24$, etc., the difference in successive values being 8. As

[12] We find the amounts of refraction created in water to be just as shown on the understanding that there is no sensible difference in density and rarity among [various] waters.[10]

[13] Now, if we look toward a rarer medium from a dense one, such as plain water, there will appear a considerable alteration in the difference between the angles as well as in the amount of angular deflection that occurs in the course of the ray's passage from water, which is denser, to the rarer medium. However, since it is impossible for us on the basis of the previously-described experiment to gauge the refraction produced when the ray passes from a denser to a rarer liquid, we have taken it upon ourselves to analyse the relationship of angles as follows.[11]

[14] *[EXPERIMENT V.2]* Let a clear glass semicylinder, represented by arc **TKL** [in figure V.3] be made in accordance with the semicircular section of the round plaque, but let its diameter be smaller than the diameter of the aforementioned bronze plaque. Then let its base be attached to the plaque so that both are completely joined. Let its center be **E**, and let its diameter **TL** coincide with [the plaque's] diameter **BD**, and let **AE** be perpendicular to the flat surface of the glass. Therefore, every line drawn from point **E** to arc **BGD** and to arc **TKL** will be normal [to those arcs].

[15] Accordingly, if we set up this experiment as we did before, and if we make a small mark on the midpoint of the semicylinder's sur-

---

tabulated by Ptolemy, however, the successive values are continually decreased by half a degree so that the resulting values for r become 8; $8 + 7.5$ ($= 8 - .5+$) $= 15.5$; $15.5 + 7$ $(7.5 - .5) = 22.5$, etc. This algorithm according to which a given quantity varies in relation to another by some given amount according to constant "second differences" was routinely used in Babylonian astronomy for the creation of ephemerides. For detailed analyses of Ptolemy's tabulations and the method that underlies them, see Albert Lejeune, "Les tables de réfraction de Ptolémée," *Annales de la Société Scientifique de Bruxelles*, series 1, fasc. 1 (1946): 93–101; and A. Mark Smith, "Ptolemy's Search for a Law of Refraction: A Case-Study in the Classical Methodology of 'Saving the Appearances' and its Limitations," *Archive for History of Exact Sciences* 26 (1982): 221–240.

[10]According to Lejeune, *L'Optique*, p. 230, n. 16, and *Recherches*, differences in density among various "waters" (e.g., seawater vs. rainwater) had been recognized long before Ptolemy's time. Thus, Ptolemy was presumably testing various waters to see whether these differences in density had any significant effect on refractivity.

[11]In *L'Optique*, p. 230, n. 17, Lejeune notes the practical problems involved with trying to measure refraction from water to air: i.e., either the eye must be placed directly into the water or the *baptistir* must be manufactured to unfeasible specifications; cf., however, n. 13 below. Moreover, as Lejeune later observes (p. 233, n. 22,), even if Ptolemy could have overcome these problems, he would still have been faced with the problem of critical angle—i.e., internal reflection from the surface of refraction when looking through the interface from an optically denser to an optically rarer medium. Accordingly, the range of possible tabulations according to 10-degree intervals for i would have been severely limited.

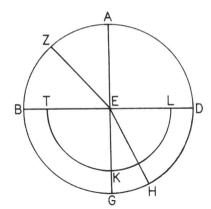

Figure V.3

face where it[s axis] touches point **E**, and if we look with either eye along line **AE** toward the edge of the glass and move a marker on the arc [**BGD**] opposite this arc [**BAD**] until it appears in front of it, it will be found to lie on **G** itself. For line **AEG** is normal to both **TEL** and **TKL**. And if we move our eye until it lies directly in line with this position, and if we look along line **GE** so that the marker that has been moved along the arc lies directly in line with **GE**, then that marker will be situated on line **EA**. For the same reason, moreover, there will be no refraction in the passing of the ray [orthogonally] from glass into air.

[16] But if we take some given arc **AZ** from point **A** and draw line **ZE**, coloring it black, and then if we sight along this line until the marker, which is moved behind the glass, appears to fall in line with it, and if we mark the place—e.g., point **H**—where we found it so that the black color coincides with **EH**, then we will also find in this case that angle **AEZ** > angle **GEH**. We will also find that the angular difference is greater than the angular difference in water, where the arc [measuring incidence] was the same.

[17] If, moreover, we station our eye at point **H**, which is opposite point **E**, and sight from point **H** along **HE**, [both of the points] **E** and **Z** will appear to coincide on one and the same line of sight. And since there appears to be a refraction of the ray in this situation, it is necessary that, whether the ray passes from air into glass, as represented by **ZE**, and is refracted along **EH**, or whether it passes from glass to air, as represented by **HE**, and is refracted along **ZE**, the refraction takes place toward **T**.[12] And since the normals dropped

---

[12]Here, for the first time, Ptolemy explicitly articulates the principle of reciprocity: i.e., no matter the direction of passage in refraction (or, for that matter, reflection), the ray-couple remains fixed, with the incident and refracted rays interchangeable. See V, 32, below, for a restatement.

from **E** to **TKL** are the same, rays [that pass along them] are not refracted, whether they pass from **E** to **K** or from **K** to **E**.

[18] In addition, if we now analyze the amount of refraction for
each of [the previous angular] positions, we will find that, when the
eye is placed at the same angular distances as before and when the
angle measured from point **E** (i.e., the angle [of incidence] formed
by normal **AE** and ray **EZ**) is 10 degrees of the 90 ascribed to the
circle's quadrant, then the [resulting] angle [of refraction] **GEH** will
measure nearly 7 degrees. When the first angle is 20 degrees, the
second will be 13.5. When the first is 30 [degrees], the second will
be 19.5. When the first is 40 [degrees], the second will be 25. When
the first is 50 [degrees], the second will be 30. When the first is 60
[degrees], the second will be 34.5. When the first is 70 [degrees], the
second will be 38.5. And when the first is 80 [degrees], the second
will be 42 [see Table V.2 below for a synopsis].

TABLE V.2

AIR TO GLASS	
incidence	refraction
10	7.0
20	13.5
30	19.5
40	25.0
50	30.0
60	34.5
70	38.5
80	42.0

[19] But if the glass is placed in contact with water, the refractions will turn out to be smaller, because the difference in refractive effects between these media is not large. In fact, the difference
in density between water and glass is less than that between air
and either of these two media. But it is possible in this case as well
for us to ascertain the amount of refraction in the way we have
already discussed.

[20] *[EXPERIMENT V.3]* Let the glass semicylinder [in figure
V.4] be attached to the bronze plaque and let it be set up so that its
center coincides with the center of the plaque. Let point **E** be colored
somewhat, and let the plaque be set up in the basin at right angles
to the water's surface, which coincides with the plaque's [horizon-
tal] diameter, and let the curved surface **TKL** of the glass [semi-
cylinder] be arranged such that it lies above [the water's surface].
Then, let water be poured into the basin just until line **TEL** of the
semicylinder coincides with the water's surface. Within the rarer
medium (i.e., water) let some arc **GH** be marked off, and again let it
be 10 degrees. Let some thin colored marker be placed at **H**, and

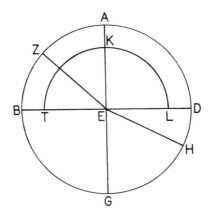

Figure V.4

sight with either eye [along **HE**] until marker **Z**, which is moved on arc **AB**, appears to lie in line with point **H** and the colored spot at **E**. With things so disposed, let the two lines **EH** and **EZ** be drawn.[13]

[21] If, then, we take the angle formed in the denser medium (i.e., in glass), that angle being designated on arc **AB**, we will find by means of this experiment that, when the angle in water as measured from the normal—namely, angle **GEH**—is 10 degrees of the 90 ascribed to a right angle, the [resulting] angle in the glass—namely, angle **AEZ**—is nearly 9.5. When the angle in water is 20 [degrees], the angle in glass will be 18.5. When the former is 30 [degrees], the latter will be 27. When the former is 40 [degrees], the latter will be 35. When the former is 50 [degrees], the latter will be 42.5. When the former is 60 [degrees], the latter will be 49.5. When the former is 70 [degrees], the latter will be 56. And when the former is 80 [degrees], the latter will be 62 [see Table V.3 below for a synopsis].

[22] At this point, then, we should state a general principle, just as we did in the case of direct vision: namely, objects that are said to appear along a single line are seen by one and the same visual ray. But we must understand that this is ideally, not [always] actually, the case. Indeed, objects that initially block the visual flux prohibit it from reaching objects that lie behind them. For, because the

---

[13]Lejeune assumes that, given the material constraints discussed in n. 11 above, Ptolemy must actually have reversed his observational procedure in this case: i.e., he must have sighted from the glass into the water; see *L'Optique*, p. 236, n. 24, and *Recherches*, pp. 163–166. Nevertheless, the fashioning of the requisite semicylindrical glass container may have been within the competence of Alexandrian glassmakers. Among the techniques they had mastered were moulding and shaping, carving and incising, and free-blowing and mold-blowing. In this case, of course, some form of blowing, whether free or into a mold, would have been necessary; for further details of these techniques, see Jennifer Price, "Glass," in Donald Strong and David Brown, eds., *Roman Crafts*, (N.Y.: NYU Press, 1976), pp. 110–125.

TABLE V.3

WATER TO GLASS

incidence	refraction
10	9.5
20	18.5
30	27.0
40	35.0
50	42.5
60	49.5
70	56.0
80	62.0

visual flux will not have penetrated the first objects, those objects that lie behind them are not seen by the aforesaid rays. It is therefore evident that they are not seen by any of the neighboring rays. It will also be evident that, since those [subsequent objects] are not seen by the same visual flux that apprehends the first object, assuming it is not blocked, the [screening and screened] objects will be in line with one another. For all of them are aligned along the visual ray itself.[14]

[23] Furthermore, it is possible for us to realize that, at the surface between air and ether, there is a refraction of the visual ray according to the difference in density between these two media. The resulting phenomena are as follows:

[24] We notice that [celestial] bodies that rise and set tend to incline toward the north when they are near the horizon and are measured by an instrument for measuring the stars.[15] For, when they lie to the east or west [i.e., at rising or setting], the circles drawn through them parallel to the equator are nearer to the north than the circles drawn through them when they are in the middle of the sky, and the more they approach the horizon, the more they are inclined to the north. Moreover, the distance from the north pole of the stars that are always visible [i.e., that do not set] will always be less when they lie on the meridian toward the horizon. For when they are on the meridian in a location that is closer to the zenith, the circle at that location that is parallel to the equator becomes larger, whereas in the previous position it becomes smaller.[16] This is a result of the re-

---

[14]This paragraph seems to be misplaced, having little or no apparent connection with the foregoing analysis of refraction. In fact, its evident purport is to justify the assumption that visual flux radiates in perfectly straight lines.

[15]Presumably what is being referred to here (by Ptolemy or one of his translators, whether Arabic or Latin), is an astrolabe, or something akin to it, that measures the latitude of celestial bodies; see Lejeune, *L'Optique*, p. 238, n. 27.

[16]According to Ptolemy, then, the apparent latitude of the stars changes constantly as they rise higher in the sky toward the zenith in their nocturnal round, even though their actual latitude remains constant. For instance, let line **EW** in figure 8 represent the horizon-line, with **E** marking the eastward side and **W** the westward side. Let the dotted line pointing toward

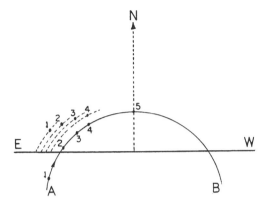

Figure 8

fraction of the visual flux at the surface that separates the air and the ether, a surface that must be spherical, its center being the common center of all the elements, which is the center of the earth.

[25] *[THEOREM V.1]* First, let **E** [in figure V.5] represent the viewer's zenith, and let one of the great circles on the spheres of the aforesaid elements cut the earth along **AB**. Let **GD** be a great circle on the interface between the air and ether, and let **EZ** be a great circle passing through a given star, and let the center of all the circles be point **H**. Let line **EAH** be drawn. Let point **A** be the eye and line **ADZ** a line coinciding with the common section of the [plane of the] horizon and circle **GD**. In addition, let **DT** be normal to the circle[s]. Let us suppose that **ADK** is a visual ray refracted along **KD** at point **D**, and let the star lie at point **K**.

[26] Since the visual ray is refracted at the interface toward a position away from point **E** according to the normal dropped at equal angles to the refracting surface, angle **KDT** [of refraction], which lies in the subtler medium, will be greater [than angle **ADB** of incidence]. Thus, the star will be seen from point **A** along line **ADZ**, and

---

N (= North) represent the meridian line, and let arc **AB** represent the path of a given star. Thus, according to Ptolemy's account, while the star is still below the horizon at position 1, it will appear above the horizon at counter-position 1 because of atmospheric refraction, which causes the star, being in a rarer medium (i.e., ether), to appear to "float" above its actual position (see V, 5 above). Thus, the projection of its apparent latitude or orbit (along the dotted line) describes a circle lying above—i.e., to the north of—its actual circle of latitude. Likewise, as it progresses along its actual orbit from 2 to 5, refraction causes the apparent position to lie farther north than the real one, but the amount of refraction diminishes continually as it rises higher in the sky. Thus, the dotted circles of apparent latitude or orbit constantly approach the actual circle of latitude or orbit until the star reaches zenith (point 5), where no refraction occurs and where the apparent orbit has reached as far south as possible. Then, and only then, will the projected circle of latitude or orbit coincide with the actual one. Moreover, the farther north the circles of latitude lie, the smaller they necessarily are in comparison to the great circle of the equator.

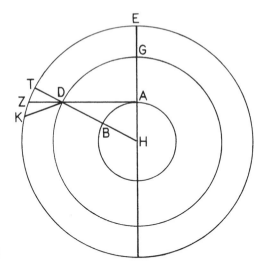

Figure V.5

its [apparent] distance from the zenith will be less than its true distance, for it will be seen along arcal distance **EZ** instead of arcal distance **KE**. Therefore, the higher its distance [toward the zenith], the
smaller the difference between the star's apparent and true location.
And if the star is at **E**, there will be no refraction, because there is no
breaking of the visual ray that passes from point **A** to point **E**, for in
that case it will be normal to the surface of refraction.

[27] *[THEOREM V.2]* With these points established, let **ABG** [in
figure V.6] represent the circle of the horizon and **AEZG** the semicircular arc of the meridian that lies above the earth. Let point **E** represent the zenith and point **Z** the apparent pole of the heavenly
sphere [of fixed stars]. Let **BHD** be the arc of a line above the earth's
surface that is parallel to the equator and that passes through certain stars.[17] Let **T** be a star that lies on this line near the horizon, and
let **KETL** be the semicircular arc of the circle that passes above the
earth through the zenith and through star **T**.

[28] Accordingly, since the star appears to lie nearer the zenith than
it truly is when it is near the horizon, and since the divergence in its
apparent from its true position is measured on the great circles passing through the points on the horizon, the point where the star that
is [actually] at **T** appears will lie between **E** and **T**, such that it appears at point **M**. Moreover, the line parallel to the equator and passing through point **M** will lie higher to the north than the line parallel
to the equator and passing through point **T**, which at our particular

---

[17]In short, **BHD** represents a line of latitude.

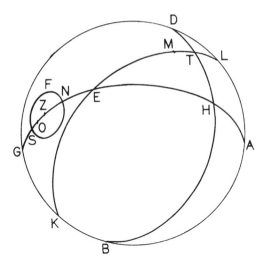

Figure V.6

latitude is inclined toward the north. And when the star rises to position **H**, it reaches a point where the visual ray is refracted without any perceptible difference between apparent and true location.[18]

[29] Likewise, if we suppose **Z** to represent the north pole and if we draw one of the circles parallel to the equator that is always visible [at our latitude], e.g., circle **NSF**, then, when the star lies at point **S** on this circle, it will appear closer to point **E**, which lies at the zenith, and it will seem to lie at point **O**. But when the star is at point **N**, then there is no difference, or only an imperceptible one, between apparent and true location.[19] And therefore, when a star approaches the horizon in its revolution, its distance from the north pole of the [celestial] sphere appears to be smaller [than it really is]; but when it approaches the zenith in the course of its revolution, that apparent distance seems larger, for arc **ZN** will be larger than arc **OZ**.

[30] It has thus been demonstrated how stellar observation must be affected by the refraction of the visual ray. It would also be possible for us not only to examine the degree of such refractions, but also to analyze such refraction in the case of certain [celestial] bodies whose distance is given—e.g., the sun and the moon—and to determine the degrees [of refraction] toward the horizon as well as the amount by which the refraction of the visual ray shifts the apparent

---

[18]In other words, point **M** represents one of the intermediate positions (2–4) of the star in figure 6, note 16 above, whereas **H** represents position 5 in that figure.

[19]That is, when the star is at **S** and therefore closest to the horizon in relation to zenith-point **E**, the ray that reaches it will be maximally refracted. Consequently, the apparent location of the star will lie as far northward as possible. On the other hand, when the star is at **N**, it is closest to zenith-point **E** and, therefore, to the point at which refraction is minimal. Hence, its northward shift is as minimal as possible.

position upward if the distance of the interface between the two media [i.e., air and ether] were known.[20] But, although this distance lies
nearer than the earth to the lunar sphere, where the ether stops, it is
not known whether the [refractive] interface lies at the same distance as the aforesaid surface, or whether it lies nearer the earth, or
whether it lies farther from [the aforesaid] surface.[21] Therefore, it is
impossible to provide a method for determining the size of the angles of deviation that occur in this sort of refraction.[22]

[31] It is, however, possible to formulate in such a way a general
claim about refraction on the basis of previously established points.[23]
We put it thus: the amount of refraction is the same whichever the
direction of passage; the difference is one of kind [rather than of de-
gree]. For in passing from the rarer to the denser medium, the ray inclines toward the normal, whereas in passing from the denser to the
rarer medium, it inclines away from the normal.[24]

[32] *[EXPERIMENT V.4]* In fact, if we set up the plaque as before
and assume that diameter **BD** [in figure V.7] lies on the interface
between the two different media, and if we draw normal **AEG** as
well as the refracted ray-couple **ZEH** inclined toward the normal,

---

[20]The amount of refractive displacement in this case will depend upon two basic factors:
1) the actual refractive power of the denser medium (air) relative to that of the rarer one
(ether), and 2) the depth of the denser medium. Ptolemy seems to be citing the distance of the
bodies in question as another factor, but in fact, the actual eye-to-object distance has virtually
no effect on refractive displacement (perhaps by "distance" he means actual latitudinal distance with respect to the equator at any given point along the known orbit of the body). All
of this can be graphically explained by recourse to figure 9, which is adapted from figure V.5.
Let arc **ABB'** represent an arc on the earth's surface, with **H** as its center and **A** as the viewpoint. Let the two arcs through **D** and **D'** represent two possible interfaces between air and
ether, and let **HBDT** and **HB'D'T'** be the respective normals to those interfaces. Finally, let
ADD'Z represent the horizontal line-of-sight. From the diagram, then, it is clear that, for the
nearer interface, a given body **K** will be displaced upward by refraction farther than a body
K' viewed through the farther interface. It is also clear that any body along line **DK** will have
the same angular displacement, no matter its actual distance along that line. Note, finally, that
the farther out the refractive interface lies, the smaller the amount of refraction; indeed, there
is a point at which the refraction is minimized to imperceptibility.
[21]As Lejeune notes in *L'Optique*, p. 242, n. 28, this passage is confusing, because it seems to
imply the possibility that the air-ether interface might lie beyond the lunar sphere, which is
a patent absurdity given his earlier recognition that the moon is one of the bodies of known
distance whose apparent location is affected by refraction. According to Aristotelian cosmology, however, the air-ether interface cannot possibly lie below the lunar sphere, since that
sphere defines the lowest limit for ether. Also, according to that same cosmology, there is a
sphere of fire separating air from ether, yet that sphere goes unmentioned.
[22]Knorr, "Archimedes," pp. 97–99, suggests that, despite the cited imponderables, Ptolemy
not only could have, but should have, found a way to estimate atmospheric refraction with
reasonable accuracy.
[23]The next seven lines of text in the critical Latin edition have been transposed to V, 75 below where they seem to belong; see Lejeune, *L'Optique*, p. 243, n. 30, for an explanation.
[24]Here, and in the following paragraph, Ptolemy is merely generalizing the claim about reciprocity made in V, 17 above.

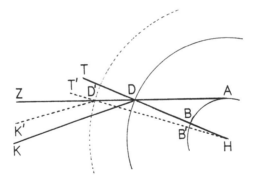

Figure 9

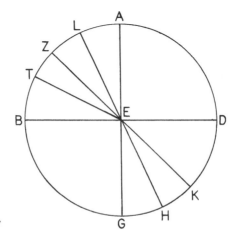

Figure V.7

with which it forms angle **GEH**, then the path of refraction remains one and the same. In fact, when the visual ray passes through point **E**, and the eye is stationed at point **Z**, the line [of sight] after refraction—i.e., line **EK**—inclines toward the normal according to its continued passage [along **EH**] while the visible object is seen along the rectilinear continuation [**EK**]. But if the eye lies at point **H**, and **EZ** lies within the rarer medium delimited by **ABD**, then, after refraction, line **EL** will take an opposite tack from that previously specified, inclining away from normal **AE** [along **EZ**] in such a way that it lies farther out [from the normal] than would the visual ray if it were to continue in a straight line.

[33] Furthermore, when the media and the angles differ from one another by a significant amount, the difference [between the angles of incidence and refraction] increases as the density of either of the media grows. Indeed, if we assume that arc **BAD** lies in the rarer medium and arc **BGD** in the denser, and if we take angle **AEZ** as it is represented, then, when the medium within section **BGD**

becomes denser than it previously was, the difference between [angle **AEZ** and] angle **GEH** will vary with the difference in density between the two media. In fact, when angle **AEZ** is 30 degrees in air, angle **GEH** in water will be nearly 22.5 degrees, whereas in glass it will be nearly 19.5 degrees.[25] And in this latter case the refraction and difference in angles measured from the point at the top [of the normal] will be greater, because the substance of glass is denser than the substance of water.

[34] So too, if we suppose that the refraction of another of the visual rays takes place at some other arcal distance [than **AZ**] from normal **AE**—e.g., along ray-couple **TEK**—then $AT : AZ > GK : HG$. By alternation, $AT : GK > AZ : GH$. By separation, $TZ : AZ > KH : GH$. And [by alternation,] $TZ : KH > AZ : GH$. Furthermore, we can determine particular cases on the basis of the refractions as we measured them if we take the resultant numbers and, on their basis, investigate particular measurements of this sort, substituting the numbers so derived for the two arcs **AZ** and **AT**.[26]

[35] But someone might object by demanding an explanation for why, in light of the first principles laid out concerning normals and the appearance of the visible object along the straight continuation of the [incident] visual ray, the sort of breaking that we have discussed is similar to the breaking that occurs in mirrors, whereas the angular relationships are not, for in the case of refraction the equality of angles is not conserved. It will be seen in response that this must be the case according to what we have explained, on which basis an even more marvelous fact will be apparent: namely, the course of nature in conserving the exercise of power.[27]

[36] In fact, the distinction [between refraction and reflection] consists in the fact that, in the case of refraction the visual ray continues in the direction of the surface at which the breaking occurs, whereas in the case of reflection, the visual ray continues away from [the direction of] that surface. It therefore follows, it seems to

---

[25]All of these angles are expressed in the Latin text as fractions of a right angle. Thus, 30 degrees = "a third part of a right angle"; 22.5 degrees = "a fourth part of a right angle"; and 19.5 degrees = "a fifth part plus a sixtieth part of a right angle."

[26]Although there is a specific mathematical law implicit in Ptolemy's tabulations, its proper formulation (in algebraic terms) would have been beyond Ptolemy given the mathematical techniques of his day. Accordingly, he was forced to fall back on the weak generalization expressed here: that the greater the angle of incidence, the greater the ratio of i to r and, thus, the greater the difference between i and r. Or, to put it another way, i and r approach one another as the incident ray approaches the normal. For a detailed analysis, see Smith, "Ptolemy's Search," especially pp. 235–238.

[27]Unfortunately, this extraordinarily suggestive point is not followed up in the remainder of the text as we now have it. It is tempting to think that Ptolemy has in mind some sort of principle of conservation, either dynamic or kinematic, that governs the angular relationship between i and r.

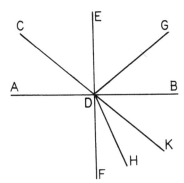

Figure 10

us, that refraction involves a certain breaking with respect to the normals, whereas reflection entails a more pronounced breaking. In both cases, each of the visual rays must continue to move along a straight line according to the portion of a right angle it cuts off [with respect to the normal]. But in such cases, a significant breaking is involved so that the ray does not preserve its [original] disposition. In the case of reflections, it is possible for the rays to continue on their course as they move upward [from the reflecting surface], but in refraction this is not the case, because it is impossible for the breaking to preserve the same relative disposition [between incident and refracted rays]. It necessarily follows, then, that such breaking in refraction is greater when the arcal distances from the normals are greater, whereas in reflection it is smaller under the same conditions.[28]

[37] Furthermore, in the case of reflection, the size of the angle formed by the normal and the line of incidence is the same as that of the angle formed by the normal and the line of reflection. It follows, therefore, that in the case of refraction the angle formed by the normal and the line of incidence is unequal to the angle formed by the normal and the line of refraction passing through the interface [of the refracting media].

[38] *[THEOREM V.3]* This point is illustrated as follows: let us take a surface at which refraction takes place, and let us suppose common section **ABG** [in figure V.8] formed by it and the plane

---

[28]The following interpretation of this passage is adapted from Lejeune's account in *L'Optique*, pp. 246–247, n. 42. Let **ADB** in figure 10 represent the interface of reflection/refraction, **EDF** the normal to the point of reflection/refraction, and **CD** the incident ray. Thus, in reflection, the breakage (along **DG**) at equal angles is more radical than that in refraction (along **DH**) at unequal angles. However, with respect to the continuation of the line-of-sight along **DK**, the divergence in reflection (according to angle **KDG**) decreases as the angle of incidence (**CDE**) increases. On the other hand, in refraction, as the angle of incidence increases, the divergence with respect to the line of sight (according to angle **HDK**) increases.

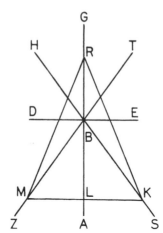

Figure V.8

within which the visual ray is refracted. Let **DBE** be normal to that common section, let **ZB** be an [incident] visual ray, and let ray-couple **ZB**, **BH** be reflected at equal angles, with **DB** representing the normal and **ZB** the oblique. Thus, angle **DBZ** = angle **DBH**, and angle **ABD** = angle **GBD**.

[39] In the case of refraction, if a given normal to **AB** is drawn along perpendicular **DE**, then angle **ZBD** is formed at point **B** from incident ray **ZB** (assuming it is oblique) and line **DBE**. Thus, **TB** will lie along the rectilinear extension of **ZB**, while **BS** will lie along the rectilinear extension of **BH**, and only under this condition will each of the angles **TBE** and **EBS** be equal to angle **ZBD**.

[40] But if the [visual ray's] passage were along **ZBT**, there would be no refraction whatever, since the opposite angles are equal.

[41] If, however, the passage were to occur along **BS**, then, in the first place, there would be no difference between the refraction that takes place within a rarer medium and that which takes place within a denser one. In fact, such [a symmetrical] refraction occurs only along the perpendicular, or as if what lies orthogonally above were inclined in both cases, which is not so by necessity.[29]

[42] In addition, the intersection of the catheti dropped from each of those [points on **BS**] with the continuation of the [incident] rays by means of which we locate the image of the visible object would always lie outside that [refracting] medium; but we have

---

[29]Latin = *sive quasi sursum recte sit declinans in utraque specie, quod non est necessario*; given the incoherence of this phrase within the context, Lejeune opines in *L'Optique*, p. 249, that it was corrupted in meaning by the Arabic translator. According to Lejeune's reconstruction, the phrase ought to read something like "the broken ray would always incline in the same direction in both kinds of breaking, which is false."

found that the [refracting] medium contains the image. For the cathetus dropped from any of the points on **BS** to **BA**—e.g., cathetus **KLM**—intersects **BZ**, as is always the case in the propagation of a visual ray.[30]

[43] Besides, it would also happen that every visual ray emanating from point **M** would invariably reach a single point whose distance from the surface of refraction is equal to the distance of the eye along the normal dropped from it [to that surface]—e.g., distance **KL**—when the eye is placed at point **M** along the visual ray. Indeed, the lines emanating from point **M** and from point **K** pass through one and the same point on line **AG**, as, for instance, do **MR** and **KR**, so it follows that angles **MRL** and **KRL** are equal, because **ML** and **KL** are equal while the angles at **L** are right. And if refractions were like this, the [refracted rays] would be affected by several magnitudes that lie in that location [at **K**]; and even when the images lie behind the eye, they will appear no different in terms of the nature of the sensation.[31]

[44] Even if refraction did not actually maintain these equal angles but, rather, the relationship were conserved in terms of the equality of angle **ZBD** to angle **DBH**, a change would be immediately apparent. While neither **DB** nor **AB** is refracted, the subsequent rays soon undergo such significant refraction that they divide angle **ABD** in half and arrange these oblique dispositions to be rectilinear and unbroken [throughout] with respect to the normal.

[45] So too, let us assume that the passage of all the visual rays takes place at the point of refraction along the normal—e.g., along **ZB** and **BE**—and that whatever [ray] strikes point **B** at a different angle [refracts] at an equal one.[32] This assumption will easily be found untenable. For it would then be the case that perpendicular rays and rays that are inclined at any angle [to the surface] would maintain one and the same position when refracted at the impeding surface. Also, the image-locations would always lie in a single plane where the normals dropped from the visible objects necessarily

---

[30]In other words, if the refraction were to take place along line **BS** and the object-point were to lie at **K**, then the image-location would lie at intersection-point **M** of the cathetus of refraction **KLM** and the incident ray **ZB**. But that intersection-point lies above the refracting medium, which is impossible in the case of refraction.

[31]The sense of this passage seems to be that, if refraction did occur according to the pattern of ray-couple **MB**, **BK**, then all possible refracted ray-couples, such as **MR**, **RK** reaching **K** would have a common image-location at **M**. Consequently, no matter what the line of incidence, **K** will always appear at the same spot according to the common image-location, even if that location were to lie behind the center of sight itself.

[32]Evidently, the supposition at play here is that, after refraction at point **B**, the ray will follow normal path **BE**, no matter the angle of incidence.

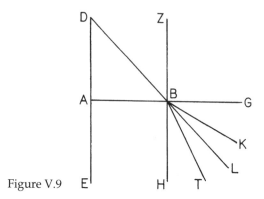

Figure V.9

intersect the visual ray that meets with them, as is the case with those perpendiculars that are dropped to **B** from **SH** and **ZT**.[33]

[46] It has now been shown that there cannot possibly be an equality of angles in the case of refractions. But, on the basis of what we have already established as well as what we have yet to say, it can be understood that the types of angular relationships we have determined [previously by experiment] are subject to none of the problems just set forth. Since we now grasp the requisite principles that govern refraction, there is no need to prolong our discussion. We ought to talk now about the intersection of the refracted ray emitted from the eye and the cathetus dropped from the visible object to the refracting surface, that intersection defining the actual location of each of the images. We ought also to explain what happens in the case of each of the plane figures under examination;[34] and, in order to make what we want to show clear, we ought to begin with objects apprehended by the eye as it is located in the preceding discussion.

[47] *[THEOREM V.4]* To start with, let **ABG** [in figure V.9] represent the common section of the interface of the two media and the plane containing the refracted ray-couple. Let **D** be the eye, **DAE** the cathetus of incidence dropped from the eye [to the interface],

---

[33]The point here seems to be that, if, for example, **MB** were the incident ray and refraction were to follow the normal, then the image-location would necessarily lie at point **B** on the plane of refraction. Precisely what, in this case, it would mean to drop perpendiculars to **B** from **SH** and **ZT** escapes me. Indeed, the entire "theorem" is so marred by obscure or even incoherent assertions and lines of reasoning that I am inclined to believe either that it is not original to Ptolemy or that it has been badly mutilated in the process of translation and/or transmission.

[34]What is intended here by "plane figures" is unclear at best; subsequent context suggests that the figures in question are the basic ones involved in the study of reflection but in this case applied to refractive surfaces: i.e., plane, spherical convex, and spherical concave.

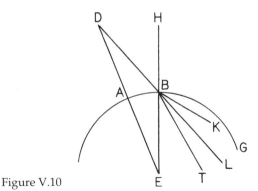

Figure V.10

and **DB** an oblique ray. Let normal **ZBH** be dropped from point **B**. Then, let **DB** be refracted in one of two ways: 1) toward the normal, as is the case when the eye lies within the rarer medium, so that it is refracted along **BT**; or 2) away from the normal, as happens when we look through a denser medium [into a rarer one], so that it is refracted along **KB**.

[48] According to our presuppositions, then, angle **DBZ** will be greater than angle **HBT** and smaller than angle **HBK**. And it is clear that cathetus of incidence **DAE** intersects neither [ray] **BT** nor [ray] **KB** on the side of points **E** and **T**, since the sum of angles **BAE** and **ABT** is greater than two right angles. It is thus even more clearly the case that lines **BK** and **D[A]E** must not meet [on the side of **E** and **T**].[35]

[49] *[THEOREM V.5]* If the interface between the two media is spherical, then let us start by assuming that its convexity faces toward the eye, and let arc **ABG** [in figure V.10] of a circle with center **E** represent the common section of this interface and the plane that contains the refracted ray-couple. Let **D** be the eye, **DAE** the cathetus dropped from the eye to point **E**, and **EBH** the normal to the point of refraction. Let both [of these perpendiculars] be connected by [ray-]line **DB**, and let it be refracted either toward the normal—e.g., [along] **TB**—or away from the normal—e.g., [along] **KB**.

[50] Consequently, the farther **KB** continues, the more it diverges from cathetus **EAD**, whereas **BT** will sometimes be parallel to

---

[35]This theorem offers an elaboration of the point already made in V.42 above that refraction cannot possibly take place within the refracting medium toward the side of the incident ray—i.e., within the area bounded by normal **BH** and cathetus **AE**. This theorem also initiates a series of theorems (V.4–V.8, paragraphs 47–54) to examine whether, and under what conditions, the ray after refraction will meet the cathetus of incidence in the case of plane, spherically convex, and spherically concave interfaces.

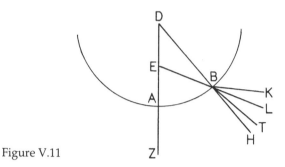

Figure V.11

cathetus **EAD**, sometimes intersect it on the side of **E** and **T**, and sometimes diverge from it. For at times angle **AEB** can be equal to angle **EBT**, so that both lines will be parallel; at times, however, it is larger, so that the two lines intersect; but at times it is smaller, so that they continually diverge.

[51] *[THEOREM V.6]* Let us continue by describing concave interface[36] **ABG** [in figure V.11], whose center is **D** and whose concavity faces the eye. Let the eye at **E** first be placed between the center and the interface's surface. Let cathetus of incidence **EAZ** be drawn, and let **EB** be an oblique ray. Let normal **DB** be dropped [from **D**] to **H**, and let ray **EB** be refracted either toward the normal along **BT** or away from the normal along **BK**.

[52] It is therefore obvious that the more the two rays **BT** and **BK** continue in the direction of **T** and **K**, the more they diverge from **H**, so their distance from cathetus **DAZ** will be greater than that of **BH** [from **DAZ**]. Furthermore, the farther **BH** is continued, the more it diverges from **DAZ**.

[53] *[THEOREM V.7]* Now, let the center [**D**] lie between [the eye at] **E** and arc **AG** [in figure V.12]. Let [ray] **EB** be refracted toward the normal along **BT**, or let it be refracted away from the normal along **BK**.

[54] Thus, the farther **BT** continues in the direction of **T**, the more it diverges from [cathetus] **EDZ**, for angle **DBE** > angle **HBT**. Line **BK**, however, is sometimes parallel to line **EAZ**, sometimes intersects it on the side of points **K** and **Z**, and sometimes diverges away from it. For angle **KBH**, which is greater than angle **DBE**, can sometimes be equal to angle **ADB**, as is the case when both lines [**BK** and

---

[36]Latin = *speculum*; Lejeune, *L'Optique*, p. 253, n. 54, suggests two possibilities for this idiosyncratic use of the term: either Ptolemy himself meant its Greek equivalent (*katoptron*) in the general sense of "surface that causes breaking," or the Arabic translator confused *dioptron* (= refracting body) and *katoptron*.

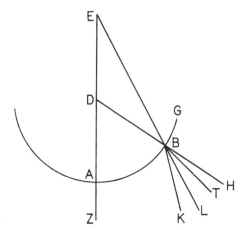

Figure V.12

EAZ] are parallel; sometimes, however, it can be greater, as is the case when the extensions of both lines intersect [toward **Z** and **K**]; and sometimes it can be smaller, as is the case when, the farther both lines are extended, the more they diverge.

[55] It is possible, then, for the cathetus dropped from the eye to the surface of refraction not to meet the refracted ray. This is the case either when the [refracting] surface is convex and its convexity lies, along with the rarer medium, on the side of the eye or when the concave side faces the eye while its center lies between the eye and the [refracting] surface, and the denser medium lies on the side of the eye. Otherwise, [such failure to intersect] can hardly be the case.[37]

[56] At this point, we want to demonstrate how the cathetus dropped from the visible object to the aforesaid [refracting] surface intersects the refracted ray.[38]

[57] *[THEOREM V.8]* To start with, let the [refracting] surface be plane, and let straight line **ABG** [in figure V.13] represent the common section of that surface with the plane containing the refracted ray-couple. Let **D** be the eye and **DBE** an oblique ray. Draw normal **ZBH** to point **B** on **AB**, and let **DB** be refracted toward the normal along **BT**, on the one hand, and away from the normal along **KB**, on the other. Let the visible objects lie at **T** and **K**, and from them let the

---

[37]The fact that the cathetus of incidence and the refracted ray can meet under certain conditions and cannot under others has no practical bearing whatever on the subsequent analysis of image-location in refraction. Other than the desire to be comprehensive, Ptolemy's motivation in demonstrating it so punctiliously is unclear.

[38]In this case, the term "refracted ray" (= *radius flexus*) refers not to the refracted branch of the entire broken ray but, rather, to the line of incidence extended below the refracting surface; see also V, 46 above.

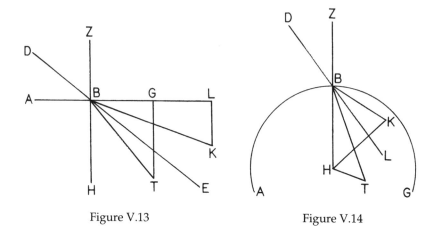

Figure V.13   Figure V.14

two catheti **KL** and **TG** be dropped to **BG**. They invariably intersect line **DB**, for angle **ABE** is larger than a right angle, while the angles at **G** and **L** are right.

[58] *[THEOREM V.9]* Now, let the surface of refraction [in figure V.14] be spherical, [let its center be **H**,] and let its convex surface face the eye. Let **D** be the eye, and let [incident ray] **DBL** be drawn.

[59] Accordingly, catheti, such as **KH**, that are drawn to any of the visible objects placed on **KB** between **Z** and **H**[39] invariably intersect **BL**. However, catheti, such as **TH**, that are drawn to such objects and that intersect **BT** and **ZH**, will sometimes be parallel to **DBL**, sometimes intersect it in the direction of **T** and **L**, and sometimes diverge from it. For it is possible for angle **HTB** to be at times equal to angle **LBT**, at times greater than it, and at times smaller than it.

[60] *[THEOREM V.10]* Now, let the concave surface face the eye, which lies at any one of the two positions [between the center of curvature and the surface or beyond the center of curvature]. Let ray **DBL** [in figures V.15a and V.15b] be drawn, and let the visible objects be assumed to lie on **BT** and **KB**.

[61] According to both positions for the eye, then, the cathetus **TG** dropped from the visible object on **BT** will intersect ray **BL**. But the cathetus **KM** dropped from any of the visible objects on **KB** will sometimes be parallel to line **DBL**, sometimes intersect it on the side of **K** and **L**, and sometimes diverge from it. For angle **KBL** can be sometimes equal to angle **BKE**, sometimes larger than it, and some-

---

[39]This specification, "between **Z** and **H**," is at best irrelevant and at worst incoherent.

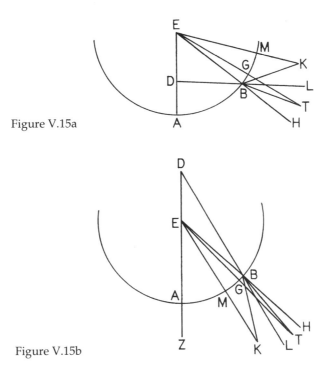

Figure V.15a

Figure V.15b

times smaller than it, depending on the distance [along **BK** from the point of refraction] of the object seen at **K**.

[62] In this case, moreover, it is possible that the aforementioned lines—i.e., the refracted ray and the cathetus dropped from the visible object to the refracting surface—do not meet and [therefore] that there will be no determinate image-location. This is possible when the refracting surface is spherical and its convex surface faces the eye, which lies in the rarer medium, or when the concave surface faces the eye, which lies in the denser medium.

[63] In every other case, though, the lines we have designated always intersect. But when they do not intersect, the visual faculty is affected in the same way that it is in the case of mirrors in regard to the intersection except that the location where the image is formed will not be determinate but is shifted to the common intersection of the normal and the refracting surface, and it takes on the image of the visible object and will coincide with it in location and in the transparent medium.[40]

---

[40]Thus, as with reflection in concave mirrors, so with refraction, when the image-location is indeterminate, the visual faculty transposes the image to the refracting surface itself, where it blends in with it; see IV, 25 and 69–74 above.

[64] Having, then, determined these points, we must now analyze distortions among images in each case [of refraction] and find out if this will accord with the principles we have already established for objects seen according to direct vision.

[65] What we actually see in such a situation is understood more easily if our eye lies within the rarer medium. In fact, those bodies that lie in the heavens are seen at a great distance in comparison with the angle subtended by the normal and the refracted ray; accordingly the refraction is very slight. This is why, and also because, the difference in density between the air in which we stand and the ether in which the stars lie is not great, so that it is with difficulty that the distortions in their images are perceived. The same holds for the difficulty in perceiving [such distortions] when someone opens his eye under water and observes. But if we wish to place a transparent body against the eye so that there is no space between them—whether or not this can actually be done—no object will be seen.[41] For, in that case, before the visual ray is emitted from the eye to follow its proper path, it is weakened by the pressure of that body, particularly if the path it follows is not short.

[66] And since it is impossible [in this case] to explain the distortions of images in the way that they can be explained in the case of direct vision,[42] we have deemed it proper to speak of those cases—with the eye placed inside the rarer medium—in which such phenomena are more evident. On that basis, what must follow in the opposite situation [with the eye in the denser medium] will become evident. Hence, when we have demonstrated the former cases, the latter will be clear, because, once we have shown in one case what ought to happen, the same demonstration will definitely hold for the other case.

[67] *[EXPERIMENT V.5]* In order to make a general analysis on this basis, we must construct three containers made of glass that is as thin and pure as possible so that they are transparent. Let one of

---

[41]As Lejeune remarks in *L'Optique*, p. 159, there is a logical gap in this passage between the initial claim—that image-distortion is more easily observed when the eye is located in the rarer medium—and the subsequent illustrations of the difficulties involved in observing such distortion when the eye is located in the denser medium (i.e., observing the distortion of planets and stars when looking through air, or attempting to observe through a transparent body placed directly upon the eye). Evidently, then, there is a missing transitional statement pointing out that it is precisely because it is so difficult to observe anything when the eye is located in the denser medium that observing image-distortion is easier when the eye is placed in the denser medium.

[42]"Distortion" is meant generally here to include not just mutations of shape but also divergences in apparent location, such divergences applying to direct vision as well as to vision by reflection or refraction.

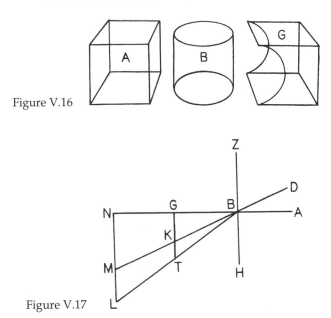

Figure V.16

Figure V.17

them, as represented below by figure **A**, be cubical in shape. Let the second, represented by figure **B**, be cylindrical. And let the third, as represented by figure **G**, be cubical except for one side, which faces the viewer; let that side be concave cylindrical in form, and let it be indented to the depth of a hemisphere.[43]

[68] Therefore, if we wish to understand individual cases involving distortions of images when the surface of refraction is plane, we must first fill the cubical container with perfectly clear water; we must then position the eye directly facing one of its sides and immerse a ruler of reasonable width vertically within the container so that part of it lies above the water's surface. Accordingly, when we have stood the ruler up vertically, the image will lie directly in line with the part of the ruler above water, but it will appear nearer and larger than the actual object, and it will have the same shape.

[69] On the basis of what we will demonstrate, it will be seen that this must hold for every particular case.

[70] *[THEOREM V.11]* Let straight line **ABG** [in figure V.17] be the common section of the water's surface and the plane containing the refracted ray-couple. Let **D** be the eye, let ray **DB** be drawn, and let normal **ZBH** pass through point **B**. Let ray **DB** be refracted along

---

[43]In short, the cavity should be semicylindrical.

BT toward the normal, as happens when the eye lies in the rarer medium. Then, let [cathetus] **TKG** be dropped normal to **AG** from point **T**.

[71] Accordingly, an object that lies at point **T** will be seen at **K**. And since lines **DB** and **BT** taken together are longer than line **DBK** (for angle **BKT** is obtuse), the distance of the visible object's image will be less than that of the object itself.

[72] If, moreover, we draw **BTL** and place the visible object at **L**, and if from that point we draw [cathetus] **LMN** normal to **AG**, the image of **L** will lie at **M**, and the image that appears at point **M** will lie farther away than the image that appears at point **K**. But it will not just lie farther away; it will do so according to a particular mathematical relationship that is proper to it. Thus, since **TG** and **LN** are parallel and lie on a plane, **MN : KG = MB : KB = LB : TB**. And so, **MN : KG = LB : BT**.

[73] Therefore, what happens in the case of something seen when the eye lies within the rarer medium has been demonstrated. When the eye is placed within the denser medium, however, the ratio of the one distance [i.e., true] to the other [i.e., apparent] will be identical according to what we have said; but the distance of the images will be greater than the distance of the actual objects.

[74] *[THEOREM V.12]* For if we assume refracted ray **BTL** [in figure V.18], which forms an angle [of refraction] **LBH** greater than angle [of incidence] **DBZ**, then the ratios of the distances will remain constant to one another. For the ratio of distance **L[B]** to distance **T[B]** will be the same as the ratio of distance **M[B]** to distance **K[B]**. But in each case, the relationship is the inverse [of that in the previous demonstration]. For the location of **T**, which represents the visible object itself, will be nearer than [that of] its image **K**, just as the location of **L** will be nearer than [that of] its image **M**.

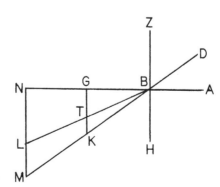

Figure V.18

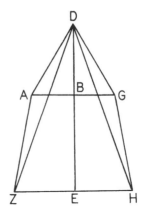

Figure V.19

[75] Furthermore, objects whose images lie at the point of inter-section between the refracted ray and the cathetus dropped from any [of their] points to the surface of refraction between two differ-ent media, if they lie within the denser medium, appear larger than they do when they are placed in a rarer medium, assuming that they maintain the same actual spatial disposition in both cases. In such a case, the visual ray will pass from the rarer into the denser medium, whereas the opposite happens when it passes from the denser into the rarer medium.[44]

[76] *[THEOREM V.13]* Let the two rays **DA** and **DG** [in figure V.19] be drawn from the eye at **D** to line **GA**, and let them flank the normal **DBE**. Then let them be refracted so as to flex apart from the normal along **AZ** and **HG**, and let them comprehend some magni-tude represented by the line connecting the two endpoints of mag-nitude **ZEH**. Finally, let lines **DZ** and **DH** be joined. It is therefore evident that angle **ADG** > angle **ZDH**, because the refracted rays lie farther from the normal [than the continuation of the incident rays along **DZ** and **DH**]. And [so], **ZH** will be seen under a larger angle [than otherwise], assuming that the distance and disposition [of the magnitude] remain constant.

[77] For this reason, then, objects that are submerged in water must invariably appear larger than they would if they were ob-served according to direct vision at the same distance and in the same disposition.

[78] *[THEOREM V.14]* But, assuming that the situation is the re-verse of that just specified, let us suppose that rays **DA** and **DG** [in

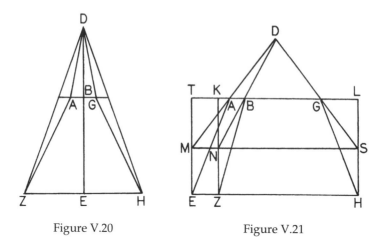

Figure V.20   Figure V.21

figure V.20] are refracted away from the normal, as happens when the eye is situated in the denser medium. Thus, if we join lines **DZ** and **DH**, angle **ZDH** > angle **ADG**. And so the actual object must appear larger than its image.

[79] Furthermore, the shape of the images will be similar to the shapes of the actual objects.

[80] *[THEOREM V.15]* . . . Otherwise, as is suitable:[45] namely, that if ray-couple **DAE** [in figure V.21], ray-couple **DBZ**, and ray-couple **DGH** are each refracted toward magnitude **EZH**, and if catheti of refraction **ET**, **ZK**, and **LH** are dropped, while lines **DAM**, **DBN**, and **DGS** are connected, then the image of line **EZH** will lie on the line passing through points **M**, **N**, and **S**. Also, **TM** : **KN** = **ET** : **ZK**, while **KN** : **SL** = **KZ** : **LH**, and **TM** : **SL** = **ET** : **LH**.[46]

[81] The same will hold as well for every type of disposition. For if segments **TE**, **KZ**, and **LH** define the observed line **EZH** as a straight line, then segments **MT**, **KN**, and **SL** define the observed [image]-line as a straight line.[47] On the other hand, if the former seg-

---

[45]Given this opening phrase and the flawed analysis that follows, Lejeune speculates in *L'Optique*, p. 265, n. 68, that this theorem might represent an interpolation added to an original theorem, itself now lost in the process of textual transmission from Greek to Arabic.

[46]As Lejeune observes in *Recherches*, p. 171, such exact proportionalities do not follow from the vague "law" of refraction articulated in V, 34 above.

[47]That this claim is patently false can be determined by the merest glance into a moderately large flat-bottomed container filled with water; the bottom appears somewhat concave, not flat. Moreover, as Knorr shows in "Archimedes," pp. 67–68, if Ptolemy had applied his own experimental data for refraction from air to water, even his faulty procedure for determining image-location would have led him to conclude that **M** is projected slightly higher than **N**, which in turn is projected slightly higher than the image of another point on **EH** somewhat

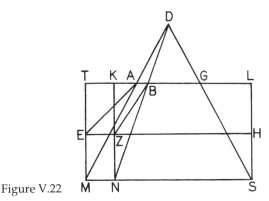

Figure V.22

ments define the observed line as convex, the latter segments define the observed image as convex. Finally, if the former segments define a concave line, the latter define the observed image as concave.

[82] As is represented by the figure [V.22], moreover, this is the case whether we suppose the refraction to occur in the direction of, or away from, the normal.

[83] But if we analyze the proposed distortions in the same way for the cylindrical container with the eye facing the convex surface, and if the ruler immersed in it is stood along the diameter of the visual field [perpendicular to the axis of the visual cone], then we will find that, when the ruler is moved, the image appears to move in the same direction as the upper section of the ruler [that is not submerged]. If the ruler lies between the eye and the [cylinder's] axis, it will appear closer [than it actually is], whereas if it lies beyond that axis, it will appear farther away [than it actually is]. Moreover, magnitudes will always appear larger [than they actually are]; also, their images will appear more convex when they lie beyond the axis and more concave when they lie in front of it. But none of these things will be clearly observable because of a coalescence of the whole image's point-image locations, a coalescence that is due to a weakness of the visual flux.[48] That this must be the case is explained as follows:

---

to the right of **Z**, and so forth. In other words, the image of the section of **EH** to the left of the normal dropped from **D** will incline constantly upward as the line of sight moves ever outward, and the same will hold for the section to the right of that normal. Quite clearly, then, either Ptolemy or Lejeune's supposed interpolator followed deductive rather than inductive imperatives in this case.

[48]Indeed, many of these observational claims about images seen in the water-filled cylinder are misleading, dictated more by the imperatives of the geometrical model than by those of actual observation. Thus, to the eye, the image of an object submerged in the water beyond

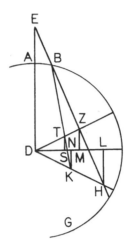

Figure V.23

[84] *[THEOREM V.16]* Let circular segment **ABG** [in figure V.23] be the common section of the convex surface of the water and the plane containing the refracted ray. Let its center be **D**, let the eye be at **E**, and let line **EAD** be connected. Let line **EBZ** be drawn, and let line **DZ** be connected. Let **EB** be refracted toward the normal along line **BTK**, let lines **DL** and **DH** be drawn, and let **DL** stand at right angles to **EAD**. Finally, to that line [**DL**], let perpendiculars **ZM**, **TS**, **KN**, and **HL** be drawn.[49]

[85] Accordingly, angle **DEZ** forms obtuse angle **ZBD**, and angle **HBK** forms obtuse angle **BKH**.[50] Therefore, according to what pertains to displacements, the image of **T** will lie at **Z**, and the image of **K** will lie at **H**. Also, the image of **T** is nearer, because **TS** < **ZM**, and the image of **K** is farther away, because **LH** > **NK**.[51]

---

the axis does not appear farther away than the object itself, although its displacement toward the side as the object itself is moved toward the side from the axis does increase dramatically. Also, the claim that the object appears more concave in the region between eye and axis and more convex in the region beyond the axis does not really square with observation—hence Ptolemy's qualification about image-coalescence and weakening of the flux.

[49]Thus, **DL** represents the line demarcating the region between the eye and the cylinder's axis and the region lying beyond that axis, **T** represents a point-object within the first region, its cathetus of refraction being **DTZ**, and **K** represents a point-object within the second region, its cathetus of refraction being **DKH**.

[50]It makes no sense, of course, to talk of angles forming angles. Moreover, there is no angle **ZBD** delineated in the diagram, and even if it were, it would be acute, not obtuse. Perhaps this represents a misguided effort on the part of either a Greek or an Arabic interpolator to reconstruct a line of argument that had become broken over the course of textual transmission—a possibility that is heightened by the fact that book 5 itself ends abruptly just a few lines later.

[51]Properly speaking, the "proof" hinges on the fact that **EBZ** < **EB** + **BT**, whereas **EBH** > **EB** + **BK**. Indeed, properly speaking, this is no proof at all as it stands, because **DL** does not necessarily separate the region at which the object appears closer from the region at which it appears farther.

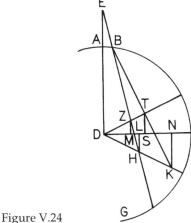

Figure V.24

[86] *[THEOREM V.17]* Supposing that the eye lies within the denser medium, let the refraction take place away from the normal, as represented by the similar figure [V.24]. Actually, the distortion is similar to that represented in the preceding case, but the displacements will be reversed. For the form that lies within [the region between eye and axis] will be farther away, and the form that lies beyond [that region] will be nearer, since **TS > ZM**, and **LH < KN**, as is obvious from the figure.

[87] Now, let perpendicular **AZ** be drawn. . . . [52]

[88] The rest of this book has not been found.

---

[52]There is, of course, no way of knowing precisely how much and what is missing from the rest of the treatise; but, as Lejeune points out in *Recherches*, pp. 174–175, we can be fairly certain that, upon completing his analysis of image-distortion and displacement for convex refracting surfaces, Ptolemy would have turned to the same sort of analysis for concave refracting surfaces, starting with observations based on receptacle **G** in figure IV.16.

BIBLIOGRAPHY

Alhazen. See Risner.

Archimedes. See Heath, Heiberg, Lejeune.

Bacon, Roger. See Bridges, Lindberg.

Barnes, Johnathan. See Lloyd.

Beare, J. I. *Greek Theories of Elementary Cognition from Alcmaeon to Aristotle.* Oxford: Clarendon Press, 1906.

Berggren, J. L. See Hamilton.

Björnbo, Axel and Vogl, Sebestian, eds. *Alkindi, Tideus und Pseudo-Euklid: Drei optische Werke.* Leipzig: Teubner, 1912.

Boll, Franz. "Studien über Claudius Ptolemäus," *Jahrbücher für classische Philologie,* supplementband 21 (1894): 53–244.

Boncompagni, B. "Intorno ad una traduzione latina dell'Ottica di Tolomeo," *Bullettino di bibliografia e di storia delle scienze matematiche e fisiche* 4 (1871): 470–492.

Boyer, Carl B. *The Rainbow: from Myth to Mathematics.* New York: Thomas Yoseloff, 1959.

Bridges, John H., ed. *The Opus Maius of Roger Bacon,* 3 vols. Oxford: Clarendon Press, 1897–1900.

Brown, David. See Price.

Bruno, Vincent. *Form and Color in Greek Painting.* New York: Norton, 1977.

Caussin de Perceval, J. J. A. "Mémoire sur L'Optique de Ptolémée," in *Mémoires de l'Institut Royal de France, Académie des Inscriptions et Belles-Lettres,* pp. 1–43. Paris, 1822.

Cohen, Morris R. and Drabkin, I. E. *A Source Book in Greek Science.* Cambridge: Harvard University Press, 1966.

Coulton, J. J. *Ancient Greek Architects at Work.* Ithaca, NY: Cornell University Press, 1977.

Dambska, Izydora. "La théorie de la science dans les oeuvres de Claude Ptolémée," *Organon* 8 (1971): 109–122.

Damianos. See Schöne.

De Lacy, Phillip. See Galen.

Delambre, J. B. "Sur l'Optique de Ptolémée comparée à celle qui porte le nom d'Euclide et à celles d'Alhazen et de Vitellion," in *Connaissance des temps pour l'an 1816.* Paris, 1813.

Delambre, J. B. *Histoire de l'astronomie ancienne*. Paris, 1817.

Doland, Edmund. See Duhem.

Drabkin, I. E. See Cohen.

Duhem, Pierre. *To Save the Phenomena: An Essay on the Idea of Physical Theory from Plato to Galileo*, trans. Edmund Doland and Chaninah Maschler. Chicago: University of Chicago Press, 1969.

Euclid. See Heath, Heiberg, Ver Eecke.

Fazzo, Silvio. "Alessandro d'Afrodisia e Tolomeo: Aristotelismo e astrologia fra il II e il III secolo D. C.," *Rivista di Storia della Filosofia* 43 (1988): 627–649.

Forbes, R. J. *Studies in Ancient Technology*, 9 vols. Leiden: E. J. Brill, 1964–1972.

Fraser, Peter M. *Ptolemaic Alexandria*, 3 vols. Oxford: Clarendon Press, 1972.

Fritz, Kurt von. "Democritus' Theory of Vision," in *Science, Medicine and History*, 2 vols., ed. E. A. Underwood. London: Oxford University Press, 1953.

Galen. *De usu partium*, in *Galen on the Usefulness of the Parts of the Body*, 2 vols., trans. Margaret Talmadge May. Ithaca, NY: Cornell University Press, 1968.

Galen. *De placitis Hippocratis et Platonis*, in *Corpus medicorum graecorum*, vol. 4, 1, 2, ed. and trans. Phillip De Lacy. Berlin: Akademie-Verlag, 1980.

Gillespie, Charles Coulston. See Toomer.

Goldstein, B. R. "The Arabic Version of Ptolemy's *Planetary Hypotheses*," *Transactions of the American Philosophical Society*, vol. 57.4. Philadelphia, 1967.

Goldstein, B. R. See Hamilton.

Gould, Josiah B. *The Philosophy of Chrysippus*. Leiden: E. J. Brill, 1970.

Govi, Gilberto. *L'Ottica di Claudio Tolomeo, da Eugenio, ammiraglio di Sicilia, scrittore del secolo XII, ridotta in latino sovra la traduzione araba di un testo greco imperfetto, ora per la prima volta, conforme a un codice della Biblioteca Ambrosiana, per deliberazione della R. Accademia delle Scienze di Torino*. Torino, 1885.

Granger, Frank. *Vitruvius I: De Architectura Books I–V*. Cambridge: Harvard University Press, 1933.

Grant, Edward. See Sabra.

Hahm, David E. "Early Hellenistic Theories of Vision and the Perception of Color," in *Studies in Perception*, ed. Peter K. Machamer and Robert G. Turnbull. Columbus, Ohio: Ohio State University Press, 1978.

Hamilton, N. T., Swerdlow, N. M. and Toomer, G. J. "The Canobic Inscription: Ptolemy's Earliest Work," in *From Ancient Omens to Statistical*

Mechanics: Essays on the Exact Sciences Presented to Asger Aaboe, ed. J. L. Berggren and B. R. Goldstein. Copenhagen: University Library, 1987.

Haskins, C. H. *Studies in the History of Mediaeval Science*. Cambridge: Harvard University Press, 1924.

Heath, T. L., trans. *The Works of Archimedes*. Cambridge: Cambridge University Press, 1897.

Heath, T. L., trans. *The Thirteen Books of Euclid's* Elements, 3 vols. Second ed., 1925; reprint ed., New York: Dover, 1956.

Heiberg, J. L., ed. *Euclidis opera omnia*, 7 vols. Leipzig: Teubner, 1883–1888.

Heiberg, J. L., ed. *Archimedis opera omnia*, vol. 1. Leipzig: Teubner, 1910.

Hero of Alexandria. *Heronis Alexandrini Opera Quae Supersunt Omnia*, 5 vols. Ed. L. Nix and W. Schmidt. Leipzig: Teubner, 1899.

Huby, P. See Long.

Jamison, Evelyn. *Admiral Eugenius of Sicily: His Life and Work*. London: Oxford University Press, 1957.

Kattsoff, Louis O. "Ptolemy's Scientific Method," *Isis* 38 (1947): 18–22.

Kaufman, L. and Rock, I. "The Moon Illusion," *Scientific American*, vol. 207 (1962): 120–130.

Al-Kindi. See Björnbo.

Knorr, Wilbur. "Archimedes and the Pseudo-Euclidean *Catoptrics*: Early Stages in the Ancient Geometric Theory of Mirrors," *Archives internationales d'histoire des sciences* 35 (1985): 27–105.

Knorr, Wilbur. "Pseudo-Euclidean Reflections in Ancient Optics," *Physis* 31 (1994): 1–45.

Lammert, Friedrich. "Eine neue Quelle für die Philosophie der mittleren Stoa I," *Wiener Studien* 41 (1919): 113–121.

Lammert, Friedrich. "Eine neue Quelle . . . II," *Wiener Studien* 42 (1920/21): 34–46.

Lammert, Friedrich, ed. "On the Criterion," in *Ptolemaei Opera Omnia*, vol. 3. Leipzig: Teubner, 1952.

Lejeune, Albert. "Les Tables de Réfraction de Ptolémée," *Annales de la Société Scientifique de Bruxelles* 60 (1946): 93–101.

Lejeune, Albert. "Le dioptre d'Archimède," *Annales de la Société Scientifique de Bruxelles* 61 (1947): 27–47.

Lejeune, Albert. *Euclide et Ptolémée, deux stades de l'optique géométrique grecque*. Louvain: Bibliothèque de l'Université, 1948.

Lejeune, Albert. *Recherches sur la catoptrique grecque*. Mémoires de l'Académie Royale de Belgique: Classe des lettres et des sciences morales et politiques, vol. 52.2. Brussels: Palais des Académies, 1957.

Lejeune, Albert. *L'Optique de Claude Ptolémée*. Leiden: E. J. Brill, 1989.

Levere, T. H. See Smith.

Lindberg, David C., ed. and trans. *John Pecham and the Science of Optics:* Per-spectiva communis, *Edited with an Introduction, English Translation, and Critical Notes.* Madison: University of Wisconsin Press, 1970.

Lindberg, David C. *A Catalogue of Medieval and Renaissance Optical Manu-scripts.* Toronto: Pontifical Institute of Mediaeval Studies Press, 1975.

Lindberg, David C. *Theories of Vision from Al-Kindi to Kepler.* Chicago: Uni-versity of Chicago Press, 1976.

Lindberg, David C., ed. and trans. *Roger Bacon's Philosophy of Nature.* Ox-ford: Clarendon Press, 1983.

Lloyd, G. E. R. "Observational error in later Greek science," in *Science and Speculation: Studies in Hellenistic Theory and Practice,* ed. Johnathan Barnes, et al. Cambridge: Cambridge University Press, 1983.

Long, A. A. "Ptolemy on the Criterion: An Epistemology for the Practising Scientist," in *The Criterion of Truth: Essays Written in Honour of George Ker-ferd,* ed. P. Huby and G. Neal. Liverpool: Liverpool University Press, 1989.

Lucci, Gualberto. "Criterio e metodologia in Sesto Empirico i Tolomeo," *Annali dell'Istituto de Filosofia di Firenze* 2 (1980): 23–52.

Luckiesh, M. *Visual Illusions: Their Causes, Characteristics and Applications,* 1928; rpt. New York: Dover, 1965.

Machamer, Peter K. See Hahm.

Martin, Th. M. "Ptolémée, auteur de l'Optique . . . est-il le même que Cl. Ptolémée, auteur de l'Almageste?" *Bullettino di bibliografia e di storia delle scienze matematiche e fisiche* 4 (1871): 466–469.

Maschler, Chaninah. See Duhem.

May, Margaret Talmadge. See Galen.

Montucla, J. F. *Histoire des Mathématiques,* 4 vols. First ed. 1758; reprint ed., Paris: A. Blanchard, 1960.

Morrow, Glenn R., trans. *Proclus: A Commentary on the First Book of Euclid's Elements.* Princeton: Princeton University Press, 1970.

Murdoch, John. See Sabra.

Nallino, Carlo. *Raccolta di scritti editi e inediti,* 6 vols. Ed. Maria Nallino. Rome: Istituto per l'Oriente, 1944.

Nallino, Maria. See Carlo Nallino.

Neal, G. See Long.

Neugebauer, Otto. *The Exact Sciences in Antiquity.* Second ed., 1957; reprint ed., New York: Dover, 1969.

Neugebauer, Otto. *A History of Ancient Mathematical Astronomy.* Berlin, Heidelberg, New York: Springer, 1975.

Nix, L. See Hero of Alexandria.

Pace, Anna de. "Elementi Aristotelici nell'*Ottica* di Claudio Tolomeo," *Rivista critica di storia della filosofia* 36 (1981): 123–138 and 37 (1982): 243–276.

Pecham, John. See Lindberg.

Price, Jennifer. "Glass," in *Roman Crafts,* ed. Donald Strong and David Brown. New York: New York University Press, 1976.

Proclus. See Morrow.

Ptolemy. See Lejeune, Toomer, Robinson.

Rashed, Roshdi. "A Pioneer in Anaclastics: Ibn Sahl On Burning Mirrors," *Isis* 81 (1990): 464–491.

Rashed, Roshdi. "Fithitos (?) et al-Kindi sur l'illusion lunaire," *Collection des Etudes Augustiniennes: Série Antiquité* 131 (1992): 533–559.

Risner, Friedrich, ed. *Opticae Thesaurus: Alhazeni arabis libri septem nunc primum editi, eiusdem liber De crepusculis et nubium ascensionibus, item Vitellonis thuringopoloni libri X.* 1572; reprint ed., New York: Johnson Reprint, 1972.

Robinson, F. E., trans. *Ptolemy's* Tetrabiblos. Cambridge, MA: Harvard University Press, 1940.

Robinson, J. O. *The Psychology of Visual Illusion.* London: Hutchinson University Library, 1972.

Rock, I. See Kaufman.

Ross, G. M. See Ross, H. E.

Ross, H. E. and G. M. "Did Ptolemy Understand the Moon Illusion?" *Perception* 5 (1976): 377–385.

Sabra, A. I. "Ibn al-Haytham's Criticisms of Ptolemy's *Optics,*" *Journal of the History of Philosophy* 4 (1966): 145–149.

Sabra, A. I. "Ibn al-Haytham's Lemmas for Solving Alhazen's Problem," *Archive for History of Exact Sciences* 26 (1982): 299–324.

Sabra, A. I. "Psychology versus Mathematics: Ptolemy and Alhazen on the Moon Illusion," in *Mathematics and its Applications to Science and Natural Philosophy in the Middle Ages: essays in Honor of Marshall Clagett,* ed. Edward Grant and John Murdoch. Cambridge: Cambridge University Press, 1987.

Sabra, A. I., trans. *The Optics of Ibn Al-Haytham: Books I-III: On Direct Vision,* 2 vols. London: Warburg Institute, 1989.

Saint-Pierre, Bernard. "La physique de la vision dans l'antiquité: Contribution à l'établissement des sources anciennes de l'optique médiévale," PhD diss., University of Montréal, 1972.

Sambursky, Samuel. *Physics of the Stoics.* London: Routledge & Kegan Paul, 1987.

Sarton, George. *Introduction to the History of Science,* 3 vols. Baltimore: Williams & Wilkins, 1927.

Schmidt, W. See Hero Of Alexandria.

Schöne, Richard, ed. and trans. *Damianos Schrift über Optik, mit Auszügen aus Geminos.* Berlin, 1897.

Shea, W. R. See Smith.

Simon, Gérard. *Le regard, l'être et l'apparence dans l'Optique de l'antiquité.* Paris: Seuil, 1988.

Simon, Gérard. "L'*Optique* d'Ibn al-Haytham et la tradition Ptoléméenne," *Arabic Sciences and Philosophy* 2 (1992): 203–235.

Smith, A. Mark. "Getting the Big Picture in Perspectivist Optics," *Isis* 72 (1981): 568–589.

Smith, A. Mark. "Saving the Appearances of the Appearances: The Foundations of Classical Geometrical Optics," *Archive for History of Exact Sciences* 24 (1981): 73–99.

Smith, A. Mark. "Ptolemy's Search for a Law of Refraction: A Case-Study in the Classical Methodology of Saving the Appearances and its Limitations," *Archive for History of Exact Sciences* 26 (1982): 221–240.

Smith, A. Mark. *Witelonis* Perspectivae *liber quintus (Book V of Witelo's* Perspectiva*): An English Translation with Introduction and Commentary and Latin Edition of the First Catoptrical Book of Witelo's* Perspectiva. Warsaw: Ossolineum (Polish Academy of Sciences Press), 1983.

Smith, A. Mark. "The Psychology of Visual Perception in Ptolemy's *Optics*," *Isis* 79 (1988): 189–206.

Smith, A. Mark. "Alhazen's Debt to Ptolemy's *Optics*," in *Nature, Experiment, and the Sciences*, T. H. Levere and W. R. Shea, eds. Dordrecht; Kluwer, 1990, pp. 147–164.

Smith, A. Mark. "Extremal Principles in Ancient and Medieval Optics," *Physics* 31 (1994): 113–140.

Stahl, George. *Roman Science.* Madison: University of Wisconsin Press, 1962.

Stratton, George M., trans. *Theophrastus and the Greek Physiological Psychology before Aristotle.* London: Allen and Unwin, 1917.

Strong, Donald. See Price.

Swerdlow, N. M. See Hamilton.

Takahashi, Ken'ichi. *The Medieval Latin Traditions of Euclid's* Catoptrica. Fukuoka: Kyushu University Press, 1992.

Taub, Liba. *Ptolemy's Universe: The Natural Philosophical and Ethical Foundations of Ptolemy's Astronomy.* Chicago: Open Court, 1993.

Toomer, G. J. *Diocles on Burning Mirrors.* Berlin, Heidelberg, New York: Springer, 1976.

Toomer, G. J. "Ptolemy," *Dictionary of Scientific Biography*, vol. 11, ed. Charles Coulston Gillispie, pp. 186–206. New York: Scribner's, 1976.

Toomer, G. J., trans. *Ptolemy's Almagest*. Berlin, Heidelberg, New York: Springer, 1984.

Toomer, G. J. See Hamilton.

Turnbull, Robert. See Hahm.

Underwood, E. A. See Fritz.

Unguru, Sabetai, ed. *Witelonis* Perspectivae *liber secundus et liber tertius (Books II and III of Witelo's* Perspectiva): *A Critical Latin Edition and English Translation with Introduction, Notes and Commentaries*. Warsaw: Ossolineum (Polish Academy of Sciences Press), 1991.

Venturi, G. B. *Commentari sopra la storia e le teorie dell'ottica*, vol. 1. Bologna, 1814.

Ver Eecke, Paul. *Euclide: L'Optique et la Catoptrique*. Paris: Blanchard, 1959.

Vitruvius. See Granger.

Vogl, Sebastian. See Björnbo.

Waerden, B. L. van der. "Klaudios Ptolemaios," *Paulys Realencyclopädie der classischen Altertumswissenschaft*, vol. 23.2. Stuttgart: J. B. Metzler, 1959.

Wagner, David L. *The Seven Liberal Arts in the Middle Ages*. Bloomington: Indiana University Press, 1983.

Witelo. See Risner, Smith, Unguru.

GENERAL INDEX

accidental coloring: *See* coloring.

accidents 63n. *See also* **accidens** *in* Latin-English index.

act/actuality 22n, 27, 29n, 59n, 76n, 77n. *See also* **actus** *in* Latin-English index.

afterimage 32, 114

Alcmaeon 22n

Alexandria 1, 3, 4, 122n, 237n: Library 3; Museum 3; Serapeum 3n

Alhazen 8, 9, 59, 61. *See also* Ibn al-Haytham.

Alhazen's Problem 57n

Almagest: See Ptolemy.

Al-Andalusī 58

angle: critical 234; of incidence 20n, 36, 43, 65n, 76n, 117n, 126n, 132n, 133, 138, 139, 198n, 226, 229, 230, 233n, 236, 237, 239, 243–245, 247n, 256; of reflection 20n, 36, 65n, 115, 117n, 132n, 133, 137–139, 198n, 226, 230n, 245; of refraction 20n, 43, 65n, 229, 230, 233, 236, 237, 239, 243–245, 256. *See also* visual angle.

angular deflection 234. *See also* **flexio** *in* Latin-English index.

anterior coloring: *See* coloring.

Antoninus Pius 1

Apex 29; 152n. *See also* **principium** *in* Latin-English index.

Apollonius of Perga 4

apperception: *See* **scientia** *in* Latin-English index.

Apuleius of Madaura 49n

Archimedes 4, 14, 15n, 49n, 83n: *Catoptrics* 16; *On the Sphere and the Cylinder* 14

architecture 52, 53, 127, 128

Aristippus, Henry 7

Aristotle/Aristotelians 3, 17–19, 22n, 27, 28, 37, 49, 50, 63n, 71n, 72n, 74n, 75n, 76n, 77n, 79n, 80n, 81n, 82n, 99n, 107n, 242n: *De anima* 17, 72n; *De caelo* 50; *De sensu et sensato* 22n, 27n, 74n, 75n, 80n; *Meteorology* 22n, 49, 50n, 81n; *Parva naturalia* 17; *Physics* 17, 19n, 74n. *See also* Pseudo-Aristotle.

armillary 233n

artistic depiction 122n, 125. *See also* **ymago** *in* Latin-English index.

astrolabe 238n

astrology 5, 51

atmospheric refraction 2, 46, 56, 58, 60n, 238–242

atomists 21, 22, 27, 75n

axis: *See* common axis, proper axis, visual axis. *See also* **axis** *in* Latin-English index.

Bacon, Roger 8n, 58–61: *De multiplicatione specierum* 58, 60; *Opus maius* 58, 59n, 60; *Perspectiva* 58, 60, 61

LATIN-ENGLISH INDEX

accidens accident: 63.18, 64.13, 79.13, 79.14, 79.19, 79.27, 80.23, 98.4, 102.25, 106.16, 108.17, 108.18, 112.10, 112.17, 125.21, 125.24; accidental: 75.9; aspect: 204.37; characteristic: 122.1; feature: 112.31, 112.33, 124.32; ideal: 237.19; illusion: 119.32; phenomenon: 89.14, 106.31; quality: 74. 1; visual effect: 111.3

accidere to affect: 112.21, 118.22; to arise: 82.8, 102.21, 102.37, 106.20, 108.3, 110.20, 113.15, 113.35, 117.19, 120.10, 120.37, 121.14, 125.34, 126.10, 130.5, 153.13, 194.20; to arouse: 81.15, 120.8; to be affixed to: 74.18; to be caused: 153.6; to be created: 82.6; to be due to: 116.2, 117.19, 117.21, 118.23, 118.24, 127.20; to be observed: 154.2; to be the case: 73.18, 90.29, 118.3, 125.7, 158.6, 165.9, 216.21, 216.35, 247.29, 259.4; to befall: 112.10; to come about: 118.26, 123.13, 131.2; to convey: 64.13, 79.13, 79.14, 79.27, 102.25, 106.16, 125.21; to follow: 133.4, 152.7; to happen: 74.2, 75.27, 81.13, 82.18, 85.15, 85.17, 91.30, 100.16, 105.18, 105.27, 106.7, 106.36, 107.21(2), 108.5, 109.27, 110.17, 111.5, 111.13, 112.37, 115.16, 118.14, 119.17, 119.27, 120.1, 121.6, 121.25, 123.7, 123.20, 126.16, 127.4, 127.18, 128.4, 137.5, 138.7, 139.34, 140.34, 141.17, 148.18, 153.12, 155.11, 161.23, 161.28, 174.5, 176.5, 187.7, 192.4, 194.10, 197.1, 198.9, 202.9, 204.31, 212.9, 214.27, 217.15, 220.15, 221.18, 224.4, 225.25, 227.3, 229.11, 247.5, 254.28, 256.1, 256.16, 258.1; to hold: 150.13; to involve: 106.22, 120.13; to occur: 64.10, 65.9, 75.27, 78.22, 97.24, 121.28, 123.16, 175.3, 189.4, 190.6, 227.13, 229.13; to overtake: 75.26; to pertain to: 106-19; to provide: 112.31; to result: 185.10, 185.25, 200.21, 204.7, 238.25; to shine upon: 76.29; to take place: 204.25; to turn out: 140.16; to undergo: 70.10, 103.7, 103.10, 233.3

accipere to judge: 211.6; to measure: 213.19

actio process: 152.7

actus action: 71.9, 112.24, 113.1; actuality: 78.27, 92.6, 202.23; effect: 76.31, 113.7; event: 140.6; inclination: 137.3; operation: 76.18

aereus nebulous: 77.19; tenuous: 120.32

agere to act: 102.9; to arise: 121.2; to do: 102.3; to operate: 70.11

ambiguitas ambiguity: 92.5; problem: 108.32

amiratus admiral: 70.2

antiqui ancients: 74.17

apparere to appear: 64.15, 64.27, 65.11, 65.20, 65.25, 65.29, 66.1, 66.2, 66.4, 66.14, 73.26, 73.28, 73.31, 74.4, 74.7, 74.9, 75.14, 75.28, 77.13, 80.19, 82.9, 82.10, 82.12, 82.13, 82.19, 82.20, 82.21, 89.14, 90.26, 90.35, 91.17, 93.5, 93.14, 93.19, 93.21, 94.7, 94.8, 95.3, 96.7, 96.8, 96.10, 97.2, 97.9, 97.17, 97.30, 99.7, 99.11, 100.9, 101.10, 101.11, 101.15, 101.21, 101.27, 103.10, 104.5, 104.26, 104.35, 106.10, 108.26, 108.29, 108.30, 108.31, 109.7, 109.15, 110.3, 110.8, 110.23, 110.26, 110.33, 110.34, 111.27, 111.30, 111.33, 112.5, 114.15, 114.19, 115.18, 116.7, 116.11, 116.21, 116.24, 116.30, 116.36, 116.39, 117.5, 117.16, 118.11, 118.20, 119.6, 119.8, 119.9, 119.14, 119.20,

apparere to appear *(cont.)*
119.22, 120.15, 121.5, 121.18, 121.22, 122.2, 122.13, 122.18, 122.20, 122.22, 123.16, 123.18, 124.4, 125.32, 126.21(2), 127.10, 131.27, 131.31, 132.7, 132.8, 132.10, 132.14, 132.17, 134.16, 135.11, 135.13, 135.15, 136.2, 136.13, 136.18, 136.23, 139.35, 141.11, 141.16, 141.24, 141.34, 142.16, 143.8, 143.13, 143.18, 145.15, 145.18, 146.1, 146.4, 146.12, 146.22, 146.35, 147.13, 147.15, 147.17, 148.27, 148.28, 148.29, 151.12, 152.17, 152.32, 153.4, 153.5, 153.7, 155.13, 156.7, 156.11, 156.12, 157.15, 157.20, 157.22, 158.15, 159.30, 160.11, 160.15, 160.17, 160.18, 161.5, 161.9, 161.19, 161.26, 161.27, 162.9, 162.14, 163.26, 165.5, 165.8, 165.11, 166.14, 166.28, 167.22, 167.23, 168.17, 169.15, 171.10, 174.4, 175.3, 182.5, 194.14, 194.28, 195.12, 195.13, 195.15, 195.19, 196.14, 197.21, 198.27, 202.5, 207.23, 208.25, 208.27, 209.23, 210.5, 212.6, 212.20, 214.6, 214.23, 214.30, 215.10, 215.28, 216.2, 216.5, 216.6, 216.7, 216.8, 216.10, 216.34, 217.4, 217.6, 217.8, 217.34, 217.37, 218.1, 218.5, 218.10, 218.16, 218.22, 218.24, 218.28, 218.32, 218.34, 219.9, 219.17, 219.20, 219.35, 222.6, 223.6, 223.8, 225.28, 225.30, 226.5, 226.8, 226.20, 226.23, 226.26, 226.29(2), 232.18, 232.20, 232.23, 234.5, 235.20, 235.21, 240.21, 241.7, 241.12, 247.16, 255.13, 257.4, 259.11, 259.14; to be adduced: 139.5; to be apparent: 100.11, 105.21, 124.18, 136.21, 145.4, 145.21, 148.16, 148.31, 149.20, 204.32, 240.11, 241.3, 247.21; to be clear: 230.17; to be detectable: 117.10; to be determinate: 141.5; to be evidenced: 136.32; to be evident: 105.19, 140.10; to be manifested: 123.32; to be observable: 91.10,259.19; to be observed: 132.5, 145.10, 222.4; to be obvious: 261.8; to be perceived: 148.9; to be seen: 64.10, 64.13, 91.29, 108.5, 108.7, 115.13, 116.22, 117.28, 120.5, 121.26, 127.9, 136.11, 140.19, 141.18, 142.6, 146.13, 146.15, 146.36, 147.21, 152.28, 156.14(2), 160.26, 161.24, 167.11, 170.17, 216.32, 225.18, 227.5, 231.3, 238.2, 238.4, 256.16; to be sensed: 121.3; to be visible: 138.14, 138.15, 140.13, 141.4, 238.21, 241.5; to create a visual effect: 131.18; to look: 90.34; to seem: 63.11, 75.7, 97.27, 111.12, 120.26, 121.13, 121.24, 127.25, 136.6, 136.30, 152.15, 182.11, 241.14; to view: 167.18
apparitio appearance: 244.20; image: 136.23, 138.10
aptare to attach: 234.17
aptus apposite: 97.29; apt: 126.33; capable: 222.9; prone: 137.30
arbitrari to assume: 181.16, 201.13, 242.17, 247.25; to consider: 97.18; to determine: 144.21; to judge: 123.10, 125.13, 137.4, 161.14; to posit: 226.27; to suppose: 104.30, 239.12, 245.23
arbitratio judgment: 106.28
aspectus focus: 85.21; glance: 82.28; image: 138.19; observation: 74.3; sight: 147.11; viewing process: 150.17; viewpoint: 192.7; vision: 137.23; visual faculty: 126.5; visual grasp: 138.12; visual perception: 78.13
aspicere to direct a line of sight: 135.8, 221.22; to fix on: 82.28; to line up along a line of sight: 232.17; to look/to look at: 66.13, 76.21, 76.25, 82.22, 91.11, 91.14, 107.20, 114.6, 114.9, 114.11, 120.34, 123.27, 124.5, 126.17, 126.30, 127.4, 128.7, 132.9, 146.8, 146.31, 152.4, 155.14, 196.1, 229.13, 234.4, 235.1, 235.6, 249.5; to observe: 195.3, 207.25, 208.29, 209.22, 254.14, 257.22, 258.15, 258.16, 259.1, 259.2, 259.3; to scrutinize: 82.29; to see: 83.2, 104.33, 109.19, 124.8, 124.27, 143.1, 143.3, 149.21, 152.5, 152.27, 169.6, 169.11, 216.17; to sight: 235.11, 235.19, 237.1; to view: 64.18, 64.19, 66.12, 77.24, 105.2, 110.27, 120.4, 122.8, 122.19, 150.16, 166.30

aspiciens cornea: 137.26, 175.22, 175.23, 176.9; eye: 165.15, 168.3, 192.13, 192.17, 193.11, 204.36, 215.14; looker: 82.23, 116.16; observer: 110.38, 116.5, 116.8, 116.18, 124.28, 139.35, 161.12, 162.14, 221.2; sight: 64.23; viewer: 65.27, 90.22, 90.23, 101.14, 101.29, 106.1, 116.15, 116.18, 117.30, 123.11, 124.21, 141.25, 193.2, 196.4, 214.24, 215.8, 218.8, 222.6, 239.5
assimilari to assimilate: 63.16; to be perceived: 137.1; to look like: 196.4
assimilatio judgment: 223.15
axis axis: 64.22, 77.23, 78.1, 78.13, 82.26, 83.5, 85.1, 85.6, 85.14, 85.22, 86.10, 87.16, 88.23, 89.19, 110.8, 110.11, 113.32, 113.34, 138.31, 141.30, 142.2, 142.10, 144.1, 144.20, 144.21, 145.25, 146.30, 147.12, 147.14, 148.7, 148.14, 148.23, 149.12, 149.15, 149.17, 150.4, 150.6, 150.20, 152.7, 152.15, 176.7, 220.8, 220.9, 220.16, 220.19, 220.21, 221.12, 221.23, 221.34, 259.13, 259.15, 259.17; visual axis: 78.17, 82.2, 89.2, 124.26, 143.9, 143.12, 144.11, 144.18
axis communis common axis: 145.2, 145.6, 145.8, 145.20, 145.23, 146.6, 146.16, 146.23, 146.25, 148.22, 149.13, 152.11, 152.16, 152.19
axis proprius proper axis: 145.6, 145.23, 146.4, 146.16, 146.24, 146.26, 152.19
axis visus visual axis: 145.6, 145.14

cadere to apprehend: 112.6; to be incident: 91.8, 133.15, 135.20; to coincide: 239.10; to connect: 194.5; to converge: 83.5, 85.22; to drop: 115.20, 131.31, 132.18, 145.7, 146.6, 146.25, 160.11, 168.3, 193.15, 224.21, 225.1, 230.8, 231.4, 231.18, 247.33, 248.11, 248.21, 251.5, 251.14, 252.20, 253.4, 257.2; to emanate: 175.25, 192.17; to extend: 233.3; to fall: 64.22, 81.29, 83.13, 88.11, 91.33, 91.37, 92.25, 92.28, 92.29, 98.17, 98.34, 99.5, 99.10, 101.1, 101.3, 102.34, 109.3, 110.1, 110.29, 119.4, 133.3, 138.8, 138.29, 143.10, 151.7, 192.33, 198.22, 201.9, 206.21, 211.17, 212.5, 222.18, 225.7, 230.22; to grasp: 123.24, 128.5; to happen: 229.17; to impinge: 127.6; to lie: 165.7, 196.23; to occur: 229.16; to pass through: 202.3; to reach: 131.28, 226.22; to sense: 123.30; to shine: 76.21, 109.8; to strike: 64.25, 74.13, 90.35, 91.20, 109.7, 118.7, 119.11, 122.11, 122.13, 122.25, 123.3, 124.24, 137.12, 137.16, 139.12, 139.26, 140.12, 146.24, 156.24, 176.2, 220.2, 220.8, 222.18, 247.27; to take place: 235.25; to touch: 142.16, 142.17, 214.29, 235.1
casus impingement: 137.11, 152.12; incidence: 133.8
centrum center: 99.2, 100.20, 101.2, 110.30, 116.38, 132.3, 134.6, 138.31, 153.35, 162.5, 162.19, 163.17, 163.27, 163.32, 165.19, 166.32, 168.2, 169.17, 170.7, 170.21, 175.20, 180.20, 182.15, 183.17, 192.8, 192.16, 192.20, 192.33, 193.16, 198.17, 199.4, 199.8, 203.23, 207.12, 207.17, 220.22, 234.18, 236.25(2), 239.2, 239.3(2), 239.8, 249.16, 250.7, 250.9, 250.18, 251.9, 260.3; centerpoint: 134.8, 134.12, 137.27, 175.23, 176.19, 180.5, 180.13, 183.20, 189.14, 189.22, 191.1, 191.7, 191.11, 192.23, 192.27, 193.20, 198.23, 200.9, 201.20, 201.22, 201.25, 201.34, 202.3, 202.14, 202.21, 203.4, 204.10, 205.7, 206.13, 207.29, 208.33, 210.12, 212.16, 214.8, 215.13, 222.11, 232.4, 232.7
certitudo perfect evidence: 225.20
certus certain: 134.2, 141.16; clear: 147.4; sure: 131.20; true: 146.18, 152.18
claritas brightness: 127.12
claritas visus visual flux: 76.20
clarus brightness: 72.1; clear: 134.16, 232.10; illuminated: 77.12; unblemished: 115.3
coaptare to align: 238.7; to apply: 217.2 1; to attach: 236.24; to be congruent with: 99.16, 157.14, 226.14; to fit: 99.3, 99.6; to gather: 70.7

concavus concave (*cont.*)
212.6, 212.10, 212.11, 212.15, 213.4, 213.7, 213.12, 213.16, 213.25, 214.3, 214.8, 215.7, 215.13, 216.16, 216.29, 217.6, 217.16, 217.20, 218.7, 220.3, 250.6, 251.9, 252.14, 253.8, 255.4, 259.3(2); concavity: 122.9, 122.33; deep-set: 107.10; depressed: 122.14; inwardly curved: 153.38; raised: 122.23, 122.27, 122.29, 123.3, 123.4, 123.6; trough: 117.3

confusio confusion: 115.30, 201.13

confusus confused: 148.15, 159.18

conservare to conserve: 244.27; to keep: 135.6; to maintain: 136.14, 136.29, 138.4, 139.20, 144.18, 209.27, 247.18, 247.31, 257.6; to preserve: 194.13; to stay: 148.2

considerare to analyze: 241.18, 254.1; to be convinced: 63.1; to consider: 138.16, 141.20, 145.9; to determine: 93.2, 241.19; to discuss: 113.9; to investigate: 143.19, 182.12; to scrutinize: 120.23; to take up an issue: 141.10

consideratio analysis: 254.30; establishing a point: 242.10; inference: 126.5; interpretation: 113.12; judgment: 108.9; perception: 125.33, 182.8; scrutiny: 120.11, 120.19, 124.32

consuescere to be accustomed: 127.23; to be customary: 64.23; to be habitual: 140.2; to be used to: 131.13

consuetudo capacity: 140.4; custom: 64.24, 83.9, 113.17, 151.8, 151.14; habit: 126.12; norm: 127.25, 136.35, 137.2, 215.5; the ordinary: 94.13; the usual: 139.34, 151.14, 171.17; wont: 121.5

contemplatio analysis: 217.26; investigation: 161.23; study: 162.11(2)

continuus adjoining: 73.27; coalescing: 195.5; coincident: 196.2, 235.14, 253.16; contiguous: 109.2, 117.9, 154.5; continual: 126.13; continuous: 73.35, 91.25, 92.4, 92.8, 98.5, 103.8, 104.2, 105.23, 106.10, 107.31, 110.36, 117.7, 117.15, 119.24, 128.13, 152.28, 159.14, 159.19, 196.10, 197.11, 197.22, 197.24, 198.3, 223.3, 223.8; enveloping: 73.16; extending: 73.17; juxtaposed: 116.34; lying in a straight line with: 119.22

conus cone: 113.19

corporalitas density: 81.23

cursus course: 139.6, 244.27; motion: 110.18; moving: 123.29; way: 152.2

curvare to be curved: 134.17, 159.8, 159.9; to curve: 196.12

curvitas convex curvature: 167.23, 168.8; convex section: 217.30, 218.37; convex shape: 223.12; convex side: 210.7; convexity: 99.25, 101.5, 118.27, 122.33, 168.18, 170.10, 249.15, 251.7; curvature: 169.2

curvus circular: 98.19; concave: 169.17; convex: 65.6, 65.12, 65.15, 99.11, 99.16, 99.17, 99.18, 99.20, 99.23, 100.8, 100.12, 101.14, 116.35, 122.2, 122.7, 122.16, 131.17, 134.17, 135.4, 137.28, 150.13, 153.36, 158.2, 158.14, 158.18, 159.6, 159.8, 159.11, 159.34, 161.24, 161.29, 162.2, 162.4, 162.8, 162.11, 162.13, 162.19, 163.32, 165.14, 165.19, 166.25, 166.31, 167.21, 167.22, 167.23, 167.25, 168.1, 168.14, 169.1, 169.8, 169.9, 169.10, 169.15, 170.16, 170.20, 171.20, 172.3, 174.4, 210.9, 211.20, 212.2, 212.8, 212.12, 212.13, 213.15, 213.27, 216.21, 217.9, 217.15, 217.19, 217.32, 219.2, 219.8, 219.16, 219.23, 220.5, 251.7, 252.5, 253.7, 259.1, 259.2, 259.8, 260.2; crest: 117.3; curved: 98.33, 161.34, 161.35, 169.2, 236.28; indented: 122.21(2), 122.29, 122.30, 123.5(2), 123.6; outward curve: 153.37

deceptio deception: 110.40; illusion: 65.7, 82.6, 106.18, 106.27, 108.24, 128.17

differentia defining characteristic: 112.26; difference: 63.8, 70.9, 72.31, 72.32, 74.5, 92.13, 100.9, 125.5, 236.17, 236.18, 244.2; differentiation: 98.20; type: 217.27; variation: 125.7

dinoscere to ascertain: 142.15; to assert: 131.26, 131.30; to be apparent/evident: 244.26, 254.25; to designate: 237.5; to determine: 98.14, 242.7, 244.14; to discern: 72.27, 103.16, 130.3; to establish: 159.21; to formulate: 242.9; to perceive: 64.1, 64.6, 64.7, 72.7, 72.9, 72.21, 73.3, 73.9, 75.3, 80.26, 81.21, 92.12, 98.1, 106.3, 123.17, 124.23, 126.6; to realize: 238.9; to see: 84.11, 244.24, 255.15; to understand: 123.8, 136.33, 156.4, 204.26, 217.14, 217.26, 248.6

discernere to analyze: 229.12; to apprehend: 139.32; to differentiate: 106.26; to discern: 81.26, 101.36, 109.19, 120.22; to distinguish: 80.21; to make a judgment: 120.37; to perceive: 112.12, 121.29, 140.1

discernitivus (see **virtus discernitiva**)

dispositio case: 97.10, 226.19, 248.14; disposition: 71.8, 71.11, 79.7, 108.1; situation: 254.5

distinctio claim: 242.10; defining outline: 73.28; differentiation: 106.18; discontinuity: 136.25; discussion: 103.2; distinction: 70.15, 143.17, 204.32, 244.28; point: 103.24

distinctio communis common section: 132.26, 231.8, 239.10, 246.2, 248.19, 249.17, 251.17, 255.18, 260.2

distinguere to determine: 145.4, 145.9; to distinguish: 175.12; to investigate: 230.1; to lie between: 242.17

dubitatio difficulty: 112.9; doubt: 130.5; issue: 106.20, 141.20, 152.30; misapprehension: 102.21; problem: 90.6

effectus effect: 71.20, 76.3; empirical fact: 131.21; observed phenomenon: 159.13

efficere to cause: 216.26; to pass: 230.20; to produce: 107.9

effusio propagation: 247.3

existimare to adjudge: 111.24, 111.35; to assume: 109.4, 113.5, 119.31, 121.24, 127.27, 176.6, 188.9, 189.3, 202.27, 213.6, 252.18; to judge: 95.1, 120.30, 126.14, 215.5; to perceive: 197.24; to place: 181.25; to represent: 224.17; to suppose: 75.8, 90.32, 213.2, 213.14, 214.26, 216.4, 230.25

experientia what is actually seen: 254.5

experimentum experiment: 65.12, 65.16, 66.5, 134.3, 231.24, 234.9, 234.22, 237.6

experiri to be experimentally determined: 142.13

fallacia false perceptual inference: 74.11; illusion: 102.15, 110.39, 112.18, 112.21, 112.34, 113.35, 116.2, 117.19, 117.21, 118.21, 120.7, 120.12, 120.13, 120.26, 120.36, 121.8, 121.14, 121.28, 125.2, 125.34, 126.9, 126.15, 161.3; inferential error: 126.29; misperception: 65.9, 106.29, 112.12

fallacia visus visual illusion: 113.9, 130.2

fallere to deceive: 126.11, 128.17; to err: 108.18; to lead astray: 126.5; to misapprehend: 120.17

flectere to break: 230.7; to deflect: 231.16; to refract: 236.1, 242.19, 250.19, 250.20, 251.6, 251.15, 251.18, 251.20, 253.4, 254.8, 255.20, 256.22, 257.12, 258.9, 260.3, 260.5; to turn: 89.19, 89.21

flexio angular deflection: 234.6; breaking: 102.5, 229.25, 240.6; deflection: 229.5; refracting: 119.20, 239.16, 248.12, 253.5, 253.7, 253.15; refraction:

flexio angular deflection (*cont.*)
65.24, 66.10, 76.23, 113.27(2), 114.20, 115.9, 117.25, 118.25, 229.22, 230.4, 235.21, 238.10, 238.25, 240.5, 240.7, 242.11, 243.1, 243.7, 244.5, 244.9, 244.14, 245.9, 245.18, 246.6, 248.4, 248.9, 251.6, 252.4, 254.9, 257.16, 259.5, 261.2

forma form: 76.10, 159.16, 196.15; image: 89.7, 89.12, 116.7, 116.22, 116.25, 116.32, 116.33, 116.36, 116.39, 117.2, 117.7, 118.29, 119.1, 119.4, 119.6, 119.13, 132.7, 132.10, 132.13, 135.12, 135.15, 136.4, 136.10, 136.18, 136.28, 137.6, 137.10, 137.23, 137.30, 139.30, 140.15, 140.16, 140.19, 140.22, 140.24, 140.27, 140.28, 140.30, 141.2, 141.5, 141.9, 141.14, 141.17, 146.34, 153.5, 153.24, 154.8, 155.10, 155.15, 156.5, 156.12, 156.20, 157.5, 157.6, 157.21, 158.12, 158.14(2), 158.21, 159.14, 159.24, 159.30, 159.33, 159.34, 159.35, 159.36, 159.37, 160.6, 160.11, 160.15, 161.12, 161.13, 161.19, 161.24, 162.13, 163.16, 163.20, 163.23, 163.25, 164.10, 165.5, 165.8, 165.11, 165.13, 165.16, 165.23, 166.13, 166.20, 166.21, 166.23, 166.27, 166.29, 167.3(2), 167.5, 167.10, 168.11, 168.13, 168.14, 168.15, 168.19, 169.7, 169.9, 169.14, 170.8, 170.14, 170.16, 171.2, 171.5, 171.8, 171.10, 171.12, 171.22, 171.23, 171.25, 171.31, 171.35, 172.1, 172.3, 172.6, 174.3, 175.14, 176.8, 176.13, 181.24, 182.4, 194.8, 194.9, 194.11, 194.13, 194.21, 194.26(2), 194.28, 195.3, 195.5, 195.8, 195.11, 195.18, 196.2, 196.4, 196.7, 196.9, 196.12, 196.26, 197.3, 197.4, 197.12, 197.22, 198.2, 198.4, 198.5, 198.27, 199.1, 201.31, 202.2, 202.7, 204.30, 204.32, 204.35, 205.2, 205.3, 205.11, 206.1, 206.4, 206.6, 206.9, 206.19, 206.20(2), 206.27, 206.29, 207.1, 207.4, 207.8, 207.14, 207.15, 207.19, 207.23, 207.27, 208.8(2), 208.9, 208.12, 208.24, 208.28, 208.30, 209.4, 209.5, 209.6, 210.2, 210.3, 210.10, 211.1, 211.2, 211.5, 211.11, 211.12, 211.19, 211.25, 211.29, 212.1(2), 212.3, 212.4, 212.6, 212.7, 212.14, 212.20, 212.23, 213.2(2), 213.4, 213.10, 213.13, 213.17(2), 213.21, 213.23, 213.24, 213.26, 214.1, 214.3, 214.13, 214.14(2), 214.22, 214.27, 215.4(2), 215.7, 215.10, 215.18(2), 215.19, 215.23, 215.26, 216.5, 216.12, 216.22, 216.30, 216.39, 217.1, 217.4, 217.13, 217.27, 218.8, 218.22, 218.38, 219.14, 219.30, 220.2, 220.15, 220.18, 221.5, 221.16, 221.17, 221.33, 222.2, 222.3, 222.8, 222.21, 223.10, 223.15, 224.2, 224.4, 224.8, 225.18, 225.26, 229.9, 233.2, 246.22, 247.1, 247.32, 253.6, 253.15, 254.2, 254.12, 254.21, 255.7, 255.12, 256.6, 256.10, 256.11, 256.29, 256.30, 258.4, 258.5, 258.11, 259.11, 259.20, 260.11(2), 260.12, 260.13; shape: 110.35, 110.38

fractio bending: 229.15; breaking: 64.11, 113.16, 113.20, 113.26, 113.28, 113.30, 114.2, 114.3, 114.13, 117.22, 117.31, 118.23, 118.25, 131.2, 153.14, 229.1, 230.9, 230.12, 244.21, 244.22, 244.30, 245.1, 245.2, 245.5, 245.12; reflection: 131.36; refraction: 114.22, 114.24, 232.25, 233.4, 233.6, 234.1, 234.10, 235.9, 235.24, 236.3, 236.16, 236.21, 241.16, 241.17, 241.21, 242.8, 242.10, 243.3, 245.1, 245.23, 246.13, 246.15, 247.7, 247.13, 247.18, 247.26

frangere to bend: 119.18; to break: 117.27, 119.10, 119.23, 230.11, 247.24; to reflect: 131.34, 137.3, 167.23; to refract: 230.22, 231.9, 232.26, 235.23, 235.24, 239.12, 239.14, 248.11, 248.20, 249.2, 249.18, 249.21, 250.11

genus genus: 74.19, 76.9, 80.4, 80.10, 98.3, 98.4, 98.6, 98.7; kind: 63.5, 108.16, 175.3

habitudo case: 92.23; character: 113.6; component: 136.12; condition: 79.29, 91.12

habitus case: 101.30; characteristic: 63.20, 70.14, 71.5, 72.5, 72.29, 73.2,
    74.15, 154.6, 161.21; condition: 74.10, 76.13, 76.15, 159.17; disposition:
    137.18, 245.6; nature: 102.21; property: 74.18, 174.3; state: 109.11

iactare to throw: 78.7
iactus projection: 78.7
illuminare to light: 78.8
illuminatio illumination: 78.8, 79.28(2), 79.30
impulsio impedance: 229.21
incidere to be incident: 71.19; to define: 121.30; to fall: 72.10, 80.4, 87.16,
    91.22, 113.17, 121.1, 144.9; to grasp: 112.31, 112.32; to impinge: 76.20,
    100.19, 126.12; to lie: 159.34, 192.8; to light: 77.10; to meet: 248.1; to
    pass: 119.32, 221.12; to perceive: 111.4, 121.17; to reach: 206.23, 237.21;
    to shine: 76.6; to strike: 80.1, 80.15, 90.32, 91.6, 91.19, 126.4
incubitus force: 140.7
insensibilis imperceptible: 90.19, 90.28, 94.2, 94.11, 95.3, 100.8, 105.18,
    106.6, 109.23, 137.15, 137.17, 241.9; nullified: 97.22; unnoticeable: 105.26
insensibiliter unconsciously: 102.3
instrumentum apparatus: 231.13; instrument: 238.15
intelligere to apprehend: 79.19; to bear in mind: 82.5, 139.15; to discern:
    109.19; to gain an understanding: 195.23; to mean: 153.29, 153.30,
    153.33, 153.36, 153.38; to perceive: 73.17; to recognize: 102.27; to under-
    stand: 63.12, 66.10, 83.20, 102.4, 217.17, 230.17, 237.19
intentio intent: 63.12
interpres translator: 63.6
interpretari to translate: 63.4
investigare to investigate: 141.8
iudicare to judge: 124.33, 125.8, 125.11, 171.14, 172.7, 172.9

lucere to illuminate: 76.15
lucescere to show forth vividly: 122.11
lucidus bright: 114.5, 120.29, 122.4, 125.14; illuminated: 74.18; illumi-
    nation: 64.1, 71.3; light: 79.2; lighted: 127.7; luminosity: 72.2, 79.24,
    80.3; luminous: 71.14, 71.15, 75.19, 75.24, 76.6, 76.7, 76.28, 80.4, 91.19,
    116.12
lumen brightness: 77.6; light: 63.16, 70.7, 71.24, 71.26, 72.10, 74.13, 75.5,
    75.9, 75.11, 75.15, 75.18, 76.3, 76.5, 76.21, 76.29, 80.1, 80.2, 80.4, 80.5,
    91.6, 91.19, 91.20, 102.6, 108.27, 108.31, 109.2, 109.6, 110.17, 116.3, 116.6,
    116.7, 116.10(2), 116.11, 116.24, 122.9, 126.34, 127.3, 127.5, 127.7; lumi-
    nous object: 79.29
lux light: 102.10

medium center: 92.25, 92.28, 92.29, 148.7, 221.21; centerpoint: 78.10; cen-
    tral: 77.22; intermediate: 144.22; middle: 78.2, 83.5, 122.11, 122.15, 135.5,
    238.18; midpoint: 156.24, 156.28, 175.21, 234.23
motio motion: 106.6, 109.18, 111.2, 119.31, 124.1, 124.13, 154.6; movement:
    128.3; projection: 139.9; speed: 111.24
motus action: 70.12; activity: 71.2, 72.10, 73.9, 74.26, 74.28; change: 102.16,
    102.24; effect: 65.7; flow: 120.2; locomotion: 102.23, 103.1; motion: 72.19,
    72.21, 72.22, 103.13, 103.16, 104.22, 104.31, 105.13, 105.16, 105.17, 105.18,
    105.20, 105.26, 109.27, 109.28, 110.36, 110.39, 111.4, 118.24, 119.29, 123.8,

motus action (*cont.*)
123.30, 123.33, 124.4, 124.18, 124.22, 125.8, 128.4, 128.12, 128.14; movement: 103.25; operation: 63.17, 70.9, 152.6; speed: 104.15; velocity: 105.4
movere; to agitate: 117.7; to move: 103.19, 103.28, 104.2, 104.14, 104.16, 104.18, 104.20, 104.28, 105.2, 105.3, 105.6, 105.9, 105.11, 105.26, 105.28, 106.3, 106.5, 106.11, 111.6, 111.10, 111.15, 111.18, 111.20, 111.24, 111.32, 111.33, 112.1, 112.5, 113.3, 116.8, 119.25, 119.32, 120.2, 120.3, 120.6, 123.9, 123.20, 123.31, 124.3, 124.9, 124.14, 126.25, 126.26, 136.6, 139.16, 140.8, 161.6(2), 216.10(2), 216.11, 217.1, 245.3; to pass: 123.12; to pivot: 104.29; to revolve: 110.26, 111.4; to roil: 116.5, 117.2, 117.17; to slide: 135. 10; to travel: 111.20, 112.4; to traverse: 111.34, 112.1

natura nature: 63.2, 75.14, 76.2, 82.29, 83.1, 90.1, 91.25, 94.14, 112.17, 126.12, 126.19, 136.35, 137.28, 144.10, 150.19, 151.8, 152.25, 161.35, 230.14, 244.27, 247.16
naturalis natural: 98.2, 98.28, 101.33, 102.14, 106.31, 112.10, 113.17, 137.2, 139.6, 152.2, 152.6; plain: 234.5
naturaliter by nature: 64.22; naturally: 83.3, 126.2, 127.16, 159.14
nervosus nervous: 75.2

observare to govern: 225.19; to maintain: 138.27, 171.35; to meet (a condition): 156.13; to preserve: 162.2, 162.3
oculus eye: 64.17, 64.18, 64.25, 72.32, 73.13, 76.25, 82.22, 82.23, 83.12, 83.16, 83.18, 83.19, 83.25, 83.27, 84.1, 84.3, 84.4, 84.5, 84.7, 84.8, 84.14, 85.8(2), 85.9, 85.10, 85.15, 85.16, 85.17, 85.19, 86.2, 86.3, 86.8, 86.9, 86.12, 86.13, 86.14, 88.1, 88.4, 88.13, 88.15, 88.19, 88.27, 88.33, 89.7, 89.13, 89.16, 89.21, 90.1, 91.11, 91.13, 101.22, 103.16, 107.10, 107.11, 119.28, 133.1, 133.12, 133.14, 135.7, 141.25, 143.1, 144.20, 146.5, 146.9, 146.31, 146.34, 147.1, 147.6, 147.10, 148.3, 150.15, 150.22, 152.5(2), 152.8, 152.21, 152.28, 159.38, 160.15, 162.14, 196.1, 196.11, 196.14, 196.15, 232.18, 235.1, 237.1, 254.14; line of sight: 233.1
optica optics: 106.21, 106.36, 229.11
orizon horizon: 116.4, 127.29, 127.30, 151.13, 238.14, 238.19, 238.22, 239.11, 240.9, 240.14, 240.18, 240.20, 241.11, 241.20

passio effect: 109.9; impression: 110.20, 194.20; passion: 64.10, 64.11, 64.14, 79.11, 79.13, 79.14, 79.19, 79.25, 79.27, 80.12, 81.15, 81.16, 81.20, 102.24, 102.26, 102.37(2), 106.15, 106.16, 112.10, 112.17, 113.4, 113.10, 113.14, 113.15, 117.19, 118.22, 120.8, 120.13, 125.19, 125.21; sense-impression: 120.20; visual impression: 110.14; visual passion: 112.35
penetrabilis visui to be seen through: 232.10; transparent: 254.32
penetrare (see also **res/corpus quam visus/lumen penetrat**) to be absorbed: 126.34; to enter: 71.19; to extend beyond: 132.2; to pass through: 114.25, 115.2, 132.3; to penetrate: 91.18, 229.19, 238.1; to traverse: 114.15
penetratio passage: 71.4, 81.17, 131.11, 131.13, 229.3, 229.6; penetration: 64.2, 131.8, 131.10, 229.4
percipere to perceive: 102.8; to sense: 124.12
percussio impression: 114.8
perpendicularis cathetus: 115.24, 138.21, 138.23, 140.35, 225.1, 225.3, 225.7, 225.10, 225.12, 225.13, 225.16, 231.17, 231.18, 246.20, 247.2, 247.3,

perpendicularis cathetus (*cont.*)
248.11, 248.21, 249.9, 249.18, 250.9, 251.5, 251.13, 252.1, 252.19, 252.21,
253.4, 257.2, 258.9; normal: 131.37(2), 132.2, 132.18, 132.25, 135.19,
137.22, 137.25, 166.32, 168.2, 169.13, 169.18, 175.20, 176.11, 181.23,
192.18, 193.15, 193.20, 201.33, 206.13, 207.29, 212.16, 220.11, 224.21,
226.3, 229.28, 231.4, 232.14, 233.1, 234.21, 235.4, 235.25, 236.7, 237.7,
239.11, 240.7, 242.14, 242.15, 242.18, 242.19, 243.4, 243.9, 244.11, 244.19,
245.2, 245.13, 245.16, 245.17, 245.19, 245.20, 246.1, 246.6, 247.8, 249.1,
249.2, 249.4, 249.19, 249.21, 249.22, 250.10, 250.11, 250.12, 250.20(2),
251.19, 251.20, 251.21, 253.15, 254.8, 255.20, 256.1, 256.2, 256.9, 257.12,
257.13, 257.17, 258.1, 259.6, 260.5, 261.2; orthogonal: 77.1, 115.20, 144.20;
perpendicular: 84.16, 92.25, 92.27, 100.15, 100.19, 100.21, 101.1, 131.31,
132.21, 137.27, 140.29, 141.29, 144.9, 145.7, 145.19, 146.6, 146.14, 146.16,
146.25, 147.6, 148.24, 148.36, 149.13, 150.6, 155.19, 156.33, 157.1, 157.3,
158.5, 160.4, 160.10, 169.19, 175.25, 180.17, 191.1, 191.2, 203.5, 203.24,
206.16, 221.29, 224.16, 230.8, 234.19, 246.7, 246.18, 248.2, 260.7, 261.9
perscrutare to analyze: 232.2, 236.3, 259.7; to examine: 204.33; to investigate: 63.2
perscrutatio analysis: 90.8; care: 221.22; phenomenon: 153.10
perspicabilis perspicuous: 80.7
piramis cone: 64.22, 78.9, 78.14, 82.27, 83.6, 89.19, 90.14, 101.13, 176.7,
220.1, 220.4, 220.6, 220.22, 221.18, 221.21, 221.22, 221.35; pyramid: 221.1,
221.8; visual cone: 77.15, 78.3, 78.11, 82.25, 84.14, 84.16, 90.29, 98.34,
101.18, 104.1, 104.23, 141.28, 144.14, 148.22, 152.11, 153.29
piramis visibilis cone: 82.1; visual cone: 64.20, 73.14, 81.8, 101.9, 104.29,
110.32, 152.1, 176.6; visual field: 123.18
positio disposition: 115.16, 131.34, 167.8, 201.30, 208.31, 247.23, 257.19,
257.24, 258.14; distance: 240.3; example: 176.16; location: 72.21, 73.28,
120.12, 120.15, 136.9, 136.17, 150.1, 154.7, 171.20, 207.20, 256.28, 256.30;
orientation: 65.5, 95.16, 97.22, 98.7, 98.10, 148.8, 148.19, 156.18(2),
156.23, 156.35, 157.16, 167.5, 207.27, 208.9, 225.26, 225.27; place: 198.1;
placement: 131.29; position: 64.21, 82.25, 102.11, 104.35, 105.3, 126.16,
157.13, 192,6, 195.7, 198.14, 215.6, 216.20, 227.15, 233.5, 252.15, 252.19
potentia potential: 198.1, 202.22
preponere to analyze: 226.9; to articulate: 132.4, 161.7; to assume: 143.24;
to discuss: 89.14, 142.20; to establish: 141.12, 167.2, 217.11, 225.9, 240.8;
to give (reasons): 171.7; to lay out: 230.13; to outline: 85.24; to precede:
152.31, 161.34; to presuppose: 79.24, 249.7; to propose: 84.11, 92.11,
131.22, 167.20, 175.16, 201.11; to set up: 157.18; to show: 187.3, 187.12;
to specify: 208.26; to start: 162.10; to stipulate: 139.29
principium analysis: 141.12; apex: 78.3; Apex: 144.13, 152.9; beginning:
79.3; discussion: 152.31; generating point: 78.4; generating source: 78.5;
opening part: 63.15; origin: 75.2, 139.12, 140.8; origin-point: 136.36,
137.26, 138.19, 152.22, 152.29; outset: 221.17; primary referent: 78.9;
principle: 85.24, 131.23, 132.4, 136.32, 143.21, 161.7, 163.15, 167.2,
171.30, 174.1, 202.9, 206.19, 217.11, 222.4, 224.18, 224.23, 225.9, 225.19,
229.10, 230.14, 231.6, 244.19, 248.8, 254.3; source: 78.6, 79.21, 119.26,
119.29; source-point: 81.27; vertex: 98.13, 103.7, 132.11
probare to demonstrate: 65.13, 65.17, 66.4
pupilla eye: 83.3; pupil: 131.29, 131.34, 137.25, 137.26, 137.28, 138.3, 138.4,
138.6, 138.12

radius line of incidence: 224.9; line of sight: 232.17; line-segment: 192.23; radial branch: 133.9, 133.10, 133.16; radial line: 172.6, 231.19; radial segment: 226.3; radiation: 91.32; ray: 64.19, 64.25, 76.32, 78.15, 81.6, 81.7, 81.10, 81.31, 82.4, 82.9, 82.11, 82.12, 82.15, 83.7, 84,17, 85.2, 85.3, 85.4, 85.7, 85.8, 85.15, 85.16, 85.18, 85.19, 86.5, 87.2, 87.18, 88.28, 88.29, 88.31, 88.32, 88.34, 88.36, 89.4, 89.5, 89.6, 89.8, 89.10, 89.11, 90.3, 90.13, 90.31, 90.34, 90.35, 91.1, 91.4, 92.27, 92.29, 92.30, 98.9, 99.8, 99.10, 100.15, 101.3, 101.22, 102.6, 103.9, 103.11, 103.23, 104.12, 104.23, 104.30, 104.33, 104.34(2), 105.1, 105.2, 105.5, 105.7, 105.10, 110.1, 111.8, 115.19, 118.2, 118.4, 118.7, 118.17, 118.18, 118.20, 119.3, 119.4, 119.29, 120.38, 121.4, 121.5, 122.11, 122.12, 124.8, 126.21, 131.10, 131.12, 131.34, 133.12, 133.13, 133.15, 135.20, 137.3, 137.4, 137.8, 137.11, 137.16, 137.18, 137.20, 137.25, 138.6, 138.27, 139.19, 143.2, 143.4, 143.10, 143.15, 146.13, 148.17, 152.12, 152.13, 152.17, 153.14, 153.32, 153.35, 154.14, 154.16, 154.19, 155.13, 155.18, 156.28, 156.35, 158.3, 158.21, 160.1, 160.9, 161.5, 161.9, 161.15, 162.20, 162.21, 162.23, 166.36, 167.23, 168.4, 168.9, 169.20, 170.15, 170.23, 171.3, 171.37, 175.25, 176.2, 176.10, 177.4, 177.7, 177.12, 177.20, 179.6, 179.21, 180.13, 181.15, 181.19, 183.3, 183.9, 183.22, 183.24, 185.1, 185.4, 185.6, 185.8, 186.7, 189.21, 190.12, 191.9, 191.16, 191.23, 191.28, 191.30, 192.17, 193.17, 196.19, 197.19, 198.18, 198.20, 200.5, 200.8, 201.11, 201.16, 201.27, 202.1, 202.4, 202.19, 202.21, 202.26, 203.11, 204.5, 204.9, 204.14, 205.8, 206.14, 208.4, 209.1, 210.15, 211.7, 211.16, 212.4, 212.9, 212.24, 213.17, 214.10, 214.25, 214.29, 215.2, 215.15, 215.21, 216.2, 216.7, 216.24, 220.10, 221.28, 222.16, 223.1, 224.7, 226.13, 227.7, 227.10, 229.20, 230.6(2), 230.11, 230.21, 231.1, 231.9, 231.10, 232.26, 234.7, 234.10, 235.22, 236.7, 238.2, 238.3, 242.19, 244.11, 246.2, 246.22, 247.30, 248.11, 248.20, 249.1, 250.10, 250.11, 251.15, 251.19(2), 252.16, 252.20, 253.4, 254.8, 255.19, 257.2, 258.8(3), 260.3; visual cone: 122.33, 123.1, 125.25, 146.4, 198.10; visual flux: 123.14; visual ray: 76.23, 80.27, 81.29, 83.13, 87.15, 90.26, 119.10, 120.27, 139.32, 140.12, 146.24, 156.8, 159.3, 225.35, 226.1, 230.19, 241.2

radius visibilis line of sight: 225.8; ray: 210.7, 225.2; visual cone: 121.30, 122.25, 123.2, 124.24, 132.11, 198.10; visual flux: 91.37, 103.21, 103.25, 105.22, 122.9, 122.27, 123.30, 124.1, 220.1, 222.10, 239. 1; visual radiation: 91.25, 136.35, 140.11, 140.32; visual ray: 65.24, 77.21, 78.9, 90.30, 91.30, 92.25, 92.27, 98.17, 99.4, 99.8, 102.34, 104.1, 104.3, 109.29, 110.29, 111.7, 117.1, 117.26, 117.34, 119.24, 120.4, 120.21, 121.4, 121.17, 124.15, 124.17, 131.3, 131.8, 131.27, 132.18, 132.23, 132.27, 139.18, 151.7, 153.9, 154.13, 156.23, 157.23, 158.7, 158.20, 160.20, 165.20, 168.18, 175.8, 196.6, 202.25, 202.28, 213.8, 216.1, 216.27, 223.7, 224.21, 225.9, 225.15, 226.21, 227.4, 229.1, 229.15, 231.4, 231.13, 238.7, 238.10, 239.12, 239.14, 240.6, 241.16, 241.21, 243.2, 243.10, 244.9, 244.21, 244.29, 246.1, 246.2, 247.4, 247.5, 247.9, 247.25, 248.1, 257.7

radius visus ray: 77.22; visual ray: 77.17, 87.15, 88.7, 88.10, 91.3, 91.22, 91.27, 92.3, 98.26, 98.33, 100.14, 124.27, 133.5, 196.17, 237.18

ratio definition: 98.28; reason: 92.5; reasoned account: 197.2; rule: 139.19

ratiocinari to arrive at a conclusion: 123.1, 127.13; to conclude: 122.31, 123.5; to detect: 122.33; to gauge: 98.29; to impute: 128.14; to infer: 124.16, 125.4, 125.6; to interpret: 122.23; to judge: 121.15, 123.14; to use reasoning: 217.17

ratiocinatio analyzing: 156.21; extrapolation: 128.13; inference: 124.31, 126.15, 126.29; method: 242.7; rational interchange: 79.17; reason: 136.34; scrutiny: 102.2

reflexio refraction: 131.9

refractio reflection: 138.32, 175.15, 187.14, 187.15, 204.26, 204.31, 208.12, 209.10, 223.3, 225.12; refractive effect: 236.17

refringere to break: 118.3, 131.9, 153.32, 153.35; to reflect: 116.6, 118.16, 132.23, 133.6, 135.21, 138.8, 138.27, 140.13, 154.13, 154.15, 154.16, 154.19, 155.19, 157.1, 158.3, 160.1, 160.9, 162.21, 162.22, 163.6, 164.1, 165.20, 166.37, 168.5, 168.9, 169.21, 170.23, 171.3, 175.26, 176.11, 176.24, 177.5, 177.7, 177.12, 177.20, 178.4, 179.6, 179.22, 180.14, 181.15, 181.20, 183.9, 183.13, 183.15, 183.18, 183.23, 185.1, 185.2, 185.5, 185.6, 185.8, 186.7, 190.12, 191.9, 191.16, 191.24, 191.29, 191.31, 192.14, 192.21, 192.23, 193.8, 193.17, 196.6, 196.19, 198.19, 198.20, 198.26, 199.17, 199.23, 200.7, 200.8, 200.10, 200.11, 201.11, 201.17, 201.27, 202.4, 202.19, 202.25, 203.11, 204.5, 204.9, 204.14, 204.22, 205.8, 206.15, 208.4, 209.2, 210.7, 210.15, 211.7, 213.1, 213.8, 213.20, 214.11, 214.16, 215.15, 215.22, 221.29, 222.19, 222.21, 223.1, 224.22, 225.35, 226.13, 246.3; to refract: 231.1, 241.2, 246.1, 247.21

regitivus (see **virtus regitiva**)

repellere to impede: 77.16; to resist: 72.25, 72.32, 74.31

res/corpus quam visus/lumen penetrat surface of refraction: 118.26; transparent medium/body/object: 102.5, 102.9, 114.21, 115.7, 115.10, 119.8, 119.14, 119.20, 119.33, 122.17, 122.27, 123.3, 153.16, 153.18, 229.6, 229.14, 253.17, 254.15

res videnda visible object: 65.27, 75.13, 79.16, 82.27, 82.28, 83.5, 83.6, 85.22, 89.20, 90.11, 91.23, 92.25, 97.27, 101.4, 103.18, 103.23, 104.9, 104.10, 104.13, 104.25, 105.24, 109.7, 109.10, 109.20, 110.39, 112.21, 112.23, 112.24, 112.29, 112.35, 113.18, 113.25, 113.26, 113.32, 113.34, 115.6, 115.20, 116.9, 116.30, 116.32, 116.37, 118.17, 118.19, 118.21, 119.11, 119.22, 119.25, 119.28, 120.4, 120.6, 121.30, 124.2, 126.12, 126.17, 131.32, 131.35, 132.7, 132.8, 132.10, 132.16, 132.18, 132.24, 135.22, 136.9, 136.26, 138.3, 138.6, 138.19, 141.22, 141.30, 142.15, 143.19, 144.12, 145.8, 146.24, 146.31, 148.6, 151.7, 154.12, 155.2, 155.15, 155.18, 156.20, 156.25, 156.33, 157.14, 157.17, 158.8, 159.36, 159.40, 162.20, 163.17, 163.23, 165.6, 165.15, 165.19, 165.21, 166.34(2), 170.17, 170.21, 171.23, 181.21, 181.22, 192.8, 192.11, 192.15, 192.16, 192.18, 192.20, 192.24, 192.31, 192.32, 193.14, 193.18, 193.25, 194.18, 197.11, 199.5, 202.15, 202.27, 203.1, 204.34, 205.2, 205.8, 206.7, 207.11, 207.16(2), 207.20, 208.2, 208.25, 208.34, 210.10, 213.13, 213.16, 214.4, 214.5, 214.9, 214.21, 215.1, 215.8, 215.14, 216.1, 217.32, 218.38, 219.15, 221.27, 222.1, 224.5, 224.8, 224.12, 224.24, 225.1, 225.3, 225.33, 226.21, 229.10, 230.9, 243.5, 244.20, 247.33, 248.12, 251.14, 251.22, 252.20, 253.16, 256.6, 256.8, 256.28; visible property: 63.19, 65.8, 65.10, 71.6, 71.8, 71.11, 74.20, 74.22, 79.6, 79.8, 102.22, 106.14, 108.19, 112.16, 112.25, 113.13, 120.9, 120.15, 120.22, 130.4, 174.5, 229.11

reverberatio reflection: 65.11, 65.13, 76.22, 113.27, 113.29, 114.17, 114.19, 115.12, 115.15, 115.17, 115.23, 116.11, 116.27, 116.28, 116.33, 117.25, 118.8, 118.13, 118.25, 131.12, 131.28, 132.25, 132.27, 133.7, 133.9, 133.13, 133.15, 136.36, 137.4, 137.8, 137.14, 138.1, 138.29, 138.30, 139.5, 140.6, 140.13, 140.24, 140.26, 141.13, 153.6, 153.12, 166.30, 167.9, 175.17, 176.14, 177.23, 180.7, 180.19, 181.2, 181.5, 182.3, 182.13, 183.1, 183.12, 184.4,

reverberatio reflection (*cont.*)
    184.7, 185.4, 185.6, 185.9, 185.23, 186.2, 186.11, 187.1, 187.4, 187.9,
    187.15, 188.10, 188.13, 188.17, 188.20, 188.21, 188.24, 189.3, 189.5, 189.18,
    189.23, 189.25, 190.2, 190.4, 190.8, 190.11, 190.19, 191.19, 192.9, 192.28,
    197.14, 198.13, 199.5, 199.11, 200.5, 200.20, 200.26, 201.4, 201.9, 201.15,
    201.24, 202.5, 202.12, 202.17, 202.23, 202.24, 203.8, 203.10, 203.13, 203.18,
    204.7, 204.8, 204.11, 204.24, 206.23, 215.2, 216.2, 216.9, 216.24, 220.17,
    221.4, 221.6, 224.7, 225.17, 225.25, 225.34, 227.4, 227.8, 227.9, 227.13,
    227.18, 229.2, 229.24, 230.4, 245.7, 245.15
reversio deviation: 140.32; reversal: 227.15
revolutio diplopia: 64.12, 113.16, 113.22, 113.31, 117.22, 117.23; radial
    sweep: 118.24; scanning: 151.1; sweep: 119.24

scientia apperception: 73.5; knowledge: 131.25; perception: 73.6; science:
    63.2, 106.36, 153.21, 229. 10; scientific account: 174.2; scientific investi-
    gation: 106-20; scientific knowledge: 131.19; scientific study: 131.24
scire to bear in mind: 124.30
scotomia vertigo: 119.27
scrutare to view: 103.10
sensibilis noticeable: 75.27, 102.28, 105.14, 105.16, 105.26, 110.37; percep-
    tible: 90.13, 90.15, 92.14, 94.1, 94.10, 95.3, 95.4, 95.12, 95.13, 96.1(2),
    97.26, 100.4, 100.6, 100.8, 100.21, 100.22, 101.26, 102.28, 103.7, 103.20,
    105.13, 105.22, 105.27, 106.7, 109.23, 110.12, 110.13, 111.8, 111.17, 111.21,
    123.18, 241.3; sensible: 70.7, 73.7, 73.15, 74.30, 75.4, 75.10, 80.12, 80.18,
    82.5, 82.7, 90.11, 98.1, 106.34, 110.19, 111.11, 113.2, 153.16, 234.2; thing
    to be sensed: 74.14; visual: 106.33
sensibilitas perception: 121.17, 151.9; sensation: 78.25, 109.12, 112.37,
    152.26, 247.17; sense-impression: 98.3, 109.25; sensible impression:
    94.14; sensibility: 79.21, 144.16; sensitivity: 101.35; visual flux: 150.19;
    visual sensitivity: 182.2
sensus appearance: 85.22; observable phenomenon: 230.15; sensation:
    92.15, 124.31, 151.3; sense: 64.5, 70.10, 73.7, 74.31, 75.1, 75.3, 79.30, 80.7,
    96.11, 103.14, 105.23, 106.23, 106.27, 106.29, 107.26, 108.13, 111.21,
    112.16, 117.12, 119.30, 122.23, 123.14, 123.34, 125.31, 127.21, 128.5, 140.1,
    141.6, 151.5, 194.20, 204.30; sense-data: 106.12; sense experience: 136.34;
    sensible impression: 93.3; visual faculty: 120.18, 123.24; visual flux:
    121.1; visual sensation: 91.24; visual sense: 112.31, 126.11, 137.2
sensus visibilis sense of sight: 79.9, 122.31; visual faculty: 124.25; visual
    flux: 75.14; visual sense: 71.17, 90.30, 112.10
sentire to feel: 72.26, 72.32, 107.22, 107.23; to mean: 92.24, 156.23; to per-
    ceive: 65.6, 81.22, 90.17, 98.24, 98.33, 99.3, 99.5, 99.22, 101.5, 102.27,
    102.29, 102.32, 102.35, 106.8, 112.27, 151.8, 214.4, 214.14, 215.8; to see:
    224.24; to sense: 72.7, 106.4, 108.16, 112.36, 113.2, 126.3, 137.9, 176.1,
    176.3, 176.5
situs direction: 83.4, 138.4; disposition: 82.15, 89.18, 138.21, 139.20, 140.32,
    209.21, 257.6; location: 75.19, 81.27, 82.3, 104.25, 104.27, 117.21, 125.8,
    136.21, 138.10, 139.30(2), 144.12, 152.4, 194.22, 194.23, 194.25, 194.32,
    195.7, 195.20, 195.25, 223.15, 231.19, 241.3, 241.10, 253.16; orientation:
    92.20, 92.22, 93.8, 94.5, 94.9, 94.11, 94.14, 95.6, 95.22, 97.24, 101.29,
    138.27, 139.1, 144.18, 158.6, 166.30, 171.32, 208.9, 209.1, 209.6, 209.20,
    226.18; path: 243.1; perspective: 159.5; place: 71.2, 72.8, 72.18, 73.9,

situs direction (*cont.*)
  74.25, 81.26, 108.25; position: 82.19, 82.20, 156.26, 208.3, 216.28, 236.4,
  247.31; side: 160.27, 216.14; situation: 141.2, 159.18, 207.15
splendidus bright: 108.7, 114.8, 221.20; illuminated: 72.13; shining: 71.28
splendor brightness: 111.14, 120.25
subsistentia consistency: 229.21; subsistence: 75.9, 75.12, 75.16
substantia substance: 71.18, 74.18, 244.7, 244.8

tabula board: 147.4; panel: 124.20; tablet: 65.16
tactus touch: 64.5, 74.31, 75.2, 99.22, 103.15, 107.22
tangere to be tangent: 131.39, 139.7, 166.1, 168.7, 178.3, 180.11, 180.17,
  181.1, 205.13, 211.21, 211.23; to touch: 110.13
transgressio transition: 80.3
transire to be drawn through: 158.1, 206.15, 211.23; to brush by: 114.14; to
  intersect: 110.32; to make a passage: 106.5; to mark out: 111.29; to
  move: 103.22, 104.6, 104.7, 104.8, 104.9; to pass into: 257.8; to pass over:
  112.6, 123.17, 128.8, 230.19, 230.21; to pass through: 71.19, 77.19, 101.9,
  101.12, 101.18, 114.12, 131.32, 134.26, 137.25, 150.12, 158.18, 158.19,
  164.13, 168.21, 169.7, 169.8, 176.19, 178.6, 181.22, 184.9, 198.26, 201.34,
  204.3, 207.12, 207.17, 211.11, 211.13, 211.15, 213.3, 213.6, 213.11, 220.19,
  220.21, 220.22, 221.10, 221.12, 222.14, 239.8, 240.13, 240.15, 240.19,
  240.22, 240.24, 243.2, 245.20, 247.10, 255.20, 258.12; to pass to: 231.20; to
  reach: 231.15; to revolve: 221.36; to transpose: 198.6
transitus course: 245.8; passage: 128.15, 234.7, 242.12, 246.12, 246.14,
  247.25; passing: 242.13; path: 254.18, 254.19; travel: 111.20; way: 202.26

vertere to be led: 120.19; to incline: 182.2, 182.10; to revert: 197.9; to
  switch direction: 227.5; to tend to curve: 196.14, 196.17
videre to apprehend: 216.7, 238.5; to find out: 254.2; to look: 107.7, 119.23;
  to observe: 216.29, 257.22; to see: 64.17, 65.27, 65.28, 66.1, 71.20, 71.24,
  71.27, 73.23, 73.24, 73.34, 74.10, 75.13, 76.3, 76.5(2), 76.12, 76.19, 76.21,
  76.25, 76.29, 76.32, 77.3, 77.7, 77.10, 77.11, 77.22, 77.23, 78.20, 78.21, 79.6,
  79.11, 79.24, 79.26, 81.21, 82.3, 82.9, 82.11, 82.12, 82.17, 83.10, 83.15,
  84.2, 84.6(2), 85.5, 85.7, 85.14, 85.18, 85.19, 85.27, 86.8, 87.14, 87.17, 88.8,
  88.11, 88.19, 88.23, 88.26, 88.28, 88.31, 88.34, 88.35, 89.4, 89.6, 89.9, 89.11,
  90.22, 90.25, 90.34, 91.12, 91.23, 91.37, 91.38, 92.6, 100.12, 101.7, 101.35,
  103.18, 104.34, 105.5, 105.7, 105.10, 106.6, 107.3, 107.10, 107.18, 107.19,
  107.20, 108.22, 111.6, 114.9, 114.19, 114.24, 116.13, 116.25, 117.5, 117.34,
  118.3, 118.13, 118.30, 119.2, 119.5, 119.15, 119.19, 120.1, 124.7, 124.9,
  125.30, 126.21, 126.35, 127.2, 127.5, 127.7, 127.12, 130.2, 131.26, 131.31,
  133.2, 133.5, 135.12, 136.5, 136.11, 136.16, 137.7, 137.24, 140.12, 140.17,
  142.7, 142.10, 142.21, 143.14, 144.2, 144.7, 145.24, 146.7, 146.27, 148.30,
  148.37, 149.1, 150.10, 150.14, 151.13, 153.2, 153.14, 153.15, 154.3, 154.8,
  155.10, 155.12, 159.12, 160.5, 160.16(2), 161.5, 161.8, 161.15, 161.21,
  161.27, 161.28, 162.13, 167.3, 168.21, 170.1(2), 170.10, 171.1, 171.8,
  171.14, 194.8, 194.18, 196.1, 197.17, 202.6, 208.8, 210.3, 211.25, 213.3,
  213.17, 213.22, 216.9, 216.19, 216.31, 216.33, 216.38, 218.38, 219.14,
  221.23, 222.3, 223.5, 226.17, 227.9, 227.11, 227.14, 230.5, 230.23, 231.12,
  237.18, 238.3, 239.18, 240.2, 243.5, 253.2, 254.4, 254.7, 256.4; to view:
  126.20, 132.14, 136.31, 161.8, 171.29, 171.38; to visually grasp: 103.12; to
  visually perceive: 103.3, 120.26

visus center of sight (*cont.*)
    127.6, 144.5, 255.4; viewpoint: 75.20, 75.24, 77.13, 103.11, 103.19, 103.22,
    111.28, 121.6, 194.18, 205.26, 207.26, 211.5, 216.23(2); vision: 71.15, 80.21,
    83.2, 102.15, 107.21, 108.14, 108.18, 128.17; visual capacity: 140.5; visual
    cone: 103.7, 138.17; visual faculty: 64.23, 65.8, 65.10, 70.12, 71.1, 71.9,
    72.4, 72.8, 75.15, 78.22, 81.20, 81.22, 81.26, 90.10, 98.9, 98.16, 102.16,
    102.21, 102.26, 102.36(2), 103.5, 103.14, 106.15, 106.24, 108.20, 108.24,
    110.20, 112.19, 112.21, 113.11, 114.8, 115.15, 116.2, 117.20, 120.8, 125.19,
    253.11; visual field: 76.16, 259.9; visual flux: 63.16, 64.11, 70.7, 71.18,
    74.13, 75.9, 75.12, 75.18, 77.16, 79.12, 79.25, 79.28, 80.5, 80.15, 80.22,
    81.14, 90.21, 91.15, 101.33, 103.13, 106.4, 107.12, 107.15, 107.16, 112.6,
    113.15, 113.17, 113.35, 114.12, 114.16, 114.25, 114.26, 114.28, 115.1, 115.7,
    116.5, 118.9, 118.22, 119.26, 125.7, 126.4, 126.34, 127.2, 127.8, 138.19,
    237.20, 238.1, 238.5, 259.21; visual perception: 79.14; visual power:
    109.18, 197.5, 197.9; visual radiation: 205.4; visual ray: 76.3, 91.18,
    119.32, 133.3, 133.5, 138.8, 215.6, 220.8, 254.17; visual sense: 110.40,
    120.14, 125.13, 137.2, 137.9, 148.5, 182.9

ymaginari to see: 196.4
ymaginatio illusion: 131.6; perception: 92.16, 97.25, 112.15, 119.18, 124.34,
    125.33, 126.10, 127.21; visual perception: 101.28, 124.30
ymago depiction: 125.17; image: 73.22, 108.26, 109.5, 118.15, 124.20,
    124.22, 124.28, 159.16, 217.7; representation: 73.33; visual
    impression: 123.7

ENGLISH-LATIN INDEX

accident: **accidens**
accidental: **accidens**
to act: **agere**
action: **actus, motus**
activity: **motus**
actuality: **actus**
adjoining: **continuus**
to adjudge: **existimare**
admiral: **amiratus**
to affect: **accidere**
to agitate: **movere**
to align: **coaptare**
ambiguity: **ambiguitas**
analysis: **consideratio,contemplatio**
   perscrutatio, principium
to analyze: **considerare, discernere,**
   perscrutare, preponere
analyzing: **ratiocinatio**
ancients: **antiqui**
angular deflection: **flexio**
Apex/apex: **principium**
apparatus: **instrumentum**
to appear: **apparere, videri**
appearance: **apparitio, sensus**
apperception: **scientia**
apperceptual touch: **comprehensio**
to apply: **coaptare**
apposite: **aptus**
to apprehend: **cadere, cognoscere,**
   comprehendere, discernere, intel-
   ligere, videre
apprehension: **comprehensio**
apt: **aptus**
to arise: **accidere, agere**
to arouse: **accidere**
to arrive at a conclusion: **ratiocinari**
to articulate: **preponere**
to ascertain: **cognoscere, dinoscere**
aspect: **accidens**
to assert: **dinoscere**
to assimilate: **assimilari**
to assume: **arbitrari, existimare,**
   preponere
to attach: **aptare, coaptare**
axis: **axis.** *See also* common axis, proper
   axis, visual axis.

to be absorbed: **penetrare**
to be accustomed: **consuescere**
to be adduced: **apparere**
to be affixed to: **accidere**
to be apparent: **apparere, dinoscere**
to be caused: **accidere**
to be clear: **apparere**
to be congruent with: **coaptare**
to be convinced: **considerare**
to be created: **accidere**
to be curved: **curvare**
to be customary: **consuescere**
to be detectable: **apparere**
to be determinate: **apparere**
to be drawn through: **transire**
to be due to: **accidere**
to be evidenced: **apparere**
to be evident: **apparere, dinoscere**
to be experimentally determined:
   experiri
to be habitual: **consuescere**
to be incident: **cadere, incidere**
to be led: **vertere**
to be manifested: **apparere**
to be observable: **apparere**
to be observed: **accidere, apparere**
to be obvious: **apparere**
to be perceived: **apparere, assimilari**
to be seen: **apparere**
to be seen through: **penetrabilis visui**
to be sensed: **apparere.** *See also* thing to
   be sensed.
to be tangent: **tangere**
to be the case: **accidere**
to be used to: **consuescere**
to be visible: **apparere, videri**
to bear in mind: **intelligere, scire**
to befall: **accidere**
beginning: **principium**
to bend: **frangere**
bending: **fractio**
board: **tabula**
to break: **flectere, frangere,**
   refringere
breaking: **flexio, fractio**
bright: **lucidus, spiendidus**

brightness: **claritas, clarus, lumen, splendor**
to brush by: **transire**
by nature: **naturaliter**

capable: **aptus**
capacity: **consuetudo, virtus.**
   See also sense-capacity, visual
   capacity.
care: **perscrutatio**
case: **dispositio, habitudo, habitus.**
   See also to be the case.
cathetus: **perpendicularis**
to cause: **efricere**
center: **centrum, medium**
center of sight: **visus**
centerpoint: **centrum, medium**
central: **medium**
certain: **certus**
change: **motus**
character: **habitudo**
characteristic: **accidens, habitus**
circular: **curvus**
claim: **distinctio**
clear: **certus, clarus**
coalescing: **continuus**
to coincide: **cadere**
coincident: **continuus**
color: **color**
to color: **colorare**
coloring: **coloratio**
to come about: **accidere**
common axis: **axis communis**
common section: **distinctio
   communis**
component: **habitudo**
concave: **concavus, curvus**
concave curvature/section/shape/
   side/surface: **concavitas**
concavity: **concavitas, concavus**
to conclude: **ratiocinari**
condition: **habitudo, habitus**
cone: **conus, piramis, piramis visibilis.** *See also* visual cone.
confused: **confusus**
confusion: **confusio**
to connect: **cadere**
to conserve: **conservare**
to consider: **arbitrari, considerare**
consistency: **subsistentia**
to contain: **comprehendere**
contiguous: **continuus**
continual: **continuus**
continuous: **continuus**
to converge: **cadere**
convex: **curvus**

convex curvature/section/shape/side:
   curvitas
convexity: **curvitas**
to convey: **accidere**
cornea: **aspiciens**
course: **cursus, transitus**
to create a visual effect: **apparere**
crest: **curvus**
curvature: **curvitas**
curve: **concavitas.** *See also* outward
   curve.
to curve: **curvare.** *See also* to tend to
   curve.
curved: **curvus**
custom: **consuetudo**

to deceive: **fallere**
deception: **deceptio**
deep-set: **concavus**
to define: **incidere**
defining characteristic: **differentia**
defining outline: **distinctio**
definition: **ratio**
to deflect: **flectere**
deflection: **flexio**
to demonstrate: **probare**
density: **corporalitas**
depiction: **ymago**
depressed: **concavus**
to designate: **dinoscere**
to detect: **comprehendere, ratiocinari**
determination: **cognitio**
to determine: **arbitrari, cognoscere,
   comprehendere, considerare,
   dinoscere, distinguere**
deviation: **reversio**
difference: **differentia**
to differentiate: **discernere**
Differentiating Faculty: **virtus
   discernitiva**
differentiation: **differentia, distinctio**
difficulty: **dubitatio**
diplopia: **revolutio**
to direct a line of sight: **aspicere**
direction: **situs**
to discern: **cognoscere, dinoscere, discernere, intelligere**
discontinuity: **distinctio**
to discover: **cognoscere**
to discuss: **considerare, preponere**
discussion: **distinctio principium**
disposition: **dispositio, habitus, positio, situs**
distance: **positio**
distinction: **distinctio**
to distinguish: **discernere, distinguere**

to do: **agere**
doubt: **dubitatio**
to drop: **cadere**

effect: **actus, effectus, motus, passio.** *See also* refractive effect, sensible effect.
to emanate: **cadere**
empirical fact: **effectus**
to enter: **penetrare**
to envelope: **comprehendere**
enveloping: **continuus**
to err: **fallere**
to establish: **dinoscere, preponere**
establishing a point: **consideratio**
event: **actus**
to examine: **perscrutare**
example: **positio**
experiment: **experimentum**
to extend: **cadere**
to extend beyond: **penetrare**
extending: **continuus**
extrapolation: **ratiocinatio**
eye: **aspiciens, oculus, pupilia, visus**
eye-level: **visus**

faculty of sight: **visus**
to fall: **cadere, incidere**
false perceptual inference: **fallacia**
feature: **accidens**
to feel: **sentire**
to find out: **videre**
to fit: **coaptare**
to fix on: **aspicere**
flow: **motus**
flux: **visus**
focus: **aspectus**
to follow: **accidere**
force: **incubitus, virtus**
form: **forma**
to form: **comprehendere**
to formulate: **dinoscere**

to gain an understanding: **intelligere**
to gather: **coaptare**
to gauge: **ratiocinari**
generating point/source: **principium**
genus: **genus**
to give (reasons): **preponere**
glance: **aspectus**
to govern: **observare**
Governing Faculty: **virtus regitiva**
to grasp: **cadere, comprehendere, incidere.** *See also* perceptual grasp, to visually grasp.

habit: **consuetudo**

to happen: **accidere, cadere**
to have to do with: **accidere**
to hold: **accidere**
horizon: **orizon**

ideal: **accidens**
to illuminate: **lucere**
illuminated: **clarus, lucidus, splendidus**
illumination: **illuminatio, lucidus**
illusion: **accidens, deceptio, fallacia, ymaginatio**
image: **apparitio, aspectus, forma, ymago**
to impede: **repellere**
imperceptible: **insensibilis**
to impinge: **cadere, incidere**
impingement: **casus**
impression: **comprehension passio, percussio.** *See also* sense-impression, sensible impression, visual impression.
impulsio: **impedance**
to impute: **ratiocinari**
incidence: **casus.** *See also* line of incidence.
inclination: **actus**
to incline: **vertere**
indented: **curvus**
to infer: **ratiocinari**
inference: **consideratio, ratiocinatio.** *See also* false perceptual inference.
inferential error: **fallacia**
instrument: **instrumentum**
intent: **intentio**
to interact: **communicare**
intermediate: **medium**
to interpret: **ratiocinari**
interpretation: **consideratio**
to intersect: **transire**
to investigate: **comprehendere, considerare, distinguere, investigare, perscrutare**
investigation: **contemplatio.** *See also* scientific investigation,
to involve: **accidere**
inwardly curved: **concavus**
issue: **dubitatio.** *See also* to take up an issue.

to judge: **accipere, abritrari, existimare, iudicare, ratiocinari**
judgment: **arbitratio, assimilatio, consideratio.** *See also* to make a judgment.
juxtaposed: **continuus**

to keep: **conservare**
kind: **genus**
to know: **cognoscere**
knowledge: **scientia**. *See also* scientific knowledge.

to lay out: **preponere**
to lead astray: **fallere**
to lie: **cadere, incidere**
to lie between: **distinguere**
light: **lucidus, lumen, lux**
to light: **illuminare, incidere**
lighted: **lucidus**
line of incidence: **radius**
line of sight: **oculus, radius, radius visibilis, visus**
to line up along a line of sight: **aspicere**
line-segment: **radius**
location: **positio, situs**
locomotion: **motus**
to look: **apparere, aspicere, videre**
to look like: **assimilari**
looker: **aspiciens**
luminosity: **lucidus**
luminous: **lucidus**
luminous object: **lumen**
lying in a straight line with: **continuus**

to maintain: **conservare, observare**
to make a judgment: **discernere**
to make a passage: **transire**
to make sense of: **cognoscere**
to mark out: **comprehendere, transire**
to mean: **intelligere, sentire**
to measure: **accipere**
to meet: **incidere**
to meet (a condition): **observare**
method: **ratiocinatio**
middle: **medium**
midpoint: **medium**
to misapprehend: **fallere**
misapprehension: **dubitatio**
misperception: **comprehensio, fallacia**
motion: **cursus, motio, motus**
to move: **movere, transire**
movement: **motio, motus**
moving: **cursus**

natural: **naturalis**
naturally: **naturaliter**
nature: **habitus, natura**. *See also* by nature
nebulous: **aereus**
nervous: **nervosus**
norm: **consuetudo**

normal: **perpendicularis**
noticeable: **sensibilis**
nullified: **insensibilis**

observable phenomenon: **sensus**
observation: **aspectus**
to observe: **aspicere, videre**
observed phenomenon: **effectus**
observer: **aspiciens,**
to occupy: **comprehendere**
to occur: **accidere, cadere**
opening part: **principium**
to operate: **agere**
operation: **actus, motus**
optics: **optica**
the ordinary: **consuetudo**
orientation: **positio, situs**
origin/origin-point: **principium**
orthogonal: **perpendicularis**
to outline: **preponere**
outset: **principium**
outward curve: **curvus**
to overtake: **accidere**

panel: **tabula**
to pass: **efficere, incidere, movere**
to pass into/over/to: **transire**
to pass through: **cadere, penetrare, transire**
passage: **penetratio, transitus**. *See also* to make a passage.
passing: **transitus**
passion: **passio**
path: **situs, transitus**
to penetrate: **penetrare**
penetration: **penetratio**
to perceive: **cognoscere, comprehendere, dinoscere, discernere, existimare, incidere, intelligere, percipere, sentire**. *See also* to visually perceive.
perceptible: **sensibilis**
perception: **comprehensio, consideratio, scientia, sensibilitas, ymaginatio**. *See also* visual perception.
perceptual grasp: **comprehensio**
perfect evidence: **certitudo**
perpendicular: **perpendicularis**
perspective: **situs**
perspicuous: **perspicabilis**
to pertain to: **accidere**
phenomenon: **accidens, perscrutatio, sensus**
to pivot: **movere**
place: **positio, situs**
to place: **existimare**

placement: **positio**
plain: **naturalis**
point: **distinctio**
to posit: **arbitrari**
position: **positio, situs**
potential: **potentia**
power: **virtus**. *See also* visual power.
to precede: **preponere**
to preserve: **conservare, observare**
to presuppose: **preponere**
primary referent: **principium**
principle: **principium**
problem: **ambiguitas, dubitatio**
process: **actio**. *See also* viewing process.
to produce: **efficere**
projection: **iactus, motio**
prone: **aptus**
propagation: **effusio**
proper axis: **axis proprius**
property: **habitus**. *See also* visible property.
to propose: **preponere**
to provide: **accidere**
pupil: **pupilla**
pyramid: **piramis**

quality: **accidens**

radial branch/line/segment: **radius**
radial sweep: **revolutio**
radiation: **radius**. See also visual radiation.
raised: **concavus**
raised portion: **concavitas**
rational interchange: **ratiocinatio**
ray: **radius, radius visibilis, radius visus**. *See also* visual ray.
to reach: **cadere, incidere, transire**
to realize: **cognoscere, dinoscere**
reason: **ratio, ratiocinatio**
reasoned account: **ratio**
to recognize: **cognoscere, intelligere**
to reflect: **frangere, refringere**
reflection: **fractio, refractio, reverberatio**
to refract: **flectere, frangere, refringere**
refracting: **flexio**
refraction: **flexio, fractio, reflexio**. *See also* surface of refraction.
refractive effect: **refractio**
to relate: **communicare**
to remain: **conservare**
to represent: **existimare**
representation: **ymago**
to resist: **repellere**
to result: **accidere**

reversal: **reversio**
to revert: **vertere**
to revolve: **movere, transire**
to roil: **movere**
rule: **ratio**

scanning: **revolutio**
science: **scientia**
scientific account/investigation/ knowledge/study: **scientia**
to scrutinize: **aspicere, considerare**
scrutiny: **consideratio, ratiocinatio**
to see: **aspicere, comprehendere, dinoscere, sentire, videre, ymaginari**
to seem: **apparere, videri**
sensation: **sensibilitas, sensus**. *See also* visual sensation.
to sense: **cadere, percipere, sentire**
sense of sight: **sensus visibilis, visus**
sense-capacity: **virtus**
sense-data/experience: **sensus**
sense-impression: **passio, sensibilitas**
sensible: **sensibilis**
sensible effect: **virtus**
sensible impression: **sensibilitas, sensus**
sensibility: **sensibilitas**
sensitivity: **sensibilitas**
to set up: **preponere**
shape: **forma**
to share: **communicare**
to shine: **cadere, incidere**
to shine upon: **accidere**
shining: **splendidus**
to show: **preponere**
to show forth vividly: **lucescere**
side: **situs**
sight: **aspectus, aspiciens, visus**
to sight: **aspicere**
situation: **dispositio, situs**
to slide: **movere**
source/source-point: **principium**
to specify: **preponere**
speed: **motio, motus**
to start: **preponere**
state: **habitus**
to stay: **conservare**
to stipulate: **preponere**
to strike: **cadere, incidere**
study: **contemplatio**. *See also* scientific study.
subsistence: **subsistentia**
substance: **substantia**
to subtend: **comprehendere**
to suppose: **arbitrari, existimare**
sure: **certus**